Cambridge Studies in Biologi
Anthropology 20
Anthropology of Modern H

All humans share certain components of tooth structure, but show variation in size and morphology around this shared pattern which is studied by dental anthropologists. This book presents the first worldwide synthesis of the global variation in tooth structure in recent populations. It:

• describes the methods and assumptions used by dental anthropologists;

• discusses the genetic basis of nonmetric dental traits;

• portrays the geographic variation of tooth crown and root traits;

• analyzes dental variation on a global scale paralleling major genetic and craniometric analyses.

The book illustrates more than 30 tooth crown and root traits and reviews their biological and genetic underpinnings. This analysis links extinct and extant populations and so serves as a useful tool for elucidating population relationships and histories.

G. RICHARD SCOTT was a Professor of Anthropology at the University of Alaska Fairbanks until his retirement in 1997. He currently resides in Arizona and is an adjunct faculty member at Arizona State University. During his research career, his interests focused on dental anthropology and skeletal biology, with geographic concentration on the American Southwest, American Arctic, and North Atlantic.

CHRISTY G. TURNER II is currently Regents Professor of Anthropology at Arizona State University. His research career spanning over 40 years has delved into an eclectic mix of subjects, including dental anthropology, human taphonomy and cannibalism, and prehistoric rock art. Previous books include *The Dentition of Arctic Peoples* (1991) and *Man Corn: Cannibalism and Violence in the Prehistoric American Southwest* (1998).

Cambridge Studies in Biological Anthropology

The anthropology of modern human teeth

Cambridge Studies in Biological and Evolutionary Anthropology

Series Editors

Human ecology: C.G. Nicholas Mascie-Taylor, University of Cambridge
                Michael A. Little, State University of New York, Binghamton
Genetics: Kenneth M. Weiss, Pennsylvania State University
Human evolution: Robert A. Foley, University of Cambridge
                Nina G. Jablonski, California Academy of Science
Primatology: Karen B. Strier, University of Wisconsin, Madison

Consulting Editors

Emeritus Professor Derek F. Roberts
Emeritus Professor Gabriel W. Lasker

Selected titles in the series

11 *Genetic Variation and Human Disease* Kenneth M. Weiss 0 521 33660 0 (paperback)

12 *Primate Behaviour* Duane Quiatt & Vernon Reynolds 0 521 49832 5 (paperback)

13 *Research Strategies in Biological Anthropology* Gabriel W. Lasker & C.G.N. Mascie-Taylor (eds) 0 521 43188 3

14 *Anthropometry* Stanley J. Ulijaszek & C.G.N. Mascie-Taylor (eds) 0 521 41709 8

15 *Human Variability and Plasticity* C.G.N. Mascie-Taylor & Barry Bogin (eds) 0 521 45399 2

16 *Human Energetics in Biological Anthropology* Stanley J. Ulijaszek 0 521 43295 2

17 *Health Consequences of 'Modernisation'* Roy J. Shephard & Anders Rode 0 521 47401 9

18 *The Evolution of Modern Human Diversity* Marta M. Lahr 0 521 47393 4

19 *Variability in Human Fertility* Lyliane Rosetta & C.G.N. Mascie-Taylor (eds) 0 521 49569 5

20 *Anthropology of Modern Human Teeth* G. Richard Scott & Christy G. Turner II 0 521 45508 1 (hardback), 0 521 78453 0 (paperback)

21 *Bioarchaeology* Clark S. Larsen 0 521 49641 1 (hardback), 0 521 65834 9 (paperback)

22 *Comparative Primate Socioecology* P.C. Lee (ed.) 0 521 59336 0

23 *Patterns of Human Growth, second edition* Barry Bogin 0 521 56438 7 (paperback)

24 *Migration and Colonization in Human Microevolution* Alan G. Fix 0 521 59206 2

25 *Human Growth in the Past* Robert Hoppa & Charles Fitzgerald (eds) 0 521 63153 X

The anthropology of modern human teeth

Dental morphology and its variation in recent human populations

G. RICHARD SCOTT
Department of Anthropology
Arizona State University

AND

CHRISTY G. TURNER II
Department of Anthropology
Arizona State University

CAMBRIDGE
UNIVERSITY PRESS

PUBLISHED BY THE PRESS SYNDICATE OF THE UNIVERSITY OF CAMBRIDGE
The Pitt Building, Trumpington Street, Cambridge, United Kingdom

CAMBRIDGE UNIVERSITY PRESS
The Edinburgh Building, Cambridge CB2 2RU, UK   www.cup.cam.ac.uk
40 West 20th Street, New York, NY 10011-4211, USA   www.cup.org
10 Stamford Road, Oakleigh, Melbourne 3166, Australia
Ruiz de Alarcón 13, 28014 Madrid, Spain

First published 1997
Reprinted 1999
First paperback edition 2000

Printed in Great Britain by J. W. Arrowsmith Limited, Bristol

Typeset in Monotype Times 10/12$\frac{1}{2}$pt [VN]

A catalogue record for this book is available from the British Library

Library of Congress Cataloguing in Publication data

Scott, George Richard.
   The anthropology of modern human teeth: dental morphology and its
variation in recent human populations / G. Richard Scott and Christy
G. Turner II.
      p.   cm. – (Cambridge studies in biological anthropology; 20)
   Includes bibliographical references and index.
   ISBN 0 521 45508 1
   1. Dental anthropology.   I. Turner, Christy G.   II. Title.
III. Series.
GN209.S33 1997
573′.6314–dc20   96-46112   CIP

ISBN 0 521 45508 1 hardback
ISBN 0 521 78453 0 paperback

Dedicated to the memory of Albert A. Dahlberg
and Jacqueline A. Turner.

Contents

ix

Acknowledgments

We express our sincere thanks to Gabriel W. Lasker who encouraged the authors to write a book on dental morphology viewed from an anthropological perspective. Both authors also owe a great debt of gratitude to the late Albert A. Dahlberg whose life's work on dental variation and development helped stimulate and maintain our interest in dental morphological research. The generosity of Al and Thelma Dahlberg in sharing ideas, ideals, and materials over the past three decades will always be remembered with great fondness and appreciation.

The individuals who contributed indirectly to this work are far too numerous to list. Scores of anthropologists, museum directors, laboratory assistants, and dentists generously allowed the authors to study skeletal material and/or dental casts in their care. Dental anthropologists from all corners of the globe kindly sent copies of their latest papers to keep us abreast of a diverse and rapidly expanding literature on tooth crown and root morphology. We also thank the many granting agencies that provided funds to help defray the costs of extensive travel expenses, including the: National Science Foundation, National Geographic Society, National Academy of Sciences USA, USSR Academy of Sciences, Wenner-Gren Foundation for Anthropological Research, and IREX. The author's home universities, the University of Alaska Fairbanks and Arizona State University also provided research funds and sabbatical leaves to facilitate our travels.

We also thank those individuals who contributed directly to this work. Grant Townsend, one of the world's leading dental geneticists, provided a very useful critique of chapter 4. Russell Gould kindly performed an independent analysis of our world database to help us gauge the generality of our findings. Steven R. Street searched a diffuse literature for obscure references to intentional dental modification. Much of the information reported in the world database of CGT was obtained through the assistance and indefatigable efforts of Jacqueline A. Turner and Korri Dee Turner.

We would like to express our sincere appreciation to our editor at Cambridge University Press, Dr. Tracey Sanderson, who has shown both patience and compassion in seeing this volume through to completion. Finally, GRS thanks his parents, Robert and Kathleen Scott, and his wife, Cheryl, and three sons, Garrett, Geoffrey, and Gunnar, for their inspiration and support through the years. The parents and daughters of CGT are similarly acknowledged.

Prologue

This book is about teeth in general and tooth morphology in particular. Excluding rare cases of anodontia, where individuals never develop teeth, all humans in all places at all times had, have, or will have teeth. People use and see their teeth every day, but ordinarily take them for granted except when they are a source of pain or discomfort. Most pay even less attention to the morphology of teeth. By simply feeling the chewing surfaces of the individual teeth at the back of the mouth, one can discern elevations or bumps that are separated from one another by depressions or concavities. These teeth in the back of the mouth are molars, the bumps are cusps, and the depressions are fissures, fossae, grooves, or sulci, depending on their form and depth. More difficult to palpate are the many ridges that may be present on the cusps or along the margins of a tooth. It is elements such as cusps, ridges, fissures, and even the number of roots each tooth possesses that occupy the attention of the dental morphologist.

What are human perceptions (or misconceptions) about those hard white objects in our mouths that we (should) brush and floss everyday? Individual knowledge of teeth is highly variable within populations and between different cultures. In *Folklore of the Teeth*, Leo Kanner (1928) reviews a broad range of subjects, including the diversity of views on toothaches and dental hygiene and varied practices in artificial dental deformation. In his first chapter, on tooth number, shape, and eruption, Kanner (1928:3) discusses how many early scholars were misinformed on something as basic as the number of teeth an individual possessed: 'We not infrequently meet with individuals of sometimes more than fair intelligence who become embarrassed when questioned as to how many teeth they have.'

As dental anthropologists who have observed at least one million teeth, we sometimes take for granted what the average individual knows about their own teeth. Kanner's observations on dental misconceptions in earlier and recent times stimulated a survey of 80 university students and their friends and families who had to respond to four questions: (1) how many deciduous (baby) teeth do humans have?; (2) how many permanent teeth do humans have?; (3) how many types of teeth do humans possess?; and (4) do teeth play a role in attracting members of the opposite sex? While this small survey may not characterize the basic dental knowledge of the average American, the results were revealing. For the first question, only 19 individuals (24%) responded correctly that humans have 20 deciduous teeth, with answers ranging from 12 to 35. As most of these students have no children and have long since shed their own deciduous teeth, this result is only mildly surprising. Survey participants were more aware of the number of permanent teeth; 44 (55%) gave the correct answer of 32 (range: 21–40). Regarding types of teeth, 38 individuals (48%) correctly specified four types of teeth with incorrect answers ranging from one to 11. Part of the confusion was terminological as some thought cuspids and canines or bicuspids and premolars were different types of teeth. Only 11 individuals (14%) answered all three questions correctly (20–32–4). Regarding the role of teeth in sexual attractiveness, the response to this question was overwhelmingly positive; 75 individuals (94%) felt teeth were an important element of physical attraction. Some felt so strongly about this that they added exclamation marks after 'yes.' Jones' (1995) research on sexual selection and physical attractiveness among humans focused on eyes, nose, lips, and facial proportions, but the importance of teeth in physical appearance was not mentioned. Our survey and the near universal attempts to enhance dental beauty suggest studies of sexual selection should factor 'teeth' into their calculations along with other facial features.

The size, form, and morphology of teeth is determined primarily by genes. Teeth, however, interface with the environment following eruption. Because of their visibility and accessibility, teeth can be influenced by behaviors that are intentional, incidental, or accidental. For example, individuals in many cultures are not content with the natural morphology of their teeth but feel a cultural or idiosyncratic urge to produce an artificial morphology more in line with their value system. Such tinkering is concentrated on teeth at the front of the mouth (incisors and canines) as they are most visible in social intercourse. Such intentional manipulations of tooth form pale in comparison to the effects of unintentional alterations produced by natural forces over the life of a dentition, for example, through dental wear and pathologies. As this book is overwhelmingly concerned with the genetic aspects of tooth crown and root morphology, a prologue provides us with the opportunity to address briefly behavioral and environmental factors that influence dental morphology. Unfortunately, these factors have a secondary impact on those of us who observe dental morphology.

Intentional alterations of tooth form

There are aesthetic ideals for how teeth should 'appear' but these ideals, and how they are attained, vary greatly in different cultures. The western ideal of dental beauty, exemplified by many celebrities, is straight, white, vertically positioned anterior teeth that are all present and accounted for. Dentists and orthodontists make a living helping individuals attain this ideal through scaling, polishing, fillings, crowns, bridges, dental applications, and dentures, as needed. Except for correcting occlusal problems, individuals are less concerned with the appearance of their molars because these teeth have low visibility in social interactions. However, the mouth, as a primary social organ, does draw the viewer's eye so there is social import attached to the appearance of the incisors and canines and, to some extent, the premolars.

The western ideal of 'attractive' anterior teeth is not universal. In some cultures, straight white teeth are far from ideal. Groups from many parts of the world, especially Africa and Southeast Asia, modify their tooth morphology through artificial deformation. These practices range from the intentional removal of teeth (ablation) to modifying crown form through filing, incising, chipping, staining, banding, and insetting (Fig. P.1). The following excerpts are from ethnographers and travelers who witnessed first hand how individuals endured great pain, both literally and figuratively, to change the appearance of their teeth. While westerners might cringe at these descriptions, any individual who wore dental appliances during adolescence can identify with the pain involved in achieving a culturally prescribed dental ideal.

Tooth ablation in Central Australia

If the operation be performed on a man he lies down on his back, resting his head on the lap of a sitting man who is his tribal Oknia (elder brother), or else a man who is *Unkulla* to him (mother's brother's son). The latter pinions his arms and then another Okilia or *Unkulla* fills his mouth with furstring for the purpose, partly, they say, of absorbing the blood and partly of deadening the pain, and partly also to prevent the tooth from being swallowed. The same man then takes a piece of wood, usually the sharp hard end of a spear, in which there is a hole made, and, pressing it firmly against the tooth, strikes it sharply with a stone. When the tooth is out, he holds it up for an instant so that it can be seen by all, and while uttering a peculiar, rolling, guttural sound throws it away as far as possible in the direction of the *Mira Mia Alcheringa*, which means the camp of the man's mother in the Alcheringa.

(Spencer and Gillen, 1899:451–2)

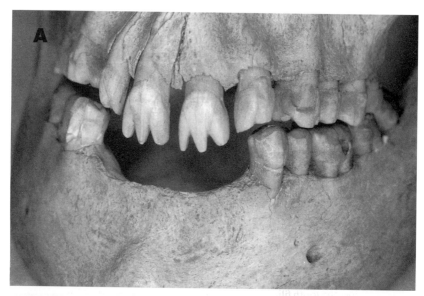

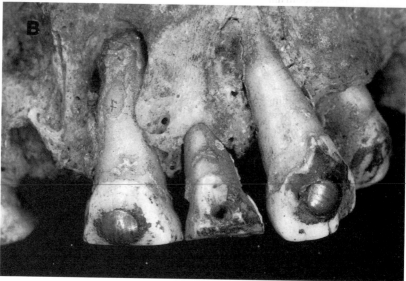

Fig. P.1.   Two examples of intentional dental modification: (A) A Jomon period male with filed upper central incisors and ablated (i.e., extracted) lower anterior teeth, a combination that produced a most interesting 'effect'; source C.G. Turner II neg. no. 5-26-84:6. (B) Brass insets in the upper central incisor, lateral incisor (inset now missing), and canine distinguish this specimen from Borneo (see section on 'brass insets among the Iban'); source C.G. Turner II neg. no. 9-19-83:1.

Tooth blackening and filing on Alor

It is during this period of adolescence that both boys and girls have their teeth blackened and filed ... The actual procedure is as follows: In July or August some young unmarried man, perhaps in his early twenties, announces that he will blacken teeth for the children of the community and designates the field house where it will be done ... He purchases from some friend in the village of Bakudatang a particular type of soil found there. .. With the earth he mixes a fruit resembling a green fig. The resulting paste is smeared on a strip of banana bark which each child cuts to fit the size of his mouth. .. For at least seven nights, and often ten, the children sleep together in a field house, with the paste held against their teeth by the flexible bark strips. .. The children all eat together, being careful to place small bits of food far back in their mouths in order not to spoil the dye. With the same objective a length of thin bamboo is used as a drinking tube during the period.

On the last day or two of the period those who are to have their teeth filed go through the ordeal. The same person who prepared the dye usually does the tooth filing. The subject's head is laid on the thigh of the operator and wedged against his side with his elbow. The jaws are propped open with a piece of corncob. The six upper and six lower front teeth are then filed to half their length with an ordinary knife blade that has been nicked to resemble a saw. Apparently experience makes it possible to avoid the root canal, which occupies only the upper half of the incisors. .... It is undoubtedly painful but, as in tattooing, it is bad form to admit it. The result of this filing means that even when the back teeth are occluded, the tongue will show pinkly between the gaping front teeth when a person smiles. *This is considered definitely attractive.* [emphasis ours]

(DuBois, 1944:83–4)

Brass insets among the Iban (Dyaks)

As among the populations of Alor, the Iban of Borneo blacken and file their front teeth. The also take dental modification one step further:

The teeth are often blackened, as black teeth are considered a sign of beauty. .. The front teeth are also frequently filed to a point, and this gives their face a curious dog-like appearance. .. Another curious way of treating the front teeth is to drill a hole in the middle of each tooth, and fix in it a brass stud. I was once present when this operation was in progress. The man lay down with a piece of soft wood between his teeth, and the 'dentist' bored a hole in one of his front teeth. The agony the patient endured must have been very great, judging by the look on his face and his occasional bodily contortions. The next thing was to insert the end of a pointed brass wire, which was then filed off, leaving a short piece in the tooth; a small hammer was used to fix this in tightly, and, lastly, a little more filing was done to smooth the surface of the brass stud. I am told the process is so painful that it is not often a man can bear to have more than one or two teeth operated on at a time.

(Gomes, 1911:38)

Dental modification among the Moi of Vietnam

At the age of puberty boys and girls alike undergo the formality of having their teeth filed down to the gums. With some kind relative sitting on the chest of the sufferer, lying on his back with his head between the legs of a primitive vise, and with a wooden bit forced into his mouth, a medicine-man breaks off the teeth with stones and hacks and chips them away. It is their idea of making themselves beautiful, and the boy or girl who has not undergone this punishment is not considered marriageable or otherwise of adult status. After a day of this frightful work the operator leaves his victim covered with blood, his gums in ribbons, his lips like hashed-beef-steak, and incapable for a fortnight of eating anything but liquids. Nor is this all, for the patient is then given a stone with which to continue the beautifying process himself, when he has a moment to spare, until not a sign of tooth remains above the level of the gums. Among some of the tribes the lower teeth are given a saw shape, so that the open mouth suggests that of an aged shark that has lost its upper plate.

(Frank, 1926:90–1)

Tooth blackening among the Annamese

. . . about marriage time, which in Annam is early in life, every Annamese, of either sex, is expected to have his teeth lacquered black by a process said to be very painful. .. Every people has its own style of beauty. .. and to the Annamese a person is handsome only if his teeth are jet-black. 'Any dog can have white teeth,' say the Annamese, looking disparagingly at Europeans. To them white teeth are not only ugly but immoral! For the congare, the Annamese girl, who has not blackened her teeth, is usually, if not always, some Frenchman's darling.

(Frank, 1926:168–9)

In some instances, artificially deformed tooth crowns provide clues to past movements of peoples. That is, when people move or are transplanted from one geographic area to another, they may continue previous cultural practices in their new locale. West Africans, forcibly transplanted to North and South America and the Caribbean in the sixteenth, seventeenth, and eighteenth centuries, maintained to some extent the tradition of filing their anterior teeth to points (Stewart 1942; Ortner 1966; Stewart and Groome

1968). Dental mutilation, widespread in Mesoamerica during pre-Columbian times (Romero 1970), has been observed in a handful of American Southwest burials. Given the extreme rarity of this practice in the Southwest, it seems likely that the few individuals with mutilated teeth are actually of Mesoamerican origin (e.g., traders). Specific patterns of tooth extraction might also indicate historical linkages between the Minatogawa population of Okinawa (*c.* 17,000 BP), the Jomon peoples of Japan (12,500 to 2200 BP), and possibly some Neolithic populations of China (Wu 1992). These are but a few examples that illustrate the potential applications of ethno-odontology to historical problems. A monograph devoted entirely to this subject, minimally exploited to date, would be a major contribution to dental anthropology.

Unintentional alterations of tooth morphology

While some groups modify their teeth intentionally to denote status and group affiliation or simply to enhance physical attractiveness, others engage in behaviors that result in unintentional alterations of crown morphology. For example, smoking pipes over a period of time, particularly clay-stemmed pipes, leaves a distinctive imprint on teeth. When the upper and lower teeth are occluded, pipe wear produces an oval-shaped opening in the region of the lateral incisors and canines that appears on either or both sides of the mouth, depending on the inclinations of the smoker. We have observed such pipe wear in prehistoric, protohistoric, and modern populations ranging from Melanesia and Siberia to the North Atlantic (Fig. P.2A). Another cultural practice that leaves a distinctive mark on the tooth crown is the use of labrets, or cheek plugs. The reasons for wearing labrets are similar to those given for intentional dental modification (i.e., group identity, status, physical attractiveness). This practice also involves pain as incisions of varying lengths had to be cut through the cheek just below the lower lip for insertion of the labret(s). Once inserted, the internal face of the labret would come in contact with the outer (buccal) surface of the lower teeth. Depending on the size of the labret(s), their placement (medial or lateral), and the length of time worn, the result would be distinct polished facets of varying size on the external (labial or buccal) surfaces of two or more teeth. From prehistoric to recent times, labret usage could be found in many New World populations with widely divergent subsistence strategies, from hunter–gatherer Eskimos in the north to agricultural Mesoamericans in the south (Fig. P.2B). In addition to pipes and labrets, other repetitive behaviors that involve the insertion of hard objects in the mouth, such as tooth-picks and bobby pins, leave permanent marks on teeth (Ubelaker *et al.*, 1969; Berryman *et al.*, 1979).

Using teeth as tools involves another set of behaviors that can result in permanent alterations to the tooth crown. Such usage was a common practice among hunter–gatherer populations prior to the Neolithic Revolution and is not unheard of even today. Who has not used their teeth to break string, tear fabric, open a container, or the like? Granted, tooth-tool use is not what it used to be, with pliers, vises, scissors, and a myriad of other tools as more efficient and less risky substitutes for teeth. However, before such tools were developed, teeth served as an intermediary in many task-related activities such as working hides, softening boots, making grass baskets, using bow-drills to create fire, and producing rope or string from sinew. These behaviors, reflected on the teeth as grooves, notches, unusual wear planes, and rounded wear, could ultimately affect crown morphology as much or more than intentional modifications of teeth (Molnar, 1972; Schulz, 1977; Hinton, 1981; Larsen, 1985).

Much to the chagrin of dental morphologists, dietary behavior alters the form of teeth more dramatically and on a much more universal scale than any other type of post-eruptive modification. In earlier human populations, the primary changes occurred through crown wear. This is a natural process brought about by the 'simple' act of bringing the upper and lower teeth in contact while chewing food (attrition). When food is contaminated with nonfood items (i.e., sand, silt, grit, etc.), the process of wear is enhanced even more (abrasion). In some instances, enamel chips are removed around the margins of the crowns when an individual bites down on foreign objects accidentally introduced into food (Turner and Cadien, 1969). Wear commences on the occlusal surfaces of the teeth and, in time, eliminates cusps, ridges, and fissures. With the introduction of highly refined foods, rates of dental wear have slowed down in many societies. This advantage is offset in many instances by increased frequencies of dental caries, which can destroy teeth more quickly than heavy crown wear. Caries are prone to develop in pits and fissures on the occlusal surface of a tooth crown, precisely the location of many morphological crown traits.

Because of the many behavioral and environmental factors that can affect teeth once they have erupted, despite their hardness, it is amazing that adults retain any crown morphology. In some skeletal series where post-eruptive modification is compounded by post-mortem loss of single-rooted teeth (especially incisors and canines), some traits might be observable in only a handful of individuals. Roots, however, are much less affected than crowns by behavior and environment so sample sizes for root

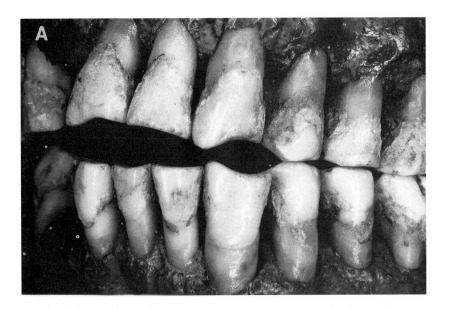

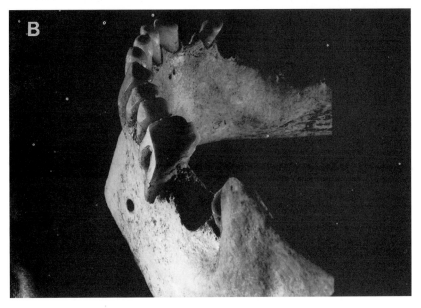

Fig. P.2. Two examples of unintentional dental modification: (A) Pipe faceting in an East Asian Goldi, affecting primarily the upper left lateral incisor and canine and the lower left canine; source C.G. Turner II neg. no. 1-19-81:2. (B) Labret faceting in this protohistoric Alaskan Eskimo was produced on the lower left canine, first premolar, second premolar, and first molar (extending onto the roots) by a large lateral labret; source G.R. Scott.

traits are often much higher than for crown traits. Although this option is rarely available in skeletal studies, many researchers who deal with living populations focus on individuals between 12 and 16 years of age. This narrow window of time is the period when an individual's teeth are least likely to be affected by the two major contributors to crown destruction, wear and caries.

Given the many genetic, developmental, behavioral, cultural, historical, pathological, and environmental insights that can be gained from teeth, it is not surprising that more and more departments of anthropology and colleges of dentistry are adding a course in dental anthropology to their curricula. A typical one semester senior-level course in dental anthropology would characteristically survey several topics, with emphasis placed on the instructor's area of research expertise, be it primate dentition, modern human dentition, or some other subject area. Dental anthropology courses in the United States usually have laboratory sessions where various practical methods are learned such as fossil identification, dental casting, thin-sectioning, and radiology. A representative course might include the following topics:

(1) Introduction. History, theoretical issues, rationale, goals and objectives, applications, scientific method.
(2) Morphological variation. Modern human crown and root anatomy.
(3) Teeth in populations. Dental characteristics of past and recent human populations.
(4) Oral pathology. Caries, periodontal disease, fluorosis, developmental anomalies, others.
(5) Teeth and behavior (comparative ethno-odontology). Use, wear, diet, hygiene, mutilation, beauty psychology, folklore.
(6) Forensic and bioarchaeological applications. Age, sex, race, individual characteristics.
(7) Growth and development. Embryology, eruption, field and clone models, occlusion, mesial drift, symmetry.
(8) Dental genetics. Normal and abnormal traits, classic, quantitative, population, and twin methods of analysis.
(9) Dental microevolution. Reduction, selection, mutation, drift, gene flow, synchronic and diachronic methods of analysis.
(10) Vertebrate dental macroevolution. Origin of teeth, major phyletic adaptations, paleontology.

(11) Primate dentition. Intra-order variation and relationships to ecology, behavior, and diet.

(12) Fossil hominid dentition. Taxonomy, evolution, reduction.

This book emphasizes topics (1), (2), (3), (7), and (8), the areas wherein the authors have conducted much of their research for the past 30 years. Both authors have traveled extensively throughout the world to collect original observations that form the core of the book's database on modern human dental variation, a database that together represents more than 30,000 living, recent, and archaeologically-derived individuals.

Students often ask instructors how they came to be interested in their area of expertise. In our case, the story begins at the University of Arizona, Tucson, where Bertram S. Kraus' research on dental morphology triggered Turner's interests in the 1950s. Turner went on to study for his PhD at the University of Wisconsin under Professor William S. Laughlin, who was leading multidisciplinary expeditions to the Aleutian and Kodiak islands of Alaska. Members of these expeditions included the prominent dental scholars Stanley Garn, Conrad Moorrees, and Albert Dahlberg whose influences and contributions are woven throughout this book. Scott completed his PhD under Turner in 1973 at Arizona State University, where dental genetics, variation, and standardization were emphasized by Professors Christy Turner and Donald H. Morris. Other students who earned their PhDs in dental anthropology at Arizona State University include Edward F. Harris, Christian R. Nichol, Joel D. Irish, and Alice Haeussler, with Diane Hawkey nearing completion. The latter three students have also handled many of the duties of officership and editorship of the Dental Anthropology Association and its newsletter.

In closing this prologue, we wish to point out that archaeologically-derived teeth and bones are the primary source for direct and diachronic evidence of human evolution. All synchronic hypotheses about human origins, affinity, and microevolution, whether based on classical serological markers, earwax, DNA, or whatever new methods that may be developed, are accepted or rejected with the actual evolutionary record provided by teeth and bones. The purpose of this book is to show the power and potential of dental morphology for these exciting tests.

1 *Dental anthropology and morphology*

Introduction

Physical anthropology focuses on human biological variation through time and space. Except for those who work in primate paleontology, the temporal bounds of the field are set by hominid origins some five to eight million years ago and extend across time through a diversity of hominid fossil species to modern members of *Homo* sapiens. The geographic bounds extend to all parts of the globe habitable by human populations. Methods employed for conducting research on human variation run the gamut from anatomical measurements and observations to physiological parameters and DNA sequencing. The subjects of study are hominid fossils and all human skeletal and living populations. Problems revolve around a multiplicity of questions involving facets of human adaptation, variation, and history.

Dental anthropology is a subfield of physical anthropology although many contributors to this area of research come from fields outside of anthropology, notably dentistry, genetics, anatomy, and paleontology. Although dental anthropology strikes outside observers as a specialized field of inquiry, it encompasses a broad range of subjects which, in turn, have invited finer levels of specialization. Some workers concentrate on developmental aspects of the dentition, from tooth germ formation to developmental defects of the crown. Others focus on post-eruptive changes such as ordinary crown wear and culturally-prescribed dental modification. The study of dental pathologies, in particular caries, periapical osteitis, patterns of tooth loss, and periodontal disease, provides yet another avenue of research. Researchers interested in those elements of the human dentition that have some underlying genetic basis study tooth size and morphology (Cadien, 1972; Bailit, 1975; Kieser, 1991; Scott and Turner, 1988; Scott, 1991a; Townsend *et al.*, 1994).

Much evidence has accumulated over the past century to indicate dental development is regulated to a significant extent by the action of genes (chapters 3 and 4). This is true not only for crown and root form in general but extends to the myriad of positive and negative structural variants of a

1

tooth. Dahlberg (1951:140) notes 'All human dentitions are basically the same. The differences between individuals are in the number and extent of the primary and secondary characters of the tooth groups, which in turn are the reflections of the genetic constitution of the individual.' From the basic blueprint, or 'master dental plan' that characterizes all human dentitions, teeth exhibit morphological and metrical traits that vary within and between populations. From an evolutionary standpoint, these traits are observable in living and fossil hominoids and hominids (Gregory and Hellman, 1926; Weidenreich, 1937; Robinson, 1956; Swindler, 1976; Wood and Abbott, 1983; Wood et al., 1983, 1988; Wood and Uytterschaut, 1987; Wood and Engleman, 1988). In recent human populations, patterned geographic variation is evident in both tooth morphology and crown size.

The enamel which covers a tooth crown is the 'hardest' part of the body, consisting primarily of calcium hydroxyapatite [$Ca_{10}(PO_4)_6(OH)_2$] (Ten Cate, 1994). Because this inorganic component is extremely durable, teeth show excellent preservation in most taphonomic contexts. In hominid fossil localities and recent archaeological sites, they are often the best represented remains. It is not uncommon to find isolated teeth when the rest of a skeleton has long since disintegrated. In addition to their qualities of preservation, teeth provide the only hard tissues of the human body directly observable in living individuals. They can be studied through direct intraoral examination ('open wide please!'), but it is usually more efficient to replicate teeth in the upper and lower jaws through negative alginate impressions that serve as a mold for pouring fine-grained plaster, thus yielding a permanent cast or study model. Some workers, especially those in Russia, make observations directly on negative wax-bite impressions. Extracted teeth from the living provide another venue for study but given their isolation from the context of the whole dentition, they are less useful than dental casts and human skeletons for a systematic analysis of dental variation.

Dental anatomy and dental morphology

Dental morphologists study the structure and form of teeth. In studies of the human dentition, there are two distinct approaches to crown and root morphology. When dental anatomists write about tooth morphology, they are concerned primarily with normative tooth form (cf. Wheeler, 1965; Carlsen, 1987). For example, the human dental formula of 2–1–2–3, shared by all catarrhine primates (Old World monkeys, apes, and humans), refers to the number of different types of teeth in each quadrant of the upper and lower jaws. In each jaw, humans have four incisors, two canines, four premolars, and six molars with paired teeth on the left and right sides (i.e., antimeres) showing mirror imagery. Human tooth crowns consist of smaller elements referred to as cusps which are augmented by regularly occurring occlusal and marginal ridges. Grooves or fissures of varying depths divide a tooth into its constituent cuspal and ridge components. Although the opponents in the two jaws (e.g., upper right first molar and lower right first molar) show size and form differences, one can characterize incisors as spatulate, canines as single-cusped (cuspids) and conical, premolars as two-cusped (bicuspids), and molars as multi-cusped. Similarly, the incisors and canines can be characterized as single-rooted while the upper and lower molars have three and two roots, respectively. Lower premolars usually have single roots, although a normative characterization of root number for upper premolars would depend on the geographic locale of the dental anatomist (e.g., one in Greenland, two in Nairobi). Although cursory in detail, this is the fundamental blueprint for the human dentition (see chapter 2 for details). Dental anatomists focus on this blueprint so their texts are designed to show students the typical, or normative, form of each individual tooth. While they illustrate variant forms of crowns and roots, such variation is of secondary importance. Dental anatomists want their students to be aware of the variety of morphological structures they might encounter in their practices (at least in 'European' dentitions), but the clinical applications of dentistry are not ordinarily hampered by subtle differences in crown and root morphology.

Two types of morphological variants are observable in the human dentition. The first type involves major deviations from the basic dental blueprint (Fig. 1.1). Adjacent teeth are sometimes fused together, or twinned. Supernumerary teeth, as additions to the 2–1–2–3 dental formula, may be normal or anomalous in form and appear either as separate structures or be fused to other teeth. While fewer than 5% of the members of a population exhibit extra teeth (i.e., hyperodontia), it is not uncommon for individuals to be missing one or more teeth (i.e., agenesis or hypodontia). Other significant departures from normative crown form include conical lateral incisors, 3-cusped upper premolars, 'mulberry' molars, and sundry anomalies. The second type of dental variation is more subtle than twinned teeth, extra or missing teeth, and anomalous crown forms. It involves minor variations in secondary cusps, fissure patterns, marginal ridges, supernumerary roots, and so forth. These minor variants are common and vary within and between populations. They are of greater evolutionary significance than rare and idiosyncratic dental anomalies that are mostly induced by environmental factors during development. As

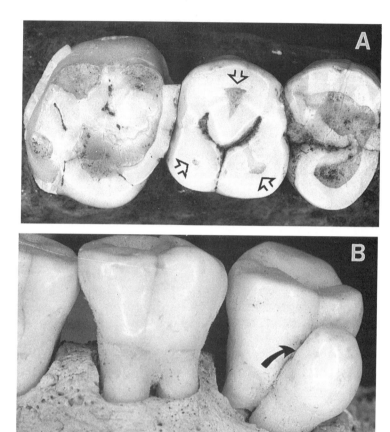

Fig. 1.1   Dental anomalies of rare occurrence: (A) This 3-cusped upper second premolar has one buccal cusp and two lingual cusps, all three of which exhibit small patches of dentine exposure (arrows); the frequency of 3-cusped upper premolars is conservatively estimated at 1/5000. (B) A peg-shaped supernumerary tooth is positioned along the buccal surface of the upper third molar; these small extra teeth are sometimes fused to adjacent teeth along the crown and/or root. (C) (*opposite*) Twinned deciduous lower incisors; developmental accidents of this sort, while interesting, provide no useful genetic information for anthropological studies. (Source: C.G. Turner II.)

Butler (1982:44) notes, 'The paleontological record indicates that dental evolution proceeds by the selection of minor variations. Presumably, large departures from the normal pattern would be functionally deleterious.' While major dental anomalies are interesting and eye-catching, it is the minor variations in human dental morphology that are useful in historic and forensic contexts.

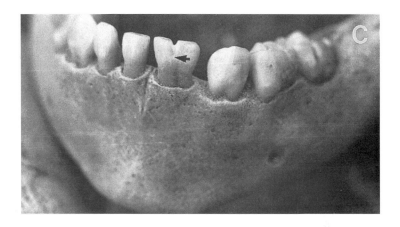

Fig. 1.1 (*cont.*)

The types of dental morphological variation we focus on are largely independent of tooth size, the subject matter of odontometrics (Kieser, 1991; chapter 3). Our interest is in the secondary structural variants of tooth crowns and roots that are manifest in two primary ways: as 'all-or-none' characters (accessory ridges, supernumerary cusps and roots, furrow patterns) or as differences in form (variation in curves and/or angles). In this volume, primary focus is on nonmetric crown and root traits that may be either present or absent within any individual dentition. Despite the presence–absence dichotomy, they are not literally all-or-none traits. Within a population, these traits show variations in degree of expression, often noted by such terms as slight, moderate, and pronounced when present. There are also major differences in trait frequency and expression between populations, and it is this variation that is of special significance to physical anthropology and allied fields.

A brief history of dental morphological studies

In the nineteenth century, dental anatomists and anthropologists described morphological variants and commented on their relative frequencies in different racial populations. Georg von Carabelli (1842) described an accessory mesiolingual cusp on the upper molars that was quite common in European dentitions. At the time, von Carabelli had no idea he would achieve a degree of immortality on the basis of this obscure accessory cusp that bears his name to this day. Early French and German anthropologists and odontologists showed that some morphological variants, particularly cusp number of the upper and lower molars, distinguished the major races of humankind. Dental anatomists like C.S. Tomes described human crown and root variants and put them in the perspective of comparative odontology. Despite such efforts, only a small foundation for the systematic study of the evolution and variability of human tooth morphology had been laid by the beginning of the twentieth century.

In 1920, Aleš Hrdlička wrote what might arguably be considered the foundation paper in the study of human tooth crown morphology in his detailed assessment of shovel-shaped incisors (so named because lingual marginal ridges enclose a fossa, giving the appearance of a 'coal shovel'). With access to many diverse archaeologically-derived human skeletal collections in the US National Museum of Natural History, Hrdlička had a distinct advantage over many of his European contemporaries. While building on the observations of earlier dental workers, Hrdlička was the first to classify the degree of expression of a morphological variant, assess this variation among several human populations, and describe its occurrence in nonhuman species. Among his findings was the close similarity between Asians and American Indians and their decided difference from European and African populations in the frequency and degree of shoveling expression. In retrospect, this might seem a small point, but this observation was made at a time when the origin of Native Americans was far from resolved. His follow-up article on 'Further studies of tooth crown morphology' (Hrdlička, 1921) extended his observations to other types of morphological variants, but this paper had less impact than the first, perhaps because he provided no methodology for observing these variants and little comparative data were provided to show differences among the major geographic races.

In his magnificent opus *The Origin and Evolution of the Human Dentition*, eminent paleontologist and comparative odontologist W.K. Gregory (1922) wrote 'apart from a few striking cases, presently to be noted, racial characters in the teeth are not very conspicuous.' Influenced in part by Hrdlička's earlier observations on human dental morphology, Gregory felt that, Europeans aside, differences in dental morphology were minor among the varied races of man. The morphological variables considered noteworthy by Gregory included shovel-shaped incisors, *tuberculum dentale* of the upper anterior teeth, upper and lower molar cusp number, lower molar groove pattern, including his *Dryopithecus* Y5 pattern, and Carabelli's cusp (chapter 2). Speaking from the vantage of the early twentieth century, Gregory characterized morphological traits as either low characters (i.e., primitive) or high characters (i.e., civilized). His low characters included central incisor shoveling, a molar cusp formula of

4–4–4/5–5–5, Carabelli's cusp on M1 and M2, and retention of the Dryopithecus pattern (Y5) in the lower molars. High characters, or those associated with a so-called modern dentition, included the absence or diminution of shoveling, the rarity of Carabelli's cusp, rounded and 3-cusped upper second molars, and lower molars with '+' rather than 'Y' patterns. In this effort, Gregory anticipated the methods of cladistics that emphasize the importance of primitive and derived traits in disentangling evolutionary history.

In 1925, T.D. Campbell published *Dentition and Palate of the Australian Aboriginal*. He noted that 'the differences between the dentitions of various types of mankind have not yet been sufficiently recognized to incite very special investigation. But this is probably due, not so much to lack of obvious differences, as to the paucity of specialized study' (Campbell, 1925:1). Although this work covered a wide array of dental topics, it did include morphological observations on upper and lower molar cusp number, upper premolar and molar root number, and Carabelli's trait. He mentions shovel-shaped incisors but provides no data, adding only that it is not characteristic of the Australian aboriginal dentition. Noting the preoccupation of his contemporaries with craniology, Campbell (1925:vii) shows prescience in his comment that 'a close and detailed study of the dentition and its associated structures does not seem to have attained the position of importance it will undoubtedly gain as time goes on.' His monograph was one turning point in this direction.

J.C.M. Shaw (1931), inspired by Campbell's work, contributed an important treatise titled *The Teeth, the Bony Palate and the Mandible in Bantu Races of South Africa*. Modeling this volume after that of Campbell, Shaw provided morphological observations on upper and lower molar cusp and root number, lower canine and premolar root number, and shovel-shaped incisors. With the publication of Shaw's volume, workers now had baseline data on South Africans as well as Australian aborigines. While lauding Shaw's contribution, Sir Arthur Keith notes in the foreword that 'from the anatomist's point of view the greater part of the world still remains in a state of dental darkness. Even in Europe and America much still remains to be done to complete a preliminary survey of the mouths of mankind.'

Despite the urging of Campbell and Keith for physical anthropologists to place more emphasis on the study of dental variation, their advice went largely unheeded at that time. Granted, other papers on human dental morphology appeared during the 1920s and 1930s, but these were limited in number and scope. Of special note are Krogman's (1927) long review article on anthropological aspects of the human dentition, Hellman's

(1928) paper on lower molar cusp number and groove pattern, and Tratman's (1938) observations on three-rooted lower first molar variation. Also published during this era were dental morphological studies of specific groups, including Hawaiians (Chappel, 1927), Finns (Hjelmman, 1929), Bushmen (Drennan, 1929), Japanese (Yamada, 1932), and American Indians (Nelson, 1938).

By 1940, we see the emergence of more intensive interest in comparative dental morphological variation. Two key workers of the time were Albert A. Dahlberg of the University of Chicago and P.O. Pedersen of the University of Copenhagen. Both of these scholars were dentists whose passion for research carried them to the study of non-European populations. In Dahlberg's case, his work on the morphology of Chicago white dentitions was expanded to American Indians with a primary emphasis on tribal groups in the American Southwest. Dahlberg's in-depth research on one of these groups, the Pima Indians of central Arizona, spanned a 35-year period. During this time, he and his wife, Thelma, amassed several thousand dental casts and associated genealogical records. Among Dahlberg's major early publications were 'The changing dentition of man' (1945a), in which he applied Butler's field concept to the human dentition (chapter 3), and 'The dentition of the American Indian' (1951), which provided valuable comparative data for this group. In 1956, Dahlberg also released a series of reference plaques to help standardize observations on morphological variables of the tooth crown. The plaques, involving ranked scales for quantifying trait expression, were distributed to workers throughout the world. The plaques for shovel-shaped incisors, Carabelli's trait, the hypocone, and the protostylid, played a significant role in stimulating further morphological studies of the human dentition.

At the time Dahlberg was systematizing research on American Indian dental morphology, P.O. Pedersen was completing field and laboratory research on living and sub-fossil Greenlandic Eskimos. His brilliant monograph *The East Greenland Eskimo Dentition* (1949) has been a primary reference for comparative data since its publication. Pedersen's vast knowledge of the European dental literature also provided a bibliographic starting point for many later students of crown and root morphology. Dental research on Arctic populations received another important contribution in 1957 with the publication of C.F.A. Moorrees' *The Aleut Dentition*. Thus, in less than a decade, Eskimo-Aleuts moved from a position of obscurity to a position of prominence in dental anthropological and morphological studies.

There was an active program of dental anthropological research in Japan during the first half of the twentieth century, but the dissemination of this research to western scholars was partly hampered by language barriers until the 1950s. Early workers, including E. Yamada, T. Fujita, and T. Sakai, helped set the stage for K. Hanihara to develop a strong tradition of dental morphological research among Japanese anthropologists that has thrived over the past four decades, as the following chapters attest.

The decade of the 1950s marks a crucial formative stage in the development of dental anthropological and morphological studies. Complementing the work of Dahlberg, Pedersen, Moorrees, and K. Hanihara, B. Kraus contributed foundation papers on the genetics and morphogenesis of tooth crown traits, G. Lasker reviewed the genetic aspects and forensic potential of dental morphology, S. Garn initiated studies on interactions in the dentition, and H. Brabant began important comparative studies of European dentitions from Upper Paleolithic to recent times.

In 1963, the study of human dental variation received a major boost with the publication of *Dental Anthropology*, edited by D.R. Brothwell. While all facets of dental anthropology were covered in this volume, from odontometrics and crown wear to morphogenesis, also included were important dental morphological papers on shoveling variation (V. Carbonell), the American Indian dentition (A.A. Dahlberg), two-rooted lower canines (V. Alexandersen), and third molar agenesis (D.R. Brothwell). The recognition of dental anthropology as a subfield of physical anthropology is essentially coincident with the publication of this work. After almost 40 years, Campbell's (1925:vii) plea that 'The subject of Dentition can no longer be considered one of only incidental interest, but must and will take its place as an important branch of the science of Physical Anthropology' was realized.

In the mid-1960s, A.A. Dahlberg and P. O. Pedersen felt the need for more international communication among members of the dental community that shared a common interest in tooth morphology. To that end, they organized, with the aid of V. Alexandersen, the first International Symposium on Dental Morphology, held in Fredensborg, Denmark, in 1965. The contributors to this symposium included anthropologists, dentists, geneticists, embryologists, and paleontologists. The overarching theme of the symposium was the structure, function, development, and evolution of teeth. To paraphrase W.K. Gregory, contributors covered the broad scope of dental variation and evolution 'from fish to man.' The original symposium proved so successful that it spawned subsequent symposia, which have since been held every three years. Proceedings of these symposia, published as edited volumes, include many valuable papers on human dental morphology (Pedersen *et al.*, 1967; Dahlberg, 1971a; Butler and Joysey, 1978; Kurtén, 1982; Russell *et al.*, 1988; Smith and Tchernov, 1992; Moggi-Cecchi, 1995; Radlanski and Renz, 1995).

In the past three decades, the number of journal articles and dissertations focusing exclusively on dental morphology far exceeds the total for the preceding 100 years. This era has witnessed the classification of many 'new' morphological traits, characterizations of numerous population samples, and more concerted efforts to understand the biological nature of crown and root traits. In the description and analysis of human dental variation, researchers from many countries are extending these efforts to all corners of the globe. Geographically, blank spaces remain (chapter 5) and much remains to be done, but the 'state of dental darkness' described by Keith over 60 years ago is slowly emerging into the light.

Dental morphology and physical anthropology

As physical anthropology embraces the study of primate, hominoid, and hominid evolution, it is impossible for authors to ignore teeth in textbooks on introductory physical or biological anthropology. Many of these texts, however, limit their comments on teeth to broad topics such as heterodont versus homodont dentitions, primate–human differences, tooth eruption in primates, Australopithecine teeth, and the like, but make no mention of dental morphological differences among recent human populations (Lasker, 1976; Bennett, 1979; Eckhardt, 1979; Harrison *et al.*, 1988; Staski and Marks, 1992). Other texts do include brief mention of morphological differences among modern humans, but these are limited to one or a few sentences on the Asian–European contrast in shovel-shaped incisors (Kelso, 1974; Brace and Montagu, 1977; Weiss and Mann, 1978), the *Dryopithecus* Y-5 pattern (Williams, 1973; Poirier *et al.*, 1994), or taurodontism (Birdsell, 1981). This situation may, however, be changing. In their sixth edition of *Physical Anthropology*, Stein and Rowe (1996) devote a side-bar to the congruence between language phyla and tooth morphology among Native Americans and discuss the implications of this finding for the peopling of the New World.

Textbooks specifically addressing human racial variation devote slightly more attention to dental morphology than introductory texts. In C.S. Coon's (1965) *The Living Races of Man*, the index has many items listed under 'Teeth' but a close perusal turns up mostly references to large-toothed or small-toothed populations. Coon does, however, refer to shovel-shaped incisors in three contexts. First, he says in his characterization of Mongoloids that 'The incisors are often, if not usually, 'shovel-shaped'; that is concave behind' (Coon, 1965:11). Second, he notes the Ainu show much less shoveling than Mongoloids, a point he uses to support the notion that this group was of Caucasoid, or 'archaic Caucasoid', descent (Coon *et al.*, 1950). Finally, he takes issue with Osman Hill (1963) who proposed the New World was peopled by Upper Paleolithic Europeans. Coon (1965:151) felt such a position was untenable, remarking that 'One objection to his hypothesis is that the American Indians have a much greater incidence of shoveled incisors than any Asiatic peoples, and in more extreme forms.' Europeans, by contrast, exhibit a much lower frequency of shoveling and even when the trait is present, it is much more muted in expression.

An author who has made significant contributions to dental genetics and anthropology, S.M. Garn (1971), devoted two pages to the 'variable dentition' in his book *Human Races*. Regarding morphology, he notes that shovel-shaped incisors are common in American Indians, and are present, though less common, in Polynesians, Finns, and fossil hominids. He says that some groups show a reduction in cusp number (e.g., Middle Easterners) while others have increased cusp numbers (e.g., Australians, Melanesians). His first reference is to elevated 4-cusped lower molar occurrence in Middle East populations, although he is not specific on how Australians and Melanesians have increased their cusp numbers. Garn also presents a histogram showing the population variation of Carabelli's trait.

In *Races, Types, and Ethnic Groups*, S. Molnar (1975) makes some reference to dental morphological variation and provides a table on shoveling variation, which is up to date. After showing how shoveling divides the world into 'haves' and 'have-nots', Molnar (1975:61) states that 'Several other features of the dentition show a great deal of variability and, in some cases, have been grouped according to race. More often, though, there is only a variability in the frequency of the occurrence of the particular trait, with all the major groups of mankind possessing it to some degree.' Given this situation, he adds that several features must be assessed in any study of population affinity. We agree with both points. Human population variation in dental morphology is mostly quantitative rather than qualitative, and it is essential to consider as many variables as possible in microevolutionary and historical studies. However, his qualification seems unnecessary given that no single biological trait or gene divides the world's many populations in an historically valid way. Why should teeth be any different?

A. M. Brues (1977), whose own research interests do not involve teeth, devoted three pages to this topic in *People and Races*. Part of her discussion focuses on crown wear, dietary behavior, and caries in modern populations, but she also addresses variation in tooth size and third molar agenesis. Regarding dental morphology, she describes broadly the variation in shovel-shaped incisors and concludes 'The shovel incisor is as clear a racial marker as numerous traits often considered to be of primary racial significance' (Brues, 1977:136). She also remarks that Carabelli's trait 'in its more marked manifestations is virtually limited to Caucasoids' (Brues, 1977:137). The view that shoveling = Asians, and Carabelli's trait = Europeans, is deeply entrenched in anthropological thinking even though it is not entirely correct (chapter 5). The most telling comment made by Brues (1977:137) is 'Other variations of cusp patterns of teeth are interesting to specialists, but we will pass them over.'

Biological anthropologists who practice in non-dental branches of the subdiscipline have long adopted the tact of 'passing over' dental morphological traits. Shovel-shaped incisors have received sufficient recognition over the years and if any dental trait is mentioned in an introductory course, it is almost assuredly shoveling. In actuality, shoveling is a very useful trait, but we will show in the following chapters that it is only one of many morphological variables that exhibits a distinctive pattern of geographic variation.

Dental morphological traits do not vary without reason across the landscape in some higglety-pigglety fashion. Tooth morphology is part of the biological heritage that humans carry with them when they migrate, much like their blood group genes, fingerprint patterns, PTC taste reactions, and other biological traits. When human groups are isolated from one another for a period of time, their crown and root trait frequencies diverge to varying degrees, depending on population size and the extent and temporal duration of isolation. When divergent populations come in contact and interbreed, the resulting populations possess convergent morphological trait frequencies. In other words, these polymorphic features of the dentition behave like other biological variables that are used to assess population history and evolutionary process. Moreover, their observability in extant populations and availability in the archaeological and fossil record give them almost unique standing among biological traits in the study of short-term and long-term hominid evolution.

Goals and organization

With the expanding application of dental morphology to anthropological and historical problems, colleagues in anthropology and allied fields have commented that they lack the expertise to evaluate the 'dental evidence' proffered for a particular hypothesis or model. This volume is partly intended to meet this need by providing some background on teeth for non-dental specialists who are working on common historical problems. For example, one classic problem which cross-cuts many disciplines is the origin(s) of Native Americans. The three wave model for the peopling of the Americas developed by C.G. Turner II on the basis of tooth crown and root trait frequencies stimulated collaboration, discussion, and debate among not only physical anthropologists but also archaeologists, geneticists, linguists, and Native Americans themselves. This model is presented in a series of articles (cf., Turner, 1971, 1983a, 1984, 1985a, 1985b, 1986a, 1989; Greenberg *et al.*, 1985, 1986), but articles have space limitations, precluding a full explication of underlying methods and assumptions. In a book devoted exclusively to an anthropological perspective on dental morphology, we have the luxury of overcoming these limitations.

At the outset, we should emphasize this book is not a manual of dental anatomy. No details are provided on how to distinguish teeth from the upper or lower jaw, from the right or left side of a jaw, or the number of a tooth within a morphological class (i.e., M1 or M2 or M3). These are very important considerations in dental morphological research, particularly when dealing with 'loose' teeth (i.e., out of socket) recovered in archaelogical contexts. However, many human osteology and dental anatomy texts address tooth identification in great detail (cf., White, 1991; Schwartz, 1995). To master the basics, these works should be consulted in a laboratory where many loose teeth, skeletons, and/or dental casts are available for observation.

We restrict our attention primarily to the morphology of permanent teeth. This limitation was a practical decision, which in no way slights researchers who have devoted energy to the study of deciduous tooth morphology. The window of time for deciduous teeth is a narrow one, much narrower than that of permanent teeth, which may (or may not in some cases) last a lifetime. In skeletal series, there are typically fewer children available for study compared to adults. Students of the living often obtain impressions from 12- to 16-year-olds after all the deciduous teeth have been shed. In this age group, the pristine crown morphology of all permanent teeth, except the third molars, can be observed before environmental factors (wear, caries, etc.) take their toll. Given the relative availability of deciduous and permanent teeth, we estimate that 90% of the papers written on crown and root morphology focus on permanent teeth. Because of this differential emphasis, Kieser's (1991) volume on odontometrics in this Series was also limited to the permanent dentition.

As dental anthropologists who specialize in dental morphology, most of our research has focused on final phenotypes directly observable on casts or skeletons. Our first task is to provide an overview of tooth crown and root morphology and describe and illustrate many of the traits that dental anthropologists employ in population studies (chapter 2). The dentition, as a whole, is an integrated system that develops with great regularity under a strict set of genetic controls. The final product of this process, the complete permanent dentition, exhibits serial structures divided into morphological classes and mirror imagery between the two sides of a jaw. Teeth also show variable degrees of sex dimorphism in and interaction between tooth morphology, size, and agenesis. These biological considerations, intrinsically interesting in themselves, significantly influence the methods of quantifying trait variation in populations (chapter 3). Although the genetics of morphological trait expression is another biological consideration, this topic has received sufficient attention to warrant separate treatment (chapter 4). The first half of this book is a preamble to our primary anthropological interest – describing and interpreting the pattern of geographic variation in crown and root trait frequencies among recent human populations. This dental variation is characterized on a global scale for 23 secondary crown and root traits (chapter 5). Many workers concede that dental variables show significant differences among the major geographic divisions of humankind, but question their utility in assessing population relationships at lower levels of differentiation. In this regard, we discuss dental differentiation and population history at three levels below global analysis, i.e., continental, regional, and local (chapter 6). Variation is also analyzed and illustrated through tree diagrams and ordinations to assess dental macrodifferentiation on a global scale. These results are compared to the pattern of relationships derived from genetic and craniometric analyses (chapter 7). This discussion has some bearing on the controversy surrounding the place of origin(s) of anatomically modern humans.

While great strides have been made in dental morphological research over the past three decades, much of this progress has been in elucidating patterns of variation among recent human populations. Crown and root traits have great but largely unrealized potential in other areas of research such as paleoanthropology and forensic anthropology. Many questions remain unresolved, but the potential avenues for research in these areas are challenging and exciting (epilogue).

2 Description and classification of permanent crown and root traits

Terminology

Dental anatomy is a common ground shared by paleontologists, odontologists, dental anthropologists, geneticists, dentists, and other specialists whose work involves teeth. Paleontologists interested in long-term dental evolution have developed nomenclatures intended to impart phylogenetic information on homologies in crown structures. Dental anatomists and dentists, minimally concerned with homologies, use positional terms for major cusps and other features of the tooth crown that are unambiguous if somewhat cumbersome (e.g., mesiolingual cusp vs. protocone). With workers coming from many disciplinary backgrounds, papers on dental anthropology contain positional terms, paleontological terms, or a combination of both. For this reason, it is necessary to describe briefly each system of nomenclature.

Positional terms

In reference to the maxilla and mandible, the midline falls directly between the central incisors. As the human jaw is parabolic, with teeth arcing toward the midline in the canine region, the terms anterior and posterior are not always appropriate to describe the 'front' and 'back' of a tooth. Instead, mesial and distal are generally preferred to denote the part of a tooth closest to or farthest removed from the midline, respectively. When the terms anterior and posterior are used, it is in the context of anterior teeth, the incisors and canines, versus posterior teeth, the premolars and molars. For the 'sides' of teeth, that aspect of a tooth facing the tongue is lingual or medial. Terms for the outer surface of a tooth are more varied. For the incisors and canines, the surface in contact with the lips is labial, while the surface adjacent to the cheeks for premolars and molars is buccal. To avoid using two terms for the same surface, some authors use buccal for both anterior and posterior teeth or substitute the term facial. As with the labial–buccal dichotomy, the chewing surfaces of the crown are referred to as incisal for anterior teeth and occlusal for posterior teeth. Occlusal is

15

favored if only one term is applied to all teeth. Approximal or interproximal are synonymous terms that refer to the mesial and distal contact surfaces between adjacent teeth.

In describing the vertical components of a tooth, dental anatomists divide a tooth into three sections: upper third (incisal/occlusal); middle third (middle); and lower third (cervical). For roots, apical indicates a position toward the tip or apex of the root. Basal (toward the base) is a general term applied to a description of either crowns (= cervical third) or roots (= apical third). Roots, like crowns, are divided into three vertical sections: lower third (apical); middle third (middle); upper third (cervical). Periapical is a positional term for a structure or disease that occurs at, or around, the root tip. Using combinations of the above terms, anatomists specify the locations of particular structures on either the chewing or vertical surfaces of crowns and roots (Fig. 2.1).

Paleontological terms and cusp numbers

E.D. Cope (1874, 1888) and H.F. Osborn (1888a, 1888b, 1897, 1907) developed the tritubercular theory of dental evolution to explain how the simple conical teeth of reptiles ultimately gave rise, through cusp budding, to the complex multi-cusped molars characteristic of mammals. Osborn (1888b) proposed names for the three main cusps of the upper molars (trigon) and lower molars (trigonid) that he thought were homologues in the upper and lower jaws. The lingually positioned main cusp of the trigon was named the protocone while the buccally positioned cusp of the trigonid was called the protoconid; as the prefix proto- implies, these cusps were considered homologous to the primitive reptilian cone. The earliest tridentate (3-cusped) teeth exhibited subocclusal denticles, or small cusplet buds, on the mesial and distal crests of the protocone (-id). According to Osborn (1907), the triangular molar form of the trigon developed when the two denticles of the upper molar rotated buccally so the protocone assumed a lingual position in the dental arcade. For the trigonid, the primary cusp (protoconid) retained its buccal position while the two denticles rotated lingually. This theory of reversed triangles resulted in a naming system that had the protocone, paracone, and metacone forming the trigon with the protoconid, paraconid, and metaconid forming the trigonid. In primate evolution, the upper molar trigon was supplemented by an additional distolingual cusp, the hypocone, a derivative of the lingual cingulum. In the lower molars, the mesially positioned paraconid was lost, leaving the original trigonid with only two cusps, the protoconid and metaconid. A subocclusal talonid was a distally positioned addition to the reduced

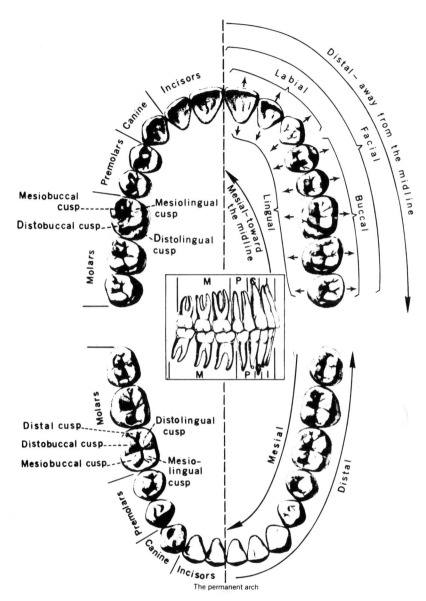

Fig. 2.1 Positional terms for the human permanent dentition. After Massler and Schour (1958), *Atlas of the Mouth*. Used with permission of the American Dental Association.

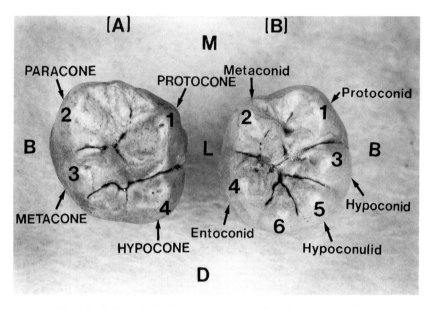

Fig. 2.2 Paleontological terms and cusp numbers for upper and lower permanent molars. (A) Upper right first molar, with four major cusps. (B) Lower right first molar, with five major cusps plus supernumerary cusp 6. M: mesial, D: distal, B: buccal, L: lingual.

trigonid; eventually, the talonid attained an occlusal position, exhibiting two main cusps, the hypoconid and entoconid, augmented in some taxa (e.g., hominoids and hominids) by a distal stylid, the hypoconulid. In 1916, Gregory numbered the cusps of the lower molar crown whereby: cusp 1 = protoconid, 2 = metaconid, 3 = hypoconid, 4 = entoconid, and 5 = hypoconulid. This numbering system corresponds to the proposed phylogenetic order of appearance of each cusp in the hominoid dentition. As the paraconid disappeared from the trigonid prior to the differentiation of the hominoids, it is not included in this numbering scheme. Numbers used to denote cusps of the upper molars are: cusp 1 = protocone, 2 = paracone, 3 = metacone, and 4 = hypocone (Fig. 2.2).

From the outset, the Osborn theory of trituberculy and associated cusp terminology was critiqued, revised and/or rejected by numerous workers, including J.L. Wortman, W.K. Gregory, G.G. Simpson, and P.M. Butler. Recently, Vandebroek (1961) and Hershkovitz (1971) proposed new terminological systems, especially for conules (-ids) and styles (-ids), that are intended to correct Osborn's misinterpretations of cusp homologies. There is now consensus among paleontologists and developmental biologists that the mesiobuccal cusp of both the upper and lower molars is homologous to the primitive reptilian cone. In recognition of this, Hershkovitz (1971:136) suggests that the paracone and protoconid be renamed the eocone and eoconid, respectively. All other major molar cusps retain their original Osborn name although the distally positioned hypoconulid is now the distostylid.

While Vandebroek (1961) and Hershkovitz (1971) make a stong case for renaming molar cusps on developmental and paleontological grounds to rectify aspects of the Cope-Osborn theory, their suggestions have had minimal impact on dental anthropologists. Osborn's terms and Gregory's numbering system for molar cusps have such a long history of usage that to change them now, especially in the context of human dental morphology, would generate more, rather than less, confusion. For example, Dahlberg (1950) used the term protostylid to describe a cingular derivative on the buccal surface of the protoconid. To adopt the new system, this trait would be 'the ectostylid of the eoconid.' Communication is a higher priority among human dental researchers than phylogenetic interpretations of cusp homologies that are generally recognized in any event, either implicitly or explicitly.

Most dental anthropologists go back and forth between positional terms, the Osbornian nomenclature, and the lower molar cusp numbers introduced by Gregory (1916) (e.g., distal cusp = hypoconulid = cusp 5). For a number of crown traits, Latin and English terms are used interchangeably for the same variable (e.g., *tuberculum dentale* = dental tubercle; *tuberculum intermedium* = cusp 7; *tuberculum sextum* = cusp 6). It would be difficult, and certainly cumbersome, to present all the terms used for the dental morphological traits we describe. Mizoguchi (1993), for example, lists 15 terms for one variable, Carabelli's trait (which is only one of the 15). Descriptions and illustrations provide sufficient detail to characterize the primary morphological features of human tooth crowns and roots without providing the entire suite of synonyms for each trait. For the sake of simplicity, we use cusp numbers for major supernumerary cusps.

Components and features of tooth crowns and roots

The dental anatomist O. Carlsen (1987) has written a textbook Dental Morphology, that diverges significantly from earlier works on dental anatomy. This book, which complements but is yet quite distinct from the present volume, provides a framework for systematizing descriptions of tooth crowns and roots. While our primary interest is in polymorphic

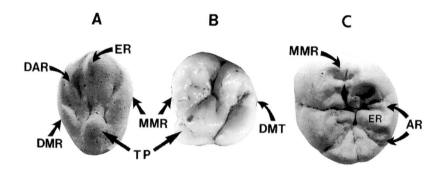

Fig. 2.3   Major morphological features of tooth crowns. (A) Upper right canine; (B) Upper left first molar; (C) Lower left first molar. Basic components: lobes and ridges (ER: essential ridge of essential lobe; AR: mesial and distal accessory ridges on cusp 4 [entoconid] of C; DAR: distal accessory ridge on A), marginal ridge complex (MMR: mesial marginal ridge on A, B, and C, and DMR: distal marginal ridge on A), and tuberculum projections (TP) or cingular derivatives on A (*tuberculum dentale*) and B (Carabelli's cusp, DMT: distal marginal tubercle).

structural variations rather than normative tooth form, these variations are best understood in the context of the recurrent crown and root features described by Carlsen.

From an external or surficial macromorphological perspective, a tooth is divided into two primary components: the crown and the root, separated at the cervical enamel line, i.e., the border between the crown and root. These primary elements have unique morphological aspects so they are described separately.

Crowns

The primary tooth crown is made up of four secondary macroscopic units: lobes, marginal ridges, cingulum derivatives, and supernumerary coronal structures (Fig. 2.3). A lobe is a constant macromorphological unit divided into two sections, a buccal and a lingual, and three segments, a median or essential lobe segment and mesial and distal accessory lobe segments. For anterior teeth, the median and accessory lobe segments exhibit mamelons along the incisal edge that are most evident in newly erupted, unworn teeth. For the posterior teeth, a cusp is that part of a lobe segment with a free apex at the occlusal edge. Incisor and canine crowns are made up of three lobe segments while each lobe of the multi-cusped premolars and molars has three lobe segments. Typically, median lobe segments exhibit lingual (anterior teeth) or occlusal (posterior teeth) ridges. The less constant ridges on accessory lobe segments are mesial and distal accessory ridges. Carlsen's descriptive system accords well with Stein's (1934:126) observation that 'The most striking polyisomeric phenomenon of man's dentition is the general tendency toward trilobulation of surfaces and cusps.'

In Carlsen's terminology, the lobes of the posterior teeth are equivalent to the major cusps defined by other workers. For example, the protocone or mesiolingual cusp of the upper molars is viewed as a lobe with lingual and occlusal lobe sections and one essential and two accessory lobe segments. The essential lobe invariably displays an occlusal cusp with a free apex, referred to as an essential cusp. To avoid confusion, we refer to the entire lobes of the posterior teeth as cusps in accordance with traditional practice.

In contrast to lobes, the marginal ridge complex may or may not be evident on any given tooth. On all teeth, elements of the marginal ridge complex occur mesially and distally. For the anterior teeth, marginal ridges can be present on either the lingual or facial surfaces. For the molars and premolars, the marginal ridge complex is occlusal and approximal. For the upper and lower molars, the marginal ridge complex is more developed mesially than distally. A structural feature exhibiting a free apex on the occlusal surface of a marginal ridge complex is a marginal tubercle. Shoveling of the incisors is the exemplar of discrete traits of the marginal ridge complex.

The cingulum is a ledge at the cervical section of a tooth crown that can be manifest either lingually or buccally. In some early hominoids, living apes, and various other mammals, the cingulum is expressed as a well defined band of enamel with a distinctive groove separating it from the vertical surface of the cusp. In modern humans, the cingulum is most evident on the cervical third of the lingual surface of the upper anterior teeth where it takes the form of a basal eminence. On the posterior teeth, where it may or may not be visible, it serves as the point of origin for Carabelli's trait and other discrete morphological features. Traits derived from the cingulum are found most often on the lingual surfaces of the upper incisors, canines, and molars and the buccal surfaces of the lower premolars and molars. Cingular derivatives, as a rule, do not occur on the labial surfaces of the incisors and canines, on the lingual surfaces of the upper premolars, or on the lingual surfaces of all teeth of the lower jaw.

Some crown traits, such as odontomes or premolar occlusal tubercles, are neither cingular derivatives nor parts of marginal ridge complexes. Moreover, they cannot be defined in the context of lobe segments and ridges. For this reason, Carlsen (1987) refers to such traits as supernumerary coronal structures that usually take the form of accessory occlusal tubercles. Most of these structures are found on the premolars and molars.

The secondary components of a tooth crown are divided from one another by grooves of varying depths, the most pronounced and regular of which are sulci. A deep rounded or angular depression on an occlusal surface where two or more grooves meet is a fossa. The groove and furrow patterns of the premolars and molars, which exhibit high levels of complexity and variability, have served as a focus for Russian dental anthropologists to define a wide range of discrete morphological traits.

The cervical enamel line, or cemento-enamel junction, is the boundary between the crown and the root complex. This line, when viewed from mesial, distal, buccal, and lingual aspects, shows different contour patterns for each tooth type. Incisors and canines, for instance, show lines that are convex on the buccal and lingual aspects but are concave mesially and distally. The premolars often have slightly concave lines viewed mesially and distally while the buccal and lingual surfaces can be convex, concave, or nearly horizontal. The molars often have lines that are approximately horizontal, with only slight convexity or concavity. One noteworthy morphological feature of the cervical enamel line is an enamel extension that is expressed on the buccal and lingual surfaces of upper and lower molars and, more rarely, on the buccal surfaces of the upper premolars. In general, enamel extensions are associated with root bifurcations and pronounced developmental grooves.

The root complex

According to Carlsen (1987), the primary elements of the root complex are root cones and supernumerary radicular structures. For consistency with other publications, we use the term 'radical' rather than 'cone' (Turner, 1991; Turner *et al.*, 1991) to refer to unseparated rootlike divisions. A root component is a combination of radicals that make up the primary roots of the molars. When a root has two or more radicals, the individual root elements may be completely or incompletely divided. In incompletely separated roots, developmental grooves delimit the boundaries of the radicals. When radicals are completely divided by a root bifurcation at some point along the total length of a root, the result is two or more separate roots (Fig. 2.4). Upper incisor roots are made up of one, two, or three radicals, with two the modal number. Lower incisors and upper and lower canines usually have roots made up of two radicals, one labial and one lingual. Except in rare instances, only developmental grooves mark the boundary of the radicals so these anterior teeth are mostly single-rooted. The lower canine is the only anterior tooth that shows completely

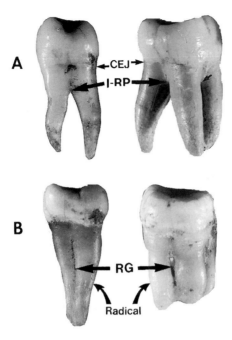

Fig. 2.4   Major morphological features of tooth roots. (A) Premolar and molar with roots separated by inter-radicular projections (I-RP), or root bifurcations, resulting in multi-rooted teeth. (B) Premolar and molar with root grooves (RG) separating radicals but no complete bifurcation, resulting in single-rooted teeth (made up of two or more radicals). CEJ: cementoenamel junction.

separated radicals, and hence two-rooted teeth, in polymorphic frequencies (i.e., > 1.0%).

Upper premolars roots are made up of two to four radicals. When there are two root radicals, they are located buccally and lingually. With three radicals, location is mesiobuccal, distobuccal, and lingual. When only developmental grooves delimit the radical boundaries, these teeth are single-rooted. When the radicals are bifurcated, these teeth may have two or three roots. For lower premolar roots, there are also two to four radicals. When there are two radicals, these are located buccally and lingually. When there are three radicals, their positions are medio-mesial, lingual and buccal. With four radicals, the buccal radical is divided into mesiobuccal and distobuccal components. As for separation structures, teeth with two radicals are usually single-rooted with only developmental grooves delimiting the radical boundaries. Roots with three or four radicals sometimes exhibit bifurcations resulting in two separate roots.

As with crowns, the roots of the molars exhibit the greatest amount of complexity. For the upper molars, the three primary root components are mesiobuccal, distobuccal, and lingual. The mesiobuccal component is made up of three radicals while the distobuccal and lingual components are generally made up of two. Lower molars have two primary root components, a mesial and a distal. Both root components are generally comprised of two radicals (buccal and lingual) although the mesial component may sometimes show a third medio-mesial radical. For both the mesial and distal root components, there may be slight separation of the radicals near the apex of the root, but this is generally not significant enough to warrant the label of three or four rooted teeth. The morphological variants of interest in the molar root complex take two forms. First, the root components may fail to differentiate completely so they are separated by only developmental grooves. This results in one- or two-rooted upper molars and one-rooted lower molars. The second type of molar root variation occurs in the form of accessory roots that are additions to, and not part of, the primary molar root components. A three-rooted lower first molar, with an accessory distolingual root, is the most common polymorphism involving an accessory molar root.

Trait descriptions

How are dental traits defined within the framework of the basic components of human tooth crowns and roots just described? As our concern is with external morphology, we define a dental trait as a positive (e.g., tubercles) or negative (e.g., grooves) structure that has the potential to be present or not present at a specific site on one or more members of a morphological tooth class (or classes). The dental traits described herein can all be classified as present or absent, but in most instances there is intermediate variation in the degree to which a tooth expresses trait presence (e.g., slight to pronounced). For example, Carabelli's trait is always located on the lingual surface of the protocone of the upper molars. In any given individual, it can be present or absent on the first, second, and/or third upper molar. If present on all three upper molars, there are not three traits each expressed on a specific tooth but one trait expressed on three teeth. The lingual surface of the protocone may be completely smooth with no positive or negative structures (i.e., trait absence), or Carabelli's trait may be present in the form of an almost imperceptible elevation associated with a small furrow. The trait can also be a large cusp that rivals the hypocone in size. Some workers define two or more traits based on the variable manifestations of a single trait, while others define individual dental traits that are composites of more than one trait. We have tried to avoid such definitions, although one classic trait, lower molar groove pattern, may partly reflect multiple interacting traits, as may *tuberculum dentale*, a very complicated trait whose morphogenesis and inheritance are poorly understood.

Researchers use a wide variety of synonyms in their descriptions of 'dental traits.' General terms employed that are considered equivalent to 'trait' include: character, characteristic, variant, variable, feature, attribute, anomaly, and polymorphism. The most commonly used adjectives to distinguish these presence–absence traits from measurable traits include: discrete, nonmetric, morphological, discontinuous, quasicontinuous, threshold, minor, and secondary. If dental, crown, or root is inserted between any of the above adjectives and nouns, a plethora of phrases can be generated (e.g., nonmetric dental trait, discrete crown attribute, minor root variant, etc.). To avoid repetition, we refer to 'dental traits' by several different phrases, but, for the sake of clarity, we limit our choices as much as possible.

More than 100 different morphological dental traits have been recognized in the human dentition. Of this number, 30 to 40 crown and root traits have been well defined, standardized, and subjected to detailed anthropological analysis. Pragmatically, workers deal only with a select subset of this smaller array of traits. At this point in time, there is no standard battery of traits described in dental morphological analyses, although the protocol adopted by the Smithsonian Institution may help rectify this situation in the future. Most commonly described are incisor shoveling, Carabelli's trait, cusp number of the upper and lower molars, and groove pattern of the lower molars. Beyond this meager list, common ground dissipates rapidly and varies by the inclination and experience of the individual observer and the nature of the study collection (e.g., root traits cannot be observed on plaster casts).

In the following sections, we describe and illustrate 35 nonmetric dental traits, most of which are used in the Arizona State University dental anthropology system (Turner *et al.*, 1991). Descriptions are organized by tooth class with upper and lower molars treated separately because of their greater morphological complexity. For each tooth class, there is a further subdivision into crown and root traits. To conserve space, the following abbreviations are used in combination to denote specific teeth in each jaw: U = upper, L = lower; I = incisor, C = canine, P = premolar, M = molar; 1 = central or first, 2 = lateral or second, 3 = third. No distinction is made between left and right antimeres.

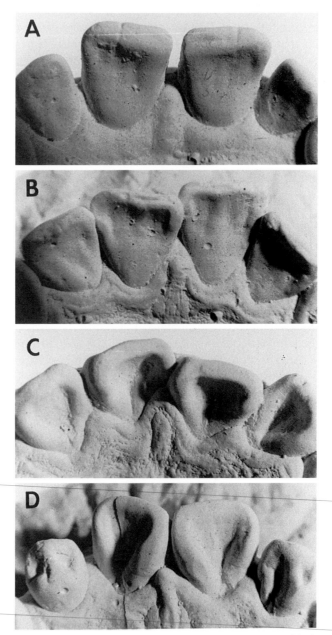

Fig. 2.5   Range of variation in shoveling of the upper central incisor. A and B show either no or trace shoveling while C and D exhibit pronounced shoveling. For A, there is a faint distal marginal ridge on the incisal third of the lingual surface but no corresponding mesial marginal ridge. (After Scott, 1991a. Used with permission from Academic Press.)

Incisors and canines

Crown traits

Shoveling (UI1, UI2, UC, LI1, LI2, LC; shovel-shaped incisors); Fig. 2.5 The distinguishing feature of this trait is the presence of mesial and distal marginal ridges on the lingual surface of the upper and lower anterior teeth. In his classic paper on shovel-shaped teeth, Hrdlička (1920) emphasized the presence of a lingual fossa surrounded by an enamel rim, but the fossa is a secondary reflection of marginal ridge development. In a shovel-shaped tooth, the marginal ridges extend from the incisal edge to the basal eminence, although, in pronounced cases, the ridges may converge on the eminence. Most workers evaluate the expression of both mesial and distal marginal ridges to score a single trait, shoveling. The two marginal ridges may, however, exhibit different degrees of expression (Mizoguchi, 1985). This fact is most critical in scoring slight expressions of shoveling as one marginal ridge may be present while the other is absent. Still, mesial and distal marginal ridge expression are so strongly correlated that little information is lost when they are considered together as one trait.

The most commonly used classification for scoring incisor shoveling has been Hrdlička's (1920) graded scale that includes the categories of absence, trace-shovel, semi-shovel, and full shovel. This classification was adopted by Dahlberg (1956) in the three-dimensional reference plaques distributed by the Zollar Dental Laboratory. Recently, some workers have employed a seven-grade scale for shoveling that provides finer subdivisions of Hrdlička's semi- and full shovel categories (Turner *et al.*, 1991). Population characterizations should focus on trait expressions of the upper incisors, especially UI1.

Double-shoveling (UI1, UI2, UC; labial marginal ridges); Fig. 2.6 Double-shoveling refers to the development of mesial and distal ridges on the labial surfaces of the upper anterior teeth (Dahlberg and Mikkelson, 1947). In most instances, the mesial labial ridge is more pronounced than the distal labial ridge. In fact, some upper incisors express a distinct mesial ridge with no corresponding distal ridge, a condition Snyder (1960:362) called 'three-quarter double shovel-shape' or 'mesial-ridged labial.' Although variably expressed on the upper anterior teeth, double-shoveling is most common and pronounced on UI1.

Tuberculum dentale (UI1, UI2, UC; lingual tubercles, cingular ridges); Fig. 2.7 These cingular derivatives are expressed on the lingual surfaces of the upper anterior teeth (rarely on lower anterior teeth) as ridges and/or

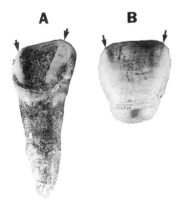

Fig. 2.6   Double shoveling of the upper central incisor. (A) Well-developed mesial and distal labial marginal ridges. (B) Weakly developed labial marginal ridges. Note that when double shoveling is faintly expressed, the labial ridges are restricted primarily to the incisal third of the crown. When the trait is pronounced, the ridges extend from the incisal third to the cervical third.

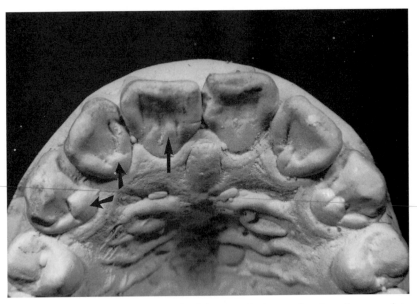

Fig. 2.7   *Turberculum dentale* of the upper incisors and canines. Arrows point to three moderate to pronounced tuberculum projections on the right upper central incisor, upper lateral incisor, and upper canine but the projections are also evident on the left antimeres (showing some asymmetry in expression).

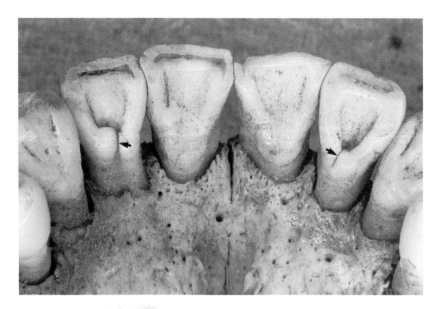

Fig. 2.8 Interruption grooves of the upper lateral incisors. Both upper lateral incisors exhibit interruption grooves (*dens invaginatus*), but expression is more pronounced on the right antimere. Such grooves, which often involve the root to a greater extent than shown here, can be manifest on the upper central incisor, but less frequently than on upper lateral incisor.

tubercles. The upper central incisor exhibits one to several ridges of variable expression (i.e., finger-like projections), rarely attaining the status of a tubercle with a free apex. By contrast, the upper lateral incisor exhibits a wide range of cingular expressions that vary from small lingual ridges to large free-standing single or multiple tubercles. Classification of lateral incisor variants has proved difficult. For the upper canine, ridge and tubercle formation shows a wide range of variation.

Although expressed in somewhat different forms, the cingular derivatives of the three upper anterior teeth are interrelated. It has been difficult, however, to pinpoint the key tooth for these lingual tubercles. For population comparisons, a case could be made for any one of the three teeth although Turner *et al.* (1991) recommend focus be placed on UI2 expressions.

Interruption grooves (UI1, UI2; corono-radicular grooves); Fig. 2.8   These developmental grooves are expressed on the cingulum and lingual aspect of the upper incisor roots. They can vertically dissect the mesial or distal marginal ridges or occur around the mid-point of the basal eminence.

Fig. 2.9   Winging of the upper central incisors shown from lingual (A) and occlusal (B) views. Note the unusual adze-shaped upper lateral incisors.

Because this variable transects both the crown and root, it is more easily scored on skeletal remains and loose teeth than plaster casts. Population studies should focus on UI2 expressions.

Winging (UI1, LI1; mesial-palatal torsion, bilaterally rotated incisors); Fig. 2.9   This variable is only indirectly a crown trait. It is characterized by bilateral rotation of the distal margins of the central incisors (primarily upper centrals) so that, from an incisal view, the incisors form a V-shape with the base of the V oriented toward the palate (Dahlberg, 1959; Enoki and Nakamura, 1959). Winging reflects orientation rather than crown or root form. Enoki and Dahlberg's (1958) classification of this trait involves combinations of unilateral and bilateral winging and counterwinging. Counterwinging, however, is often caused by anterior tooth crowding and is of little utility in population studies. Focus is ordinarily placed on bilateral winging of the upper central incisors although lower central incisors occasionally express this trait as well.

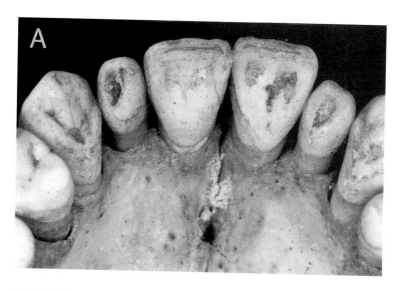

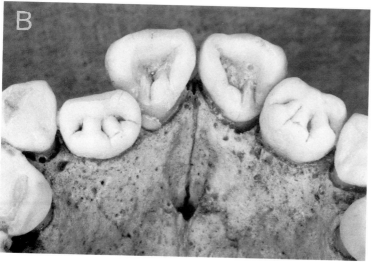

Fig. 2.10 Lateral incisor variants. Upper lateral incisors are highly variable teeth that exhibit a wide diversity of forms and morphological details. The variation may involve size (diminutive upper lateral incisor), form (conical or barrel-shaped UI2), or unusual development of the marginal ridge complex, basal cingulum, or both. Only two of many variant forms are illustrated here. (A) Diminutive UI2 (examine size relative to upper central incisor in MD diameter and crown height. (B) Exaggerated cingula of both UI2 antimeres with some involvement of the marginal ridge complex.

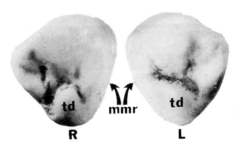

Fig. 2.11   Mesial canine ridge (Bushmen canine). The upper canines shown are antimeres. The left canine (L) exhibits a mesial marginal ridge (mmr) and *tuberculum dentale* (td) that are not divided by a developmental groove – this tooth would be scored as affected for mesial canine ridge. The right canine (R) exhibits a mesial marginal ridge and a *tuberculum dentale* separated by a developmental groove – this tooth would be scored as unaffected.

Lateral incisor variants (UI2; peg-shaped, diminutive, conical, tri-form); Fig. 2.10   Upper lateral incisors are more variable in size and morphology than any other tooth, and have the second highest frequency of congenital absence after third molars. Dahlberg (1956) divided lateral incisor variants into a series of categories with no rank order. Some of the major categories of variation include: peg-shaped or conical; diminutive (normal morphology but less than 3σ from mean mesiodistal diameters); and tri-form or T-form (narrow projection from cingulum that runs to incisal edge so incisal view gives appearance of the letter T). In addition to these variants, some lateral incisors are barrel-shaped (marginal ridges fold lingually and coalesce), a condition thought by some to be the most extreme form of shoveling. Others emphasize the cingular contribution to this phenotype. Although lateral incisor variants show an interesting pattern of geographic variation, they are typically rare in populations (0–5%) and their interrelationships remain obscure. At this point, their anthropological potential is only starting to be explored (Bailey-Schmidt, 1995).

Mesial canine ridge (UC; Bushmen canine); Fig. 2.11   The mesial lingual ridge is an almost invariant feature of the upper canines while tuberculum dentale is polymorphic. In some cases, a large *tuberculum dentale* coalesces with the mesial lingual ridge to form what Morris (1975) calls the Bushman canine. This fact is most evident when one antimere exhibits the 'Bushmen canine' while the other exhibits a large free-standing tuberculum projection. This trait is most common in African populations, especially the Bushmen, but it has been observed in populations from other geographic areas, in some cases falsely so (Irish and Morris, 1996).

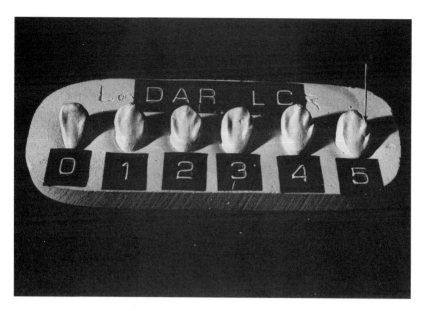

Fig. 2.12   Canine distal accessory ridge. Distal accessory ridges can be present on either the upper or lower canines. Illustrated is the 6-grade scale used to rank lower canine distal accessory ridge expression. An arrow points to the maximal form of trait expression on the lower canine.

Distal accessory ridge (UC, LC); Fig. 2.12   The lingual surface of the upper and lower canines commonly exhibits a median ridge and mesial and distal marginal ridges. Between the median ridge and distal marginal ridge is a polymorphic feature referred to as the distal accessory ridge (Morris, 1965; Scott, 1977a). This trait is more common and robust in the upper canines than the lower canines. In both teeth, it often lacks a dentinal component and is especially susceptible to crown wear. Observations should be limited to unquestionably unworn teeth. This is the most sexually dimorphic crown trait in the human dentition, with males showing significantly higher frequencies and more pronounced expressions than females. The dimorphism in the lower canine distal accessory ridge is associated, in part, with the relatively high sex dimorphism in lower canine tooth dimensions (Noss *et al.*, 1983a).

Root traits

Double-rooted lower canines (LC); Fig. 2.13   In nonhuman primates, lower canines typically have two roots. This condition is rare or absent in human populations, but it does attain a frequency of 5–10% in some

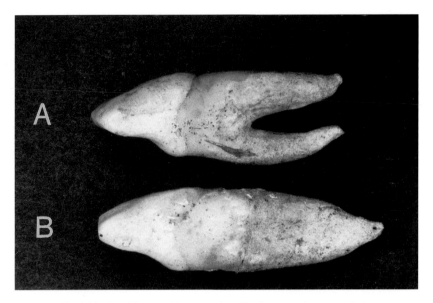

Fig. 2.13   Double-rooted lower canine. The lower canine normally has one root as illustrated by tooth B. In some instances, the radicals are completely separated, resulting in a two-rooted form, as shown by A.

groups. Viewed from the mesial or distal aspect, the two radicals of the lower canine root should be separated for at least one-fourth of total root length for a canine to be classified as two-rooted (Turner *et al.*, 1991). Ordinarily, 2-rooted lower canines show root division between one-fourth and one-half total root length, but, in rare instances, the radicals bifurcate close to the cervical enamel line (Alexandersen, 1962).

Premolars

Crown traits

Paracone/protoconid accessory ridges (UP1, UP2, LP1, LP2); Fig. 2.14   These ridges are located on the mesial and distal accessory lobe segments of both upper and lower premolars between the median occlusal ridge of the buccal cusp (paracone or protoconid) and the mesial and distal marginal ridges. Either the mesial or distal accessory ridge can be present or absent resulting in four possible phenotypes: mesial + distal accesory ridges present; mesial accessory ridge only; distal accessory ridge only; and both accessory ridges absent. The distal accessory ridge is more common than

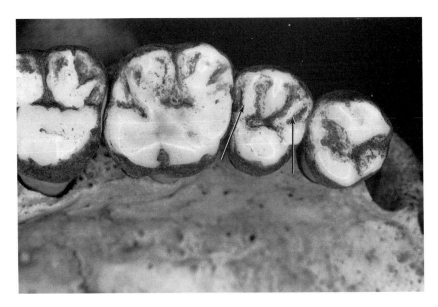

Fig. 2.14 Premolar accessory ridges. Accessory occlusal ridges occur on the buccal cusp of both upper and lower premolars. In this illustration, the upper second premolar exhibits both mesial and distal accessory ridges.

the mesial accessory ridge although the expression of the two ridges is not completely independent (Gilmore, 1968).

Premolar accessory ridges vary in degree of expression but no ranked scale has been developed to standardize this variation. In the past, scoring has been limited to presence or absence of the mesial and distal ridges. The focal tooth for this trait is unclear. Accessory ridges are more common on the upper than lower premolars and are more frequent on UP2 than UP1.

Accessory marginal tubercles (UP1, UP2); Fig. 2.15   The sagittal sulcus of the upper premolars often bifurcates as it approaches the mesial and/or distal marginal ridge complex. When the bifurcation involves two deep margino-segmental grooves, the result is a bulge or free-standing accessory tubercle on the marginal ridge. A premolar can express both mesial and distal tubercles, a mesial or distal tubercle, or no accessory marginal tubercle. No ranked scale has been developed for this trait although the tubercles vary in degree of expression.

Odontomes (UP1, UP2, LP1, LP2; tuberculated premolars, occlusal tubercles, dens evaginatus); Fig. 2.16   This trait takes the form of a cone-like

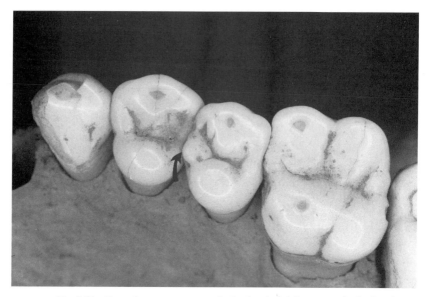

Fig. 2.15   Premolar accessory marginal tubercles. A large marginal tubercle, triangular in shape, is noted along the mesial marginal ridge of the upper second premolar.

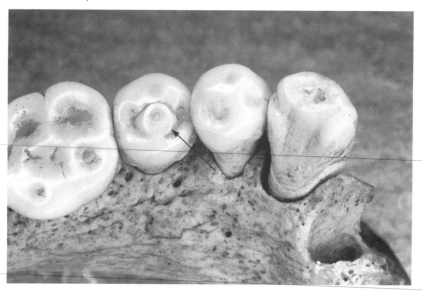

Fig. 2.16   Premolar odontomes. Occlusal tubercles, or odontomes, are found in close association with the median occlusal ridge of the premolars. On unworn teeth, odontomes are typically conical in form. The large odontome shown on the lower second premolar has been worn to the point where dentine is visible.

Fig. 2.17   Distosagittal ridge (Uto-Aztecan premolar). Only upper first premolars exhibit this rare geographically restricted trait. The distobuccal displacement of the buccal cusp and a fossa adjacent to the distal marginal ridge (arrow) are the key characteristics of the distosagittal ridge. Note also another form of upper lateral incisor variant.

projection that emanates from the median occlusal ridge of the buccal cusp of either upper or lower premolars. The tubercle has enamel and dentine components and involves a pulpal extension almost half of the time (Merrill, 1964). For that reason, breakage or rapid wear of the tubercle frequently leads to pulp exposure and periapical osteitis (Oehlers, 1956). Although the trait is most readily observed in adolescents, the base of the tubercle remains evident with a moderate amount of wear. This trait is scored as present or absent although there is some variation in size and form. It is unclear if there is a focal tooth for this trait. Given its rarity in all human populations, trait frequency can be estimated by dividing the total number of premolars exhibiting odontomes by the total number of observable premolars (upper and lower).

Distosagittal ridge (UP1; Uto-Aztecan premolar); Fig. 2.17   For most upper first premolars, the main axis of the paracone does not parallel the main axis of the protocone. Instead, its distal margin is displaced buccally relative to the protocone (Morris, 1981). The distosagittal ridge, or Uto-Aztecan premolar (Morris *et al.*, 1978), is distinguished by even more

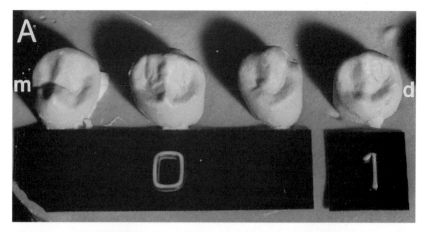

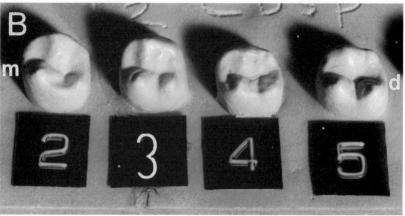

Fig. 2.18   Lower premolar multiple lingual cusps. In contrast to the stable lingual cusps of the upper premolars, the lower first and second premolars exhibit lingual cusps that vary in form and number. For lower second premolars, the teeth in row A exhibit single lingual cusps while those in B illustrate two distinct lingual cusps. In most cases, when multiple lingual cusps occur, the mesial cusp is larger than the distal cusp. m: mesial, d: distal.

exaggerated distobuccal rotation of the paracone in combination with the presence of a fossa at the intersection of the distal occlusal ridge and distal marginal ridge. This very rare trait occurs primarily in American Indian populations. It has never been observed in upper second premolars.

Multiple lingual cusps (LP1, LP2); Fig. 2.18   Upper premolar cusp number is basically invariant with one buccal and one lingual cusp. In very rare instances, the lingual cusp expands and is divided by a deep groove running in a buccolingual direction, the result being two lingual cusps with palpable apexes. Such three-cusped upper premolars have only been seen unilaterally. By contrast, multiple lingual cusps are common in the lower premolars which can exhibit one, two, three or more lingual cusps or cusplets. This trait has been difficult to quantify because the lingual cusp is expressed in such a variety of forms. The lower first premolar may have a single lingual cusp, lacking a free apex, fused to the paraconid. By convention, this fused 'cusp' is still scored as one lingual cusp. Kraus and Furr (1953) and Ludwig (1957), working with lower first and second premolars, respectively, count the number of lingual cusps that have independent apexes, no matter how slight. Following this operational definition, two small cusplets on the crest of the lingual cusp would be classified in the same category as a premolar with two large lingual cusps. An alternative classification is based on the observation that the primary lingual cusp of the lower premolars tends to have a mesial orientation relative to the buccal cusp. When accessory cusps are evident, they are usually distal to this primary lingual cusp. This additional cusp is typically smaller than the primary lingual cusp but is sometimes equal in size and, less frequently, exceeds the primary lingual cusp in dimensions. As lingual cusp variation differs on LP1 and LP2, separate ranked scales are required to classify the two teeth (Scott, 1973; Turner *et al.*, 1991). The lower second premolar is the focal tooth for multiple lingual cusps.

Root traits

Root number (UP1, UP2); Fig. 2.19   Upper premolar roots usually have one, two, or three radicals. A single-rooted upper premolar has no root bifurcation but may have developmental grooves showing the limits of the radicals. When the lingual and buccal radicals are separated by a root bifurcation, the result is a two-rooted premolar. For the rare three-rooted upper premolar, all three radicals show complete separation. Radicals may be divided close to the root apex or the bifurcation can extend as much as three-quarters of the root's total length. Turner (1981) recommends that a two- or three-rooted upper premolar should show radical division at a point no less than one-fourth of total root length. The focal tooth for this root variant is UP1.

Tomes' root (LP1); Fig. 2.20   Lower premolar roots may exhibit up to four radicals. Two-rooted lower premolars are found only in those cases where either three or four radicals make up the root complex. For premolars with three radicals, two-rooted variants arise when the lingual

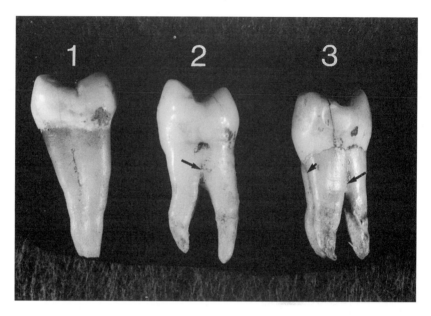

Fig. 2.19   Upper premolar root number. Arrows point to inter-radicular projections for two- and three-rooted upper first premolars. The one-rooted upper first premolar exhibits only a root groove. For the three-rooted tooth, the two radicals of the buccal root component are separated to produce two buccal roots.

radical is separated from the buccal and mediomesial root component. Two-rooted variants arise from four radicals when the mesiobuccal and mediomesial root component is divided from the distobuccal and lingual root component. Partially and completely divided roots of the lower first premolar are named after C.S. Tomes (1889) who is credited as being the first to define this trait in the human dentition. A lower first premolar with a Tomes' root does not show the distinct division of radicals that characterizes 2-rooted upper premolars. Lower second premolars rarely exhibit bifurcated radicals so they are not ordinarily scored for Tomes' root.

Upper molars

Crown traits (UM1, UM2, and UM3 unless otherwise noted)

Hypocone (cusp 4, distolingual cusp); Fig. 2.21   The hypocone is the distolingual cusp of the upper molars. Derived from the cingulum, this was the last major cuspal addition to the upper molar crown during the course

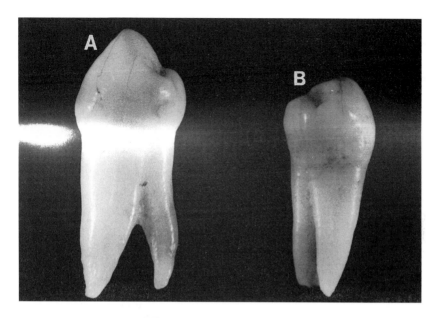

Fig. 2.20 Tomes' root. A and B are two-rooted lower first premolars. For A, an inter-radicular projection separates a small lingual root from a larger buccal root component. B exhibits the more typical expression of a Tomes' root.

of primate evolution (Swindler, 1976). It is also the upper molar cusp most often reduced in size and even lost in the later stages of hominid evolution. Most humans from all geographic areas retain this cusp on the upper first molar, but it may be reduced in size in some individuals. The second and third upper molars are decidedly polymorphic with respect to the hypocone, with groups showing highly variable frequencies of cusp size reduction and loss.

Cusp number of both the upper and lower molars has long been of interest to paleontologists and dental anthropologists. Early in the twentieth century, workers reported cusp number of the upper molars in terms such as 4–4–4, 4–4–3, 4–3–3 that characterized UM1, UM2, and UM3 for the presence or absence of the hypocone. Dahlberg (1951) refined this counting system by adding the categories of 4− and 3+ which indicated two stages of hypocone reduction between full cusp forms (4) and hypocone absence (3). Most dental morphologists have employed this four grade scheme to characterize the hypocone phenotype for the past 40 years. An alternative classification involves six grades of hypocone expression rather than the traditional four (Turner *et al.*, 1991). This classification includes a grade of expression neglected by most observers – that is, a distinct

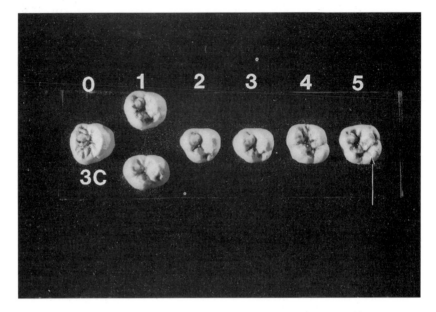

Fig. 2.21   Hypocone of the upper molars. The standard plaque for the
hypocone (Turner *et al*., 1991) shows the range of variation in the distolingual
cusp of the upper molars. An arrow points to the largest grade of hypocone
expression while 3C denotes a three-cusped upper molar lacking a hypocone.
For grade 1, the hypocone is indicated only by an outline on the protocone.
After this plaque was developed, it became apparent that the difference in
expression between grades 3 and 4 was too marked so another grade (3.5) was
added to smooth the transition.

hypocone outline on the distolingual aspect of the protocone that lacks a
free apex. This phenotype would be scored as absent following the
Dahlberg classification although the cusp is present, albeit markedly
reduced in size. As the hypocone is an almost constant feature of the upper
first molars, the focal tooth for studying hypocone variation is UM2.

Carabelli's trait (Carabelli's cusp; tubercle of Carabelli); Fig. 2.22   This
intensively studied morphological trait is a cingular derivative expressed on
the mesiolingual or lingual aspect of the protocone of the upper molars.
Carabelli's trait exhibits a wide range of variation in expression from small
ridges and associated grooves to large tubercles with free apexes. An eight
grade scale, including absence and seven degrees of trait presence, was
established by Dahlberg in 1956. This classification is the most widely used
for scoring trait expression although other authors propose ranked scales
with fewer grades (cf. Dietz, 1944; Shapiro, 1949; Kraus, 1951; Tsuji, 1958;

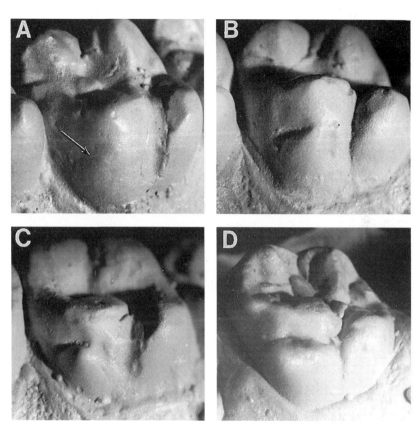

Fig. 2.22 Carabelli's trait of the upper molars. (A) The arrow points to an extremely subtle manifestation of Carabelli's trait that borders on 'threshold expression' (i.e., just barely visible); (B) a form of intermediate trait expression (grade 4); (C) a small tubercle with a free apex (grade 5); (D) a large cusp with a free apex that contacts the lingual groove separating the protocone and hypocone (grade 7). Because workers have scored intermediate grades of Carabelli's trait expression using different standards, we focus on distinct tubercle and cusp forms (grades 5–7) in our population comparisons.

Alvesalo *et al.*, 1975). All classifications take into account the gradation from minor grooves and ridges to large free-standing cusps.

On the mesiolingual surface of the protocone, an intersegmental groove separates the median lobe and accessory mesial lobe, running vertically from the occlusal edge to the cervical enamel line. Minor expressions of Carabelli's trait are often manifest as slight distal deflections of the cingulum around the midpoint of this groove (grade 1). At this same point, a pit can also be expressed (grade 2). When the grooves and ridges associated with this deflection are more pronounced, a Y-shaped formation

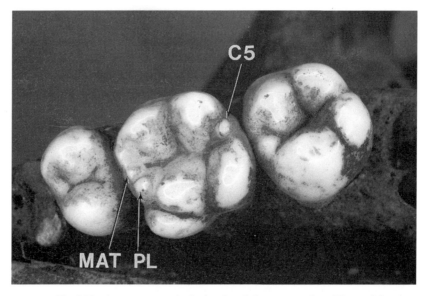

Fig. 2.23   Accessory marginal tubercles of the upper molars. The mesial and distal marginal ridge complexes of the upper molars can both exhibit accessory tubercles. On the upper first molar, there is an accessory tubercle on the distal marginal ridge (C5, or cusp 5) and two accessory tubercles on the mesial marginal ridge, a protoconule (PL) and mesial accessory tubercle (MAT). The mesial accessory tubercle is also expressed on the second molar (but not PL or C5). Two additional upper molar accessory tubercles, the mesial paracone tubercle and lingual paracone tubercle, are illustrated in Fig. 2.36.

occurs (grade 4). The three most pronounced expressions on the Dahlberg scale involve the formation of a tubercle with a distinct groove separating the structure from the protocone (grades 5, 6, and 7). The most pronounced expression, grade 7, exhibits a distinct free cusp apex, and its distal border comes in contact with the interlobal groove separating the protocone and hypocone. The upper first molar is the focal tooth for this variable.

In 1925, T.D. Campbell commented that the attention given to Carabelli's trait in countless publications went far beyond its biological significance. Despite this cautionary note, a few hundred more papers on Carabelli's trait have appeared since 1925. With this in mind, we beg the reader's forgiveness when we use (or overuse) this particular trait to illustrate general points. Carabelli's trait is to dental morphology what the ABO blood group system is to serology, the cephalic index to craniometry, or stature to anthropometry.

Cusp 5 (distal accessory tubercle, metaconule); Fig. 2.23   This trait takes the form of an occlusal tubercle on the distal marginal ridge of the upper molars. It is positioned between the metacone and the hypocone although it is more closely associated with the metacone. In appearance, it is often rounded or conical in form although it is almost triangular in more pronounced expressions. Two parallel margino-segmental grooves on the distal aspect of the upper molars delimit the buccal and lingual borders of this trait. For an accessory tubercle, this trait shows a relatively wide range of expression that was divided into absence and five grades of presence by Harris and Bailit (1980). This trait is most common on the upper first molar but often exhibits more pronounced forms of expression on the second and third molars. The focal tooth for cusp 5 is UM1.

Some terminological confusion has been associated with this trait. Cusp 5 has relatively wide usage (Turner *et al.*, 1991) but Harris and Bailit (1980) point out that Black (1902) and other dental anatomists also referred to Carabelli's cusp as the fifth cusp of the upper molars. For this reason, Harris and Bailit (1980) substitute the term metaconule for cusp 5. This term, introduced in 1892 by Osborn and Wortman, was considered preferable to the synonymous plagioconule of Vandebroek (1961) and Hershkovitz (1971). However, Hershkovitz (1971:135) illustrates the plagioconule (= metaconule) on the oblique ridge connecting the protocone and metacone, not on the distal marginal ridge. The only possible homologue to cusp 5 illustrated by Hershkovitz is the postentoconule. Metaconule, as used by paleontologists, refers to an occlusal tubercle between the protocone and metacone that is not part of the marginal ridge complex. Kanazawa *et al.* (1990, 1992) illustrate both the metaconule and cusp 5 (= distal accessory tubercle) on upper molar Moiré contourographs, and they are two different traits. Although the 'distal accessory tubercle' (Kanazawa *et al.*, 1992) or postentoconule might be just as appropriate, we view these as synonyms and retain cusp 5 to designate this trait.

Mesial marginal accessory tubercles (edge tubercles); Figs. 2.23 and 2.36 Traditionally, dental morphologists placed more emphasis on the single accessory tubercle of the distal marginal ridge of the upper molars (i.e., cusp 5) than on the multiple tubercles of the mesial marginal ridge complex. Kanazawa *et al.* (1990, 1992) have systematized observations on these tubercles which, in many populations, are more common than the distal accessory tubercle. The four primary tubercles defined by Kanazawa *et al.* (1990) involve the mesial accessory lobe segments of the paracone and protocone as well the mesial marginal ridge complex.

(1) Mesial paracone tubercle (expressed as independent part of the mesial accessory ridge of the paracone).

Fig. 2.24   Parastyle of the upper molars. Despite moderate crown wear, distinctive parastyles (arrows) are expressed on the buccal surfaces of the upper second and third molars. Forms of this magnitude are likely equivalent to paramolar tubercles.

(2) Protoconule (hypertrophied mesial accessory ridge of protocone with independent cusp tip).
(3) Mesial accessory tubercle (on mesial marginal ridge between mesial paracone tubercle and protoconule).
(4) Lingual paracone tubercle (of the four accessory tubercles, this is the only one which is not part of mesial marginal ridge complex; it is lingual and mesial to terminus of the median ridge of the paracone. Kanazawa *et al.* (1992) note this feature sometimes merges with either the mesial paracone tubercle or mesial accessory tubercle).

These four tubercles should be scored separately as independant variables. The focal tooth for observation is UM1.

Parastyle ('paramolar tubercle'); Fig. 2.24   This trait is a cingulum derivative expressed on the buccal surface of the paracone of the upper molars. In rare instances, it is expressed on the metacone of the upper molars and the buccal surface of the upper premolars. It is unclear if low grades of parastyle expression are related to Bolk's (1915) paramolar

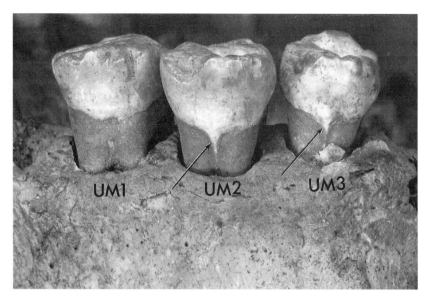

Fig. 2.25   Enamel extensions of the upper molars. These three upper molars (UM) illustrate the range of variation in enamel extension expression: UM1 exhibits no enamel extension, UM2 shows a distinct extension toward the root bifurcation, and UM3 exhibits a moderate extension.

tubercles. Some large paramolar tubercles are supernumerary conical teeth fused to the buccal surface of the upper molars. These tubercles often have their own root, are usually asymmetrical, and are most commonly found on the third molars, occasionally on the second molars, and very rarely on the first molars (Kustaloglu, 1962). The paramolar tubercles illustrated by Dahlberg (1945b) on the lower first molars are protostylids, a cingular derivative unrelated to paramolar structures (Dahlberg, 1950). Although the parastyle, or paramolar structures in general, may provide insights into dental evolution and development, the anthropological potential of this trait or complex of traits has yet to be demonstrated.

Enamel extensions (cervical enamel projections); Fig. 2.25   In contrast to other molar crown traits expressed on the occlusal surface or derived from the cingulum, enamel extensions are defined by contours of the cervical enamel line. The normative form of the cervical enamel line on the lingual and buccal surfaces of the molar crown is horizontal. An enamel extension is distinguished as a line directed apically toward the bifurcation of the roots (Pedersen, 1949; Lasker, 1950). These extensions can deviate from 1 to more than 4 mm from the horizontal axis of the cervical enamel line providing the basis for the three grades of presence (slight, moderate, pronounced) defined by Turner *et al.* (1991).

In contrast to most morphological crown and root traits, enamel extensions are expressed on both upper and lower molars and premolars. While enamel extensions can be buccal or lingual, most workers score their presence on the buccal surface of the molars. Turner *et al.* (1991) recommend scoring these lines on the upper molars although lower molar expressions should also be recorded. The focal tooth for enamel extensions is the first molar (upper or lower).

In both upper and lower molars, enamel extensions are sometimes associated with small spherical masses of enamel called enamel pearls. These are located at the point where the roots bifurcate and may or may not be in direct contact with the tip of the enamel extension. The frequency of enamel pearls in different populations is not known, but they appear to covary with enamel extensions. That is, they are most common in populations (e.g., Eskimos) who also have a high frequency of enamel extensions (Pedersen, 1949).

Root traits

One- and two-rooted upper molars; Fig. 2.26   Upper molars have three major root components. Variation from the norm for upper molar root number is not the presence of divided roots but the absence of one or two root bifurcations. A 2-rooted upper molar results from the incomplete separation of the distobuccal and lingual root components or the mesiobuccal and distobuccal root components. For 1-rooted upper molars, there is incomplete separation of all three major root components.

Most people retain three roots on their upper first molars. Upper third molars, by contrast, frequently show one- and two-rooted phenotypes. Given space restrictions toward the rear of the jaw, environmental factors can play a role in determining third molar root number. For these reasons, one- and two-rooted variants are scored on UM2.

Lower molars

Crown traits (LM1, LM2, and LM3, unless otherwise noted)

Cusp number (hypoconulid); Fig. 2.27   In living and fossil ape dentitions, it is uncommon to find lower molars with fewer than five major cusps (Gregory and Hellman, 1926; Swindler, 1976). During the later stages of hominid evolution, the primary trend in lower molar cusp reduction

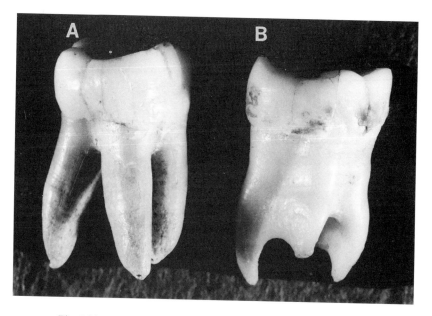

Fig. 2.26 Upper molar root number. Upper first molars (e.g., A) typically exhibit distinct inter-radicular projections in the cervical third of the root complex separating two buccal roots and one lingual root; the result is a three-rooted tooth. Upper second molars (e.g. B) have the same root components as the first molars, but inter-radicular projections may occur at the middle third or apical third of the root complex or be absent altogether; in such cases, the result is either a one- or two-rooted upper molar.

involves the loss of the distal (or distobuccal) cusp, the hypoconulid. The result is the classic 4-cusped lower molar of Gregory (1922) and Hellman (1928). Referring to a lower molar as having five or four cusps masks a range of variation from no reduction in the size of the hypoconulid through varying degrees of cusp reduction to total loss of the cusp. A six-grade ranked scale has been developed to score the degree of hypoconulid reduction more precisely (Turner *et al.*, 1991).

Most humans still exhibit five cusps on their lower first molars. Lower third molars frequently show five cusps although the form of the major cusps may be atypical. The focal tooth for the hypoconulid is LM2 which shows the greatest populational variation in cusp number. However, 4-cusped LM1, while much less common than 4-cusped LM2, also shows an interesting pattern of geographic variation.

Groove pattern (Dryopithecus pattern); Fig. 2.28 The Dryopithecus pattern of the lower molars, evident among fossil and extant hominoids

Fig. 2.27   Hypoconulid of the lower molars. The hypoconulid, or cusp 5, of the lower molars is illustrated in Figs. 2.2, 2.28, and 2.29 so the lower molars in this plate illustrate absence rather than presence of the hypoconulid. In this dentition, the lower first and second molars are both 4-cusped teeth. In our population comparisons, we focus on 4-cusped lower molars.

and retained to varying degrees in modern humans, was defined early in this century by W.K. Gregory (1916). This pattern is determined by occlusal groove configuration and the manner in which the major cusps of the lower molars come in contact at the central fossa. In describing the Dryopithecus Y pattern, Hellman (1928:159) noted 'As the two outer grooves extend from the outer toward the inner part of the tooth, they converge meeting the inner groove at about the center of the crown. The figure thus formed by the three grooves resembles quite closely that of the letter Y, the outer furrows forming the diverging limbs and the inner the stem.' As several apparent Y patterns are evident in the system of sulci on the lower molar crown, Hellman (1928) added the criteria that cusps 2 and 3 have to be in contact to form the Y pattern. By contrast, Hellman used + − pattern to denote lower molars that showed either a cruciform configuration (all major cusps contact at a single point) or contact between cusps 1 and 4.

Jørgensen (1955) distinguished the cruciform pattern (still +) from the configuration where only cusps 1 and 4 are in contact, which he specifies as an X pattern. Workers now use this modified three class system (Y, +, and

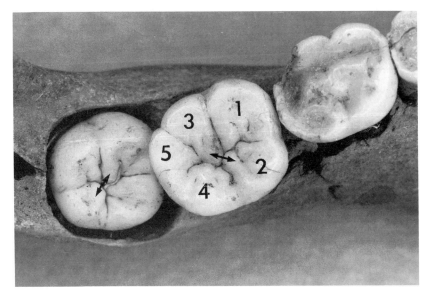

Fig. 2.28   Groove pattern of the lower molars. Gregory's famous *Dryopithecus* Y5 pattern is evident on the lower first molar. The cusps are numbered to emphasize the point that lower molar groove pattern is determined by cusp contact. When cusps 2 and 3 are in contact (as arrowed), the result is a Y pattern. Contact between cusps 1 and 4, shown by the partially erupted lower second molar, results in an X pattern.

X) to classify groove pattern expression. Most populations still retain a high frequency of the Y pattern on the lower first molar. For this reason, the focal tooth for characterizing groove pattern is LM2.

Although Gregory, Hellman, and other early workers reported frequencies for groove pattern and cusp number in terms of combinations, such as Y5, +4, etc., these variables are independent and should be reported separately. Hellman (1928) was correct to note evolutionary trends in both cusp number reduction and changes in groove pattern configuration, but they are not associated (Garn *et al.*, 1966a).

Cusp 6 (tuberculum sextum, entoconulid); Fig. 2.29   This supernumerary cusp is positioned on the distal aspect of the lower molars between the hypoconulid and entoconid. Slight expressions are indicated by a shallow occlusal groove toward the lingual boundary of the hypoconulid. In most instances, cusp 6 is only a quarter to a half as large as the hypoconulid. On occasion, it is larger than the hypoconulid. Turner's (1970) six grade classification for this trait focuses on the size of cusp 6 relative to cusp 5

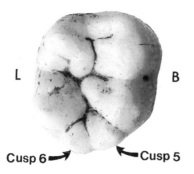

L   **B**

Cusp 6 ➡   ⬅ **Cusp 5**

Fig. 2.29   Cusp 6 of the lower molars. This lower right first molar illustrates a commonly occurring manifestation of cusp 6 (grade 2 in ASU dental anthropology scoring system). Cusp 6 is about half as large as the adjacent cusp 5. L: lingual; B: buccal.

(hypoconulid) with the higher grades on the scale showing relatively larger cusp 6 expressions.

There is some question as to whether cusp 6 can be scored when the hypoconulid is absent. This is not often a problem on LM1 where the hypoconulid is commonly present. However, for LM2, a distal cusp scored as the hypoconulid may actually be cusp 6 if it is derived from the distal lobe of the entoconid. Cusp 6 is most common on the lower first molar but is more pronounced on the second and third molars. Given the problems introduced by hypoconulid absence on the second molar, the key tooth for recording cusp 6 frequencies is LM1.

Cusp 7 (tuberculum intermedium, metaconulid); Fig. 2.30   This supernumerary cusp is expressed between the metaconid and entoconid of the lower molars. From an occlusal aspect, pronounced expressions of cusp 7 are wedge-shaped with the greatest breadth along the lingual border of the crown with the narrow tip extending to the central fossa. Distinct cusp 7 forms are easily scored, but there is some question as to what constitutes minimal trait expression. In the ranked scale developed by Turner (1970), grade 1A does not assume a wedge-shaped form but is represented by a cuspulid and associated lingual groove distal to the median occlusal cusp of the metaconid. Some authors distinguish this manifestation from the more typical cusp 7 forms, preferring the term post-metaconulid (Grine, 1981). A cuspulid referred to as the entostylid or pre-entoconulid, also associated with a lingual groove, may be found mesial to the tip of the entoconid. The entostylid is less common than the post-metaconulid and is not considered a manifestation of cusp 7.

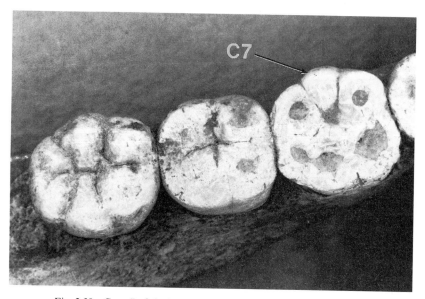

Fig. 2.30 Cusp 7 of the lower molars. A robust manifestation of cusp 7 is expressed on the lower right first molar. From an occlusal view, the form of this trait is triangular, or wedge-shaped (note: the 'smiley face' or 'jack-o-lantern' produced by the pattern of dentine exposure on the lower first molar is entirely accidental).

It is difficult to determine if cusp 7 and the post-metaconulid are separate traits or different forms of expression of a single trait. One point in favor of the former interpretation is the occasional presence of both cusp 7 and the post-metaconulid on the same tooth. However, if expression on the deciduous second molar portends that of the permanent first molar, the trait classified by K. Hanihara (1961) as cusp 7 generally takes the form of a post-metaconulid rather than a wedge-shaped supernumerary cusp. Until this problem is resolved, it is a straightforward matter to record cusp 7 on a scale that allows separation of the post-metaconulid (grade 1A) from typical cusp 7 forms (grades 1–4) when this division is desired. The focal tooth for making observations on cusp 7 is LM1.

Protostylid; Fig. 2.31 This cingulum derivative is expressed on the buccal surface of the protoconid. It is frequently, though not invariably, associated with the buccal groove separating the protoconid and hypoconid. Dahlberg (1950) was the first author to bring this variable to the attention of dental morphologists and show how it differed from Bolk's paramolar tubercle. Dahlberg (1956) later established an eight grade scale paralleling,

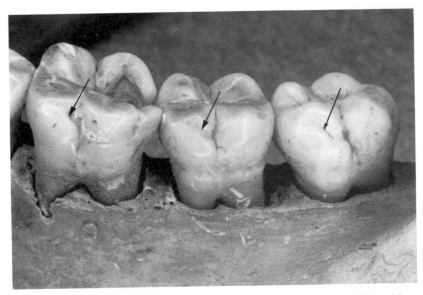

Fig. 2.31   Protostylid of the lower molars. This cingular variant, indicated by arrows, is expressed on the mesiobuccal cusps of all three lower molars. The protostylid, while showing some parallels to Carabelli's trait of the upper molars, is not expressed as commonly, nor to such pronounced degrees as its maxillary cingular counterpart.

to some extent, his standard for Carabelli's trait. He considered a pit in the cervical portion of the buccal groove to be the lowest degree of protostylid expression. His grade 2 was defined by a distal curvature of the buccal groove. Grades 3 through 7 involve forms of positive expressions up to the point where the protostylid has a free apex. The most pronounced degrees of protostylid expression, where the trait has a free-standing cusp with a palpable tip, are rare throughout the world.

The protostylid is expressed in highest frequency on the lower first molar although the degree of expression is often greater on the second and third molars. Some question remains on whether or not the buccal pit is part of the protostylid complex. In position and form, it is not comparable to the pit expression that is part of the Carabelli's trait complex. Still, using the Dahlberg standard, a worker can segregate buccal pits from phenotypes showing positive relief. LM1 is the focal tooth for observing this trait.

Deflecting wrinkle; Fig. 2.32   The median occlusal ridge of the metaconid often follows a straight course from the cusp tip to the central fossa. In some cases, however, this ridge assumes a more mesial position toward the

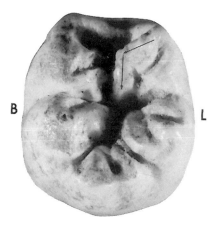

Fig. 2.32   Deflecting wrinkle of the lower first molar. Arrows indicate the course of the essential occlusal ridge from the cusp tip of the metaconid. In most lower molars, the essential ridge follows a straight path from the cusp tip to the central occlusal fossa. The deflecting wrinkle is distinguished by a change in course (deflection) of the essential ridge about halfway along its total length.

apex and then, about midway along its course, angles toward the central fossa. A straight ridge is considered normal (unaffected) while ridges that exhibit angulation are referred to as deflecting wrinkles (Weidenreich, 1937; Morris, 1970).

Turner *et al.* (1991) describe three grades of presence for the deflecting wrinkle although grade one does not involve deflection of the essential ridge and would be overlooked by most observers. Grades 2 and 3 show deflected ridges but differ in regard to how the terminus of the ridge contacts the entoconid. The deflecting wrinkle is found most commonly on LM1 so this is the focal tooth for observing this trait. It is rarely evident on second and third molars.

Anterior fovea (precuspidal fossa); Fig. 2.33   On the lower molars, the groove separating the protoconid and metaconid may run an uninterrupted course from the central fossa to the mesial marginal ridge complex. In some cases, however, there is a deep triangular depression distal to the mesial marginal ridge. The mesial marginal ridge and the mesial accessory (or median) ridges of the protoconid and metaconid form the boundaries of the triangle (Hrdlička 1924).

Although definitive expressions of the anterior fovea are common, the presence or absence of this trait may be dictated by independent features of the crown. For example, a well-developed mesial marginal tubercle may

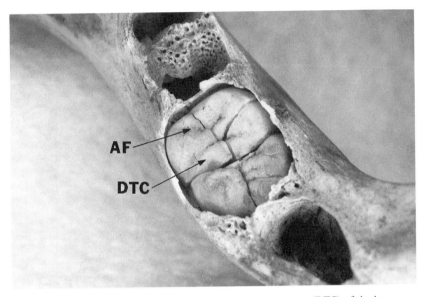

Fig. 2.33   Anterior fovea (AF) and distal trigonid crest (DTC) of the lower molars. Two lower molar crown traits are illustrated by this unerupted molar. The AF is the triangular depression distal to the mesial marginal ridge. The DTC is formed through some combination of metaconid and protoconid distal accessory ridges.

extend distally into the anterior fovea, obscuring its triangular form and the depth of the depression. Mesial accessory ridges on the protoconid and metaconid may likewise disrupt formation of a fovea. In this respect, the anterior fovea could be considered a composite morphological variable with presence dictated to some extent by the expression of variably expressed occlusal and marginal ridges. Although lower third molars sometimes exhibit large anterior fovea, the focal tooth for observing this trait is LM1.

Distal trigonid crest; Fig. 2.33   This trait is characterized by a crest or ridge that courses buccolingually along the distal aspect of the primitive trigonid, now represented by the protoconid and metaconid. It often appears as an extension of the distal accessory ridge of the protoconid although the distal accessory ridge of the metaconid may also be involved in forming the crest. This trait is most common on the deciduous lower second molar. As our emphasis is on the permanent dentition, LM1 is considered the focal tooth for scoring distal trigonid crest expression.

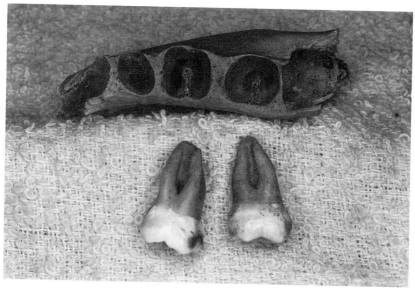

Fig. 2.34 Lower molar root number. Lower first molars usually show inter-radicular projections in the cervical third of the root complex that divide the mesial and distal roots, resulting in two-rooted teeth. When inter-radicular projections fail to differentiate the mesial and distal roots on either the buccal surface, lingual surface, or both, the result is a single-rooted tooth. This jaw has lower first and second molars with lingual inter-radicular projections but no corresponding buccal projections.

Root traits

Lower molar root number; Fig. 2.34   The normative root number for the lower molars is two. One-rooted lower molars result from incomplete separation of the mesial and distal root components. In some cases, this involves both the buccal and lingual root surfaces while, in others, either the buccal or lingual surface shows a root bifurcation while the other surface remains unseparated.

Single-rooted lower first molars are rare. Lower third molars, by contrast, are often one-rooted, with a root complex that is shortened vertically and aberrant in form due to space constrictions. For these reasons, characterizations of lower molar root number focus on LM2. To score a lower molar as two-rooted requires that the roots are divided for no less than one-fourth of total root length (Turner *et al.*, 1991).

Three-rooted lower molars (radix entomolaris); Fig. 2.35   This trait reflects the presence of a supernumerary root lingual to the distal root

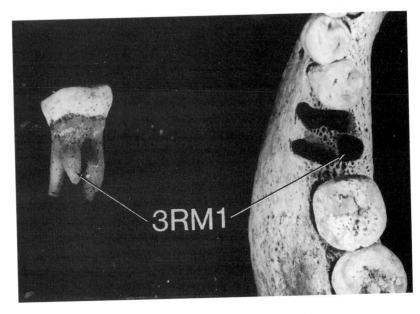

Fig. 2.35   Three-rooted lower first molars (3RMI). This supernumerary radicular structure is positioned lingually to the distal root of the lower first molar. This plate illustrates how 3RMI can be scored by examination of either the tooth or its socket.

component of the lower molars. Best known in the literature as 3-rooted lower first molars, or 3RM1, this extra root is conical in shape and generally smaller than the normally occurring radicals of the lower molar root components. This supernumerary root is most often expressed on the lower first molar but does occur occasionally on the second and third molars (Carlsen and Alexandersen, 1990). The focal tooth for population studies is LM1.

Problematic trait definitions

For morphological analysis, it is essential to define traits in such a way that they do not represent different manifestations of one trait or composites of multiple traits. Under our description of cusp 7, for example, we noted that the post-metaconulid and conventional supernumerary cusp form might represent two separate traits. In the classification we use, these phenotypic variants are distinguished from one another and can be treated together or separately (Turner *et al.*, 1991). When this question is resolved, data on the two forms of expression can be disentangled or merged. For the proto-stylid, buccal pits and positive cingular expressions may also represent different traits, but using Dahlberg's classification, these expressions can be separated as grade 1 vs. grades 2–7.

Some researchers report lower molar cusp number as either 4, 5, 6, or 7. These are composite numbers which represent three separate variables: the hypoconulid (cusp 5) and cusps 6 and 7. These three variables should be reported separately to avoid confusion. Alternatively, traits should not be defined and listed separately as 'cusp number less than 5 on molar 1' and 'cusp number more than 5 on molar 1' (Berry, 1976:261). A lower molar cusp number greater than 5 could include a 5-cusped tooth with cusp 6, cusp 7, or both cusps 6 and 7. This also applies to trait definitions that define the affected category as '5 cusps or more' on LM2 (Sofaer *et al.*, 1972a). Another composite trait definition is 'additional cusps on marginal ridge' of the upper molars, that includes extra cusps on either or both mesial and distal marginal ridges (Smith and Shegev, 1988:540). This single 'trait' could be made up of any combination of four separate traits, the protoconule, mesial paracone tubercle, marginal accessory tubercle, and cusp 5 (distal marginal tubercle).

Another methodological problem is when multiple manifestations of one trait are classified as two or more separate variables. For example, one article lists groove pattern Y (LM1), groove pattern X (LM1), groove pattern Y (LM2), and groove pattern X (LM2) as four characters. This list also includes Carabelli's pit present and Carabelli's cusp present as two separate characters (Berry, 1976). In either instance, one trait is considered as four or two separate dental variants, respectively. While researchers should provide information on the different expressions of a trait, it should be made clear that only a single trait is represented. Even when standard trait classifications are used, one should employ caution in defining what is meant by 'affected category.' For lower molar groove pattern, some workers define the affected category as 'Y or X pattern' while the non-affected is '+ pattern' (Sofaer *et al.*, 1972a:360). The sequence for groove pattern expression goes from 'Y' to '+' to 'X' so both ends of the pattern spectrum are treated as affected while the intermediate category is non-affected. This is equivalent to considering tall and short individuals as affected and medium individuals as non-affected for stature.

Methods of observation

Morphological variables of the human dentition are discrete insofar as individuals within a population express or do not express a particular trait. In another respect, these variables show quantitative variation in degree of expression – a trait may be but barely visible, or very large, or somewhere between these extremes. Attentive novices can be taught, in a short period of time, to observe traits that are distinctly or even moderately expressed. Observational difficulties are associated primarily with minimal levels of expression that fall on the boundary between trait absence and trait presence. As crown and root traits can be present or absent and variably expressed when present, different options are available to score these traits with regard to scale of measurement and recording method.

Scale of measurement

Tooth crown traits can be measured in three ways, corresponding to the three scales of measurement defined by statisticians: nominal (categorical data); ordinal (ranked data); and interval (metric data).

Nominal (absence = 0, presence = 1)

As most crown traits are, in one sense, all-or-none characteristics, a basic method of recording is by presence or absence. This was the first approach adopted in morphological studies, characterizing most data collected during the late nineteenth and early twentieth century. For example, upper molar cusp number was scored as either 4 or 3, based on the presence or absence of the hypocone, while lower molar cusp number was scored as 5 or 4, based on the presence or absence of the hypoconulid. Some workers today continue to dichotomize morphological variables, often citing their difficulties in scoring variable degrees of trait presence with consistency (cf. Sofaer *et al.*, 1972a, b).

Ordinal (absence = 0, presence = 1, 2, 3 ... n)

Hrdlička (1920) was one of the first workers to fully appreciate the limitations of identifying only the presence–absence dichotomy for dental morphological traits. His ranking procedure for shovel-shaped incisors defines four grades of expression: absence, trace-shovel, semi-shovel, and shovel. Other early dental morphologists were likewise not satisfied with limiting observations to presence or absence. For the upper molars, they were aware of the continuous gradation from a 4-cusped to a 3-cusped tooth. Some workers used multiple categories to describe the range of hypocone expression. For example, Campbell (1925) and Shaw (1931) used four categories to score upper molar cusp number (i.e., 4+, 4, 3+, 3).

Campbell's (1925) approach for most crown and root traits was to use graded categories and not just presence–absence.

Ranked standards take into account the variable expressivity exhibited by most tooth crown and root traits. These standards include trait absence but trait presence is subdivided into two or more grades ranging from slight to pronounced. To establish a scale, workers survey variation in trait expression across a broad array of populations, make a determination on minimal and maximal expression which serve as grade 1 and grade 'n', and insert intermediate expressions as grade 2 through 'n-1'. Ideally, the steps between grades should be approximately equal.

Following a ranking procedure, Dahlberg (1956) established permanent reference plaques for Hrdlička's shoveling classification and developed similar plaques for other morphological variables, including Carabelli's trait, the protostylid, and the hypocone. When these plaques were distributed internationally, this enabled workers to score trait expression more consistently. Influenced by the Dahlberg method of trait classification, K. Hanihara (1961) developed comparable plaques for morphological traits expressed on the deciduous teeth. In 1970, Turner developed ranked scales for lower molar cusps 6 and 7 which were the starting point for the ASU dental anthropology scoring system (Turner *et al.*, 1991). Over the past two decades, Turner and his students have set up ranked scales for a wide range of crown and root traits, which, in the tradition of the Dahlberg plaques, are distributed internationally. The ASU dental anthropology laboratory also distributes Dahlberg's original standard plaques.

Interval (absence = 0, presence = .10, .20, .30 ... n mm)

The majority of 'nonmetric' dental traits are not amenable to direct metric measurement because: (1) they exhibit variation in both form and size, not just size; (2) as three-dimensional objects, they often lack the kinds of landmarks that are needed to standardize measurements; and (3) they may show such slight levels of expression, there is no way to measure them in millimeters. The range and type of expression shown by Carabelli's trait illustrates all three of these problems and, for that reason, no one has successfully measured this feature on an interval scale.

Shovel-shaped incisors are, in part, an exception to the foregoing limitations of measuring trait expression. In 1947, Dahlberg and Mikkelson used a Vernier scale with a modified Boley gauge to measure the depth of the incisor lingual fossa in millimeters. In their Pima Indian sample, depth of the fossa relative to the lingual marginal ridges ranged from 0.3 to 2.6 mm for UI1 and from 0.1 to 1.5 mm for UI2. This work stimulated

K. Hanihara *et al.* (1975) and Blanco and Chakraborty (1977) to measure lingual fossa depth in Japanese and Chilean families so they could use metrical data to calculate intrafamilial correlations (chapter 4). This method is most useful when applied to Asian and Asian-derived groups with high frequencies and pronounced expressions of shoveling, although Aas and Risnes (1979a, 1979b) were able to measure shoveling in a Norwegian sample where the mean fossa depth was only 0.5 mm for UI1 and 0.3 mm for UI2.

In 1970, K. Hanihara and his colleagues developed an ingenius approach to metrically quantify the expression of the hypocone. This involved taking photographs of upper first molars at about natural size (perpendicular to the horizontal plane of the tooth) which were enlarged 5× in one-dimension or about 25× in two-dimensions. From these enlarged prints, they cut around the outline of the entire upper molar crowns and weighed them on an analytical balance accurate to 1/1000 g. Following this, the hypocone was excised from the enlarged crown and weighed. The quantitative variable used in their analysis was not 'weight' of the hypocone but the percentage of the total crown area taken up by the hypocone. Measured in this way, the distributions for the hypocone on UM1 and UM2 were approximately normal with slightly higher mean percentages for males. MZ and DZ twins were also compared for hypocone 'dimensions' in this study. Based on their results, this method of quantifying hypocone expression appears fundamentally sound but has not been widely adopted, perhaps due to the time and expense involved.

Analytical implications of measurement scale

In the analysis of dental traits, hypotheses are formulated and tested at the within- and between-group levels. Within-group analyses center on questions of sex dimorphism, degrees of asymmetry, inter-trait associations, and familial correlations (chapters 3 and 4). Between-group analysis focuses on population differences in trait frequencies and expressions, often involving the use of biological distance statistics (chapter 6). The type of data collected, whether categorical, ranked, or metrical, dictates which analytical tools are appropriate to assess a particular hypothesis.

For the analysis of geographic variation in dental traits, most authors focus on total trait frequencies that can be derived by observing presence or absence. The drawback to this method is that researchers do not always agree on what constitutes trait presence. Comparative data show that workers who score trait expression as present or absent have a tendency to discount lower degrees of expression which, in turn, lowers total trait frequencies. With dichotomous data, workers are limited largely to nominal scale statistics, principally chi-square for tests of difference and phi or tetrachoric correlations for tests of association.

If more crown traits were metrically quantifiable, this would be the best approach to data collection and analysis. However, excepting incisor shoveling and perhaps the hypocone, crown traits are not ordinarily amenable to this type of quantification. The best observational method at our disposal today is to score traits on ranked scales. These scales provide both qualitative and quantitative characterizations of tooth morphology. Observationally, the use of ranked scales stimulates greater caution in scoring low levels of trait expression that are often overlooked when traits are scored only as present or absent. With ranked scales, workers can assess the distribution of slight, moderate, and pronounced expressions and derive total trait frequencies and the frequencies of specific levels of expression. Analytically, ranked data allow the use of both nominal and ordinal scale statistics. Bending the rules slightly, workers have also achieved good results using parametric statistics.

Primary recording methods

Several media are available for scoring dental morphological traits. These traits can be observed directly on the teeth of human skeletons, on extracted teeth, and on the teeth of living subjects (intraorally). Observations once removed from actual teeth can be made on wax-bite impressions that show negative relief while observations twice removed can be made on plaster casts that show positive relief. If the impression and casting methods are followed with care, there is little loss of information from actual teeth. Skeletons and extracted teeth are the primary sources for the study of root traits. Some root variants (e.g., 3RM1) are also observable in X-rays of living subjects.

Irrespective of the materials at hand, whether real teeth or models of teeth, most observations on crown and root traits have been made through visual inspection. Given the minute size of slight trait expressions, high intensity lights and some form of magnifying glass (minimally 10 ×) are strongly recommended for visual observations. When scoring morphological traits, a worker has to take into account post-eruptive modifications to the tooth crown, particularly wear, caries, and chipping, before making a determination on trait absence or presence, and, when applicable, degree of presence. Some traits are particularly susceptible to wear not long after a tooth has erupted (e.g., canine distal accessory ridge) so a tooth should be examined from all angles under a bright light, providing various perspectives on shading, to detect the presence of wear facets. The protocone of the upper molars often wears down more quickly than the paracone and metacone and this could lead workers to score Carabelli's trait as absent when its status is unknowable due to wear. As a general rule for scoring crown traits, observations should not be made when there is any question about significant modifications to the specific location on the crown or root where a trait is expressed.

Alternative methods

The traits defined and illustrated in this chapter are a product of a dental anthropological tradition that began in western Europe in the latter half of the nineteenth century and then diffused to the Americas, Australia, Japan, South Africa, and India. The foundation has expanded during the twentieth century but the overall philosophy guiding dental morphological studies has not undergone much change. With respect to methods, dental morphologists have traditionally: (1) made observations on real teeth or models of teeth; (2) emphasized positive structures of the human tooth crown such as cusps, tubercles, and ridges; and (3) concentrated on presence–absence traits that can be reported by frequencies. Alternative approaches have been developed that diverge from one or more of these methods. While these alternatives have not been adopted widely, they do show promise as additional avenues of research in human dental morphology.

Odontoglyphics

With a primary focus on positive crown relief, relatively little attention has been paid to the complex system of grooves, fissures, sulci, and fossae that separate cusps and their components. As an alternative to the study of positive structures, Russian dental anthropologist A.A. Zubov and his colleagues adopted an approach to human dental morphological variation that places emphasis on negative relief. Zubov's position is that variation in the furrow patterns that separate the major cusps and segments of individual cusps show consistent variations in form, have a genetic basis, and exhibit patterned geographic variation. He notes 'This furrow pattern, on the whole stable but at the same time varying in details, is a valuable and interesting object of morphological study and may form a basis for a whole branch of odontology which, by analogy with dermatoglyphics, may be appropriately called odontoglyphics' (Zubov, 1977:269).

From a phylogenetic and developmental standpoint, odontoglyphics presumes that tooth crown structures have an inherent tendency to differentiate into three elements. Using Bolk's term 'odontomer,' which is any element of the crown having an independent calcification center, Zubov refers to the major cusps of the molars and premolars as primary odontomers. With sufficient differentiation, these primary odontomers divide into three parts called second-order odontomers (i.e., essential and accessory lobe segments and ridges). With maximal differentiation, secondary odontomers show a tripartite division, resulting in third-order odontomers.

In Zubov's system, it is not the odontomers but the furrows separating the odontomers that serve as the foundation for odontoglyphic trait classifications. The deep grooves that separate the major cusps of the crown are called intertubercular or first-order furrows (denoted by Roman numerals). The shallower grooves that pass over a cusp, dividing it into three segments, are second-order furrows (denoted by Arabic numerals). When second-order odontomers differentiate, the divisions are marked by shallow third-order furrows. Specific furrows and their variants are noted by a combination of Roman and Arabic numerals and abbreviations for the major cusps, following Osborn's terminology, e.g., III, 2me.

Following the principles of odontoglyphics, Zubov has defined a large number of furrow pattern types for the upper and lower molars. Of the 66 dental morphological variables Zubov and Nikityuk (1978) analyzed in MZ and DZ twins, 55 involved the presence of specific furrows and furrow triradii or diradii. In common with fingerprint patterns (i.e., whorls, loops, arches), these types are categorical or nominal, with no rank ordering when more than two variants are defined for a single furrow.

The development of odontoglyphics by Russian dental anthropologists has been dictated to some extent by their primary medium of study – wax bite impressions. These impressions are not used to make positive plaster casts but are observed directly. As wax bites reproduce the furrow system in great detail, it is understandable why the groove-fossa system of the posterior teeth became a primary focus of study. Despite its demonstrated potential, odontoglyphics has had only a minimal impact outside of Russia. There are several reasons for this: (1) furrow patterns are most evident on wax bite impressions, a medium rarely used outside of Russia; (2) the categorical nature of the trait classifications limits the kinds of statistical analyses that can be performed; (3) the sheer complexity of the

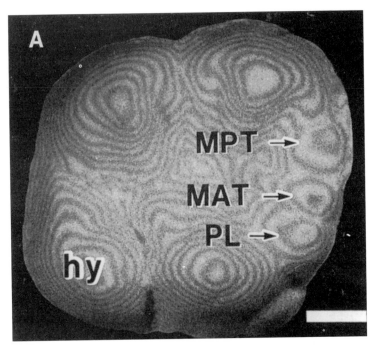

Fig. 2.36   Moiré contourograms of upper molar accessory marginal and occlusal tubercles. (A) A cusp or tubercle is scored as present when it is encircled by a continuous contour band, e.g. MPT (mesial paracone tubercle), MAT (mesial accessory tubercle), PL (protoconule). (B) (*opposite*) Upper molar occlusal tubercles include the metaconule (ML) and lingual paracone tubercle (LPT); hy: hypocone. After Kanazawa *et al.* (1990); used with permission from Wiley-Liss, Inc.

system of trait classification involving numerous types and subtypes is inherently daunting; (4) it is difficult for workers not trained in odontoglyphics to discern the relative importance of different variants; (5) most of the research on odontoglyphics has been published in Russian, so western workers have had limited exposure to this approach; and, importantly (6) its use is limited largely to the study of children with unworn teeth. Although hidden gems exist within the odontoglyphic system, wider adoption requires detailed documentation of specific crown furrow traits, involving an assessment of sexual dimorphism, asymmetry, inter-trait associations, familial correlations, and geographic variation. We should add that the development of odontoglyphics has not impeded Zubov and other Russian dental morphologists from observing thousands of individuals for positive crown structures such as shoveling, Carabelli's trait, molar cusp number, the distal trigonid crest, etc.

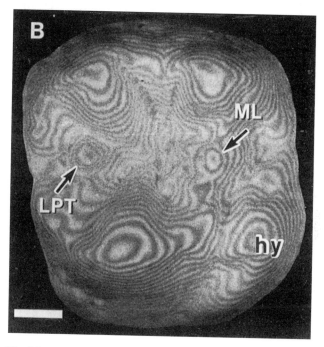

Fig. 2.36 (*cont.*)

Moiré contourography

A three-dimensional object such as a tooth should, in most cases, be observed directly rather than indirectly through a medium that reduces its structures to two dimensions. For this reason, the use of photographs to record the presence or absence of specific crown traits has met with little success. However, Japanese dental anthropologists have adapted Moiré contourography to the study of crown trait variation that circumvents ingeniously the original limitations of photographs.

The first step in Moiré contourography is to obtain a high contrast print under specified conditions for lighting, distance, and object position (e.g., tricuspal plane for molars) with the grating of a Moiré grid horizontal and perpendicular to the axis of the lens. The end product, a Moiré contourogram, resembles a topographic map (Fig. 2.36). It has regularly spaced contour bands at 0.2 mm intervals shaded alternatively black (or gray) and white which show the microrelief of the occlusal surface of a tooth crown, from its highest (cusp tips) to lowest (interlobal grooves and fossae) points. The major cusps of the molar crown show 5 to 12 enclosed contour bands extending from the cusp tip to an interlobal groove/fossa.

When applied to teeth, Moiré contourograms are useful for obtaining data on both odontometric (e.g., three-dimensional measurements) and nonmetric traits. For presence–absence variables, this method has been particularly useful for observations on small accessory tubercles of the marginal ridge complex of the upper molars. While these features have been described by previous workers through visual inspection (cf. Korenhof, 1960), their anthropological significance has never been fully demonstrated. The criterion for scoring these variables as present is precise and replicable; when there is at least one enclosed contour band at a specific trait location, a tubercle is classified as present (Kanazawa *et al.*, 1990, 1992). This method has also been applied to the analysis of traditional lower molar accessory cusps and ridges (i.e., cusps 6 and 7, deflecting wrinkle, transverse ridge) (Kanazawa *et al.*, 1989) and median and accessory cuspal ridges on the major cusps of the upper and lower molars (Sekikawa *et al.*, 1987a, 1987b).

Moiré contourography shows great promise for objectively quantifying the subtleties of tooth crown relief, including observations on discrete morphological variants. The extent to which this method will supplant the traditional approach of direct visual inspection is hard to gauge. This method, with its emphasis on microtopography, is particularly susceptible to the vagaries of crown wear – newly erupted teeth are the most suitable subjects for analysis so samples must be limited mostly to children. Even though Kanazawa *et al.* (1992) does score Carabelli's trait from contourograms, primary emphasis has been on the occlusal surfaces of the molars. Applying this method to assess relief on the lingual and buccal aspects of the tooth crown is more difficult as the points of orientation on these surfaces are not as readily standardized as those of the occlusal table of the molars. Finally, the Moiré method is more time consuming and costly than traditional methods although, under the right circumstances, the benefits would outweigh the costs.

Morphology and angles

Dental morphologists focus on structural details of tooth crowns and roots and pay less heed to other methods of characterizing tooth shape. One such method is to measure angles using major landmarks of the tooth crown. For example, upper premolars are rarely a focus of morphological studies as they exhibit a limited suite of presence–absence characteristics. However, as Morris (1981) shows, the main axis of the buccal and lingual cusps of the upper first premolars do not usually run in parallel. Relative to a straight line drawn through the sagittal sulcus separating the two cusps, the distal accessory lobe of the buccal cusp shows more buccal displacement from this line than the mesial accessory lobe. The angle between the two straight lines drawn through the sagittal sulcus and along the main axis of the buccal cusp may vary from 0 to 25 degrees. In a survey of 17 samples, Morris (1981) found this angle discriminated five American Southwest Indians samples (means: 9.4–10.8 degrees) from 12 European, African, Asian, and Pacific samples (means: 5.9–7.2 degrees). Although the geographic distinction of Southwest Indians is interesting, this angle does not discriminate among the other geographic groups.

Morris (1986) stressed both linear measurements and angles in his analysis of maxillary molar occlusal polygons. This polygon is formed by marking the cusp apexes of the paracone (A), protocone (B), metacone (C), and hypocone (D) and drawing four straight lines between AB, AC, BD, and CD. The polygon is characterized by four linear measurements and four angles. In one respect, this approach demonstrates how odontometric (measurements) and morphological (angles and shape) data can be assessed simultaneously. In the five samples Morris (1986) examined, the Papago differed from Bantu, Bushmen, Asiatic Indians, and Europeans in two of the four primary angles of the occlusal polygon. Angular differences among the latter four groups were relatively minor.

While Morris provides an alternative to the study of structural details of the tooth crown, more work on angular variation is required before this approach can be fully incorporated into dental morphological studies. Angles for both the upper first premolar and upper first molar show American Southwest Indians to be distinctive, but Europeans, Africans, and Asians do not differ among themselves. In some respects, this parallels the findings from odontometrics where absolute crown size often fails to distinguish major geographic groups. Although this approach, like Moiré contourography, is very time consuming, it nonetheless has potential for elucidating the variability of crown shapes, especially when multiple angles are analyzed in conjunction with intercuspal measurements.

Observer error

A primary methodological goal in science is to make systematic observations that can be replicated with accuracy by a single observer when the same materials are examined on two or more occasions. Moreover, the observations of one worker should coincide closely with those of another worker when both examine a common set of objects for the same variables. In other words, data should be obtained with a minimum of intraobserver and interobserver error. In studies of human dental morphology, some workers have commented that 'The scoring of morphological characters is known to be a subjective evaluation and therefore open to differences of interpretation, not only between observers, but also within a single observer from one scoring session to the next' (Sofaer *et al.*, 1972a:359). We grant there are problems in this area but also maintain that the problems have been overstated, they are not insurmountable, and their order of magnitude is no greater than that involved in blood typing where scales of agglutination are used.

Between those instances where a trait is distinctly present or distinctly absent, there are gray areas that dental morphologists do not universally agree upon. One problem is in discerning that point where trait expression is barely evident, a phenotype that has been referred to as a 'threshold expression.' Another difficulty is in evaluating cases where 'something' is manifest at a particular site, but this something may or may not represent the trait being observed. One worker could interpret this 'something' as trait presence while another views it as surficial 'noise', scoring the trait as absent. In other words, strength of trait expression is not the only problem. If it were purely a matter of size, an appropriate method of magnification could be used to settle disputes. While magnification might lead to more agreement, this would never be total. In some cases, there is no way to be absolutely certain if a trait is present or not.

In a study of three American Southwest Indian groups, Sofaer *et al.* (1972a) attempted to compensate for observer error by ranking traits relative to the percentage of concordance in scoring them consistently on two separate occasions. As they scored traits as either present or absent, cases of concordance included 00 and ++ while discordance was 0+ or +0. For ten crown traits, their concordance rates varied from 78 to 99% with a mean of 90%. The traits that posed the greatest problems in scoring were Carabelli's trait and lower molar groove pattern. Sofaer *et al.* (1972a) did not record trait presence in two or more categories, because they believed scoring traits in this manner would only increase the frequency of observer error.

In another study of Southwest Indian dental variation, Scott (1973) measured intraobserver error by scoring 50 Papago dentitions for 31 crown traits on two separate occasions. In contrast to Sofaer *et al.* (1972a), this author used ranked standards involving more than one grade of trait presence. Scott assessed intraobserver error in terms of presence–absence concordance and replicability of assigning ranks for degree of expression. His concordance values for scoring traits consistently on the basis of presence or absence ranged from 85.7 to 100% with a mean of 95.2%. For

29 graded variables, 1158 repeated observations led to 162 misclassifications by one rank (14.0%) and 20 misclassifications by two or more ranks (1.7%). Total replicability for exact rank assignment was 84.3%. Directional bias in assigning higher or lower ranks between scoring sessions was evident for only three variables. Spearman's rank order correlations between the two observational sets varied from 0.585 to 1.000 with a mean of 0.862.

The most detailed assessment of intra- and interobserver error in scoring tooth crown traits is the study of Nichol and Turner (1986). Nichol observed 50 dental casts of Kodiak Island Eskimos on three separate occasions for 45 crown traits, of which 37 were scored relative to graded reference plaques and eight were scored as present or absent. Turner observed the 45 traits on the same dental casts on a single occasion. To measure intraobserver error, Nichol's three observational sets were compared to one another (i.e., crn1–2, 1–3, 2–3). For interobserver error, each of Nichol's observational sets were compared to those of Turner (i.e., crn1–cgt; crn2–cgt, crn3–cgt).

Nichol and Turner (1986) utilized several measures to estimate the two types of observer error. First, they asked a question not often addressed in analyses of this type: what is the frequency of scoring a trait during one observational set that is not scored (same trait, same tooth) during a different scoring session? This situation occurred about 7% of the time for intraobserver scoring and 13% of the time between the two observers. This situation arises because dental morphologists are driven by two conflicting goals: maximize sample size by scoring as many individuals in a sample as possible, but be confident that trait scoring is accurate. Many teeth, unfortunately, are not in pristine condition but exhibit various sorts of environmental modifications (they can be worn or chipped and have caries or fillings); for dental casts, there is the possibility of errors being introduced by the casting process. Because of these complications, it is understandable why different decisions are sometimes made on whether to score a trait or not during two observational sessions.

For scoring differences between observational sets, Nichol and Turner (1986) examined variant scoring: (1) of any magnitude, (2) by more than one grade for ranked variables, and (3) by presence or absence. For variant scoring of any type, intraobserver error varied between 25 and 30%. Interobserver error of any magnitude was not significantly higher, averaging around 30%. Misclassification of ranked traits by more than one grade was much lower for both intraobserver (6–7%) and interobserver (6–10%) comparisons. Ignoring ranks and scoring traits just as present or absent gave intraobserver concordances of about 90% while interobserver values were slightly lower at 87%. These concordance values are similar to those of Sofaer *et al.* (1972a) and Scott (1973).

Several key points can be derived from Nichol and Turner's (1986) analysis of intraobserver and interobserver error. First, one-third of the total misclassifications occurred at the presence–absence breakpoint. As the authors note, 'it may mean that the greatest difficulties in making consistent scoring judgments occur at the trait threshold' (Nichol and Turner, 1986:307). Second, tooth crown traits are variable in terms of consistent observations; some traits are difficult to score the same way during different scoring sessions while others are observed with a high level of replicability. The traits most difficult to score were marginal accessory tubercles of UP1, the anterior fovea of LM1, *tuberculum dentale* of the upper anterior teeth, and the distal accessory ridge of the upper and lower canines. Traits with the highest scorability indexes included lower premolar lingual cusp number, cusp 6, double-shoveling, incisor shoveling, premolar odontomes, winging, the hypocone, the metacone, lower molar cusp number, cusp 5 of the upper molars, and the parastyle. Their remaining variables fell between these extremes. Their third point is the importance of experience in scoring crown traits consistently. Of course, this is true in all arenas of scientific investigation but tends to be understated or overlooked in the analysis of tooth crown and root morphology.

In making morphological observations, a worker is constantly involved in decision making. One must ponder a wide range of questions for any single observation: (1) Is the trait present or not? (2) Does the manifestation at the location where a trait is expressed represent presence, or is it something else? (3) The trait is clearly present but what rank should be assigned? (4) The site where the trait is expressed is modified to some extent by wear, filling, caries, or casting error, so should the trait be scored or not? If one errs, it should be on the side of caution. A general principle to follow is: when in doubt, don't guess!

Many articles on dental morphology have been published by relatively inexperienced workers whose specializations are not infrequently in some other area of dental, anthropological, or genetic research. For some reason, dental morphology attracts occasional enthusiasts who publish one or two papers on the subject, often abandoning their quest after concluding there is a lack of objectivity in scoring crown traits. What they fail to take into account is the experiential component of morphological studies and the many attempts to gain control over observer error.

It will probably never be possible to attain 100% concordance in replicated observations of tooth crown and root traits, either by single observers or between observers. The reference plaques developed by

Dahlberg, K. Hanihara, Turner, and others have enhanced observational precision, but they have not been a panacea for the reasons noted above (i.e., threshold expressions, post-eruptive modifications, surficial noise, varying levels of experience, etc.). Moiré contourography is a promising method that might help reduce observer error, given its precise definitions of trait presence, but its applications to date have been limited to the occlusal surfaces of unworn teeth. In his many excellent morphological studies, Y. Mizoguchi adopts the standard practice of comparing intraobserver error rates (double-observations) to measured differences between twins, sides of the jaw, and so forth. For traits observed with excessive error, he tempers his interpretations accordingly. This careful practice should be widely adopted.

Under ideal circumstances, observations on entire samples should be made on two separate occasions by two different workers. Observers could then assess the four sets of test/re-test data for any discrepancies. Those traits and specimens posing observational difficulties could be re-examined for final assignment of phenotypic expression, with one alternative being the exclusion of the trait or specimen from the sample. This approach is practical when workers focus on one or a few traits in a sample with only a few hundred individuals (e.g., Townsend *et al.*, 1992). However, either method would be wholly impractical if observations were made on many traits across a large number of samples, given the temporal and fiscal constraints of most research. Under these circumstances, the keys to gathering 'good' dental morphological data are: well defined trait definitions, the use of three dimensional reference plaques (when available), good lighting, some magnification, experience, double observations on sample subsets, and caution.

3 Biological considerations: ontogeny, asymmetry, sex dimorphism, and intertrait association

Introduction

In chapter 2, we focused on directly observable phenotypes of the tooth crowns and roots. In the next chapter, we discuss how studies of twins, families, and populations have shed light on the genetics of morphological trait expression. This intervening chapter on dental development and related topics is appropriately placed, given Butler's (1982:45) statement that the 'link between gene and character is ontogeny.' While it is beyond the scope of this work to review all the intricacies of dental ontogeny, we discuss in general terms how odontogenesis generates crowns and roots, major and minor cusps, ridges, fissure patterns, extra roots, and so forth. We also address organizing principles proposed to account for patterned morphogenetic gradients in the mammalian dentition with special reference to human crown and root traits.

Sections: 'Fluctuating asymmetry', 'Sex dimorphism', and 'Intertrait association' deal with three additional biological concerns of dental morphologists. Teeth in the two sides of the upper and lower jaws are bilateral structures that show mirror imaging (this is why isolated teeth can be determined as right or left antimeres). If the images were mirrored in all details, there would be perfect symmetry. Few anatomical structures expressed bilaterally (e.g., right and left hands; right and left ears) show complete symmetry; teeth are no exception. The asymmetry sometimes observed in crown and root trait expression has both developmental and methodological implications. Humans exhibit some degree of sexual dimorphism in all measurable body parts, including tooth size. But, do morphological traits show similar male–female differences or are they primarily monomorphic in frequency and expression? In the last section, we review interrelationships within and between crown traits, tooth size, and agenesis.

Although labeled 'biological considerations', this chapter is also concerned with methodology. The dental system, with repeated and bilateral

74

structures, allows considerable leeway in how workers score dental traits and derive population frequencies for genetic and anthropological studies. At the beginning of any study, workers have to determine how to deal with asymmetry, sex differences, and trait independence. We review the different approaches that can and have been followed by various workers and discuss our approach to dental morphological analysis. Our explicit theoretical position is that most variation in asymmetry, sex dimorphism, and other minor differences within individuals results from environmental factors.

Dental ontogeny

How do teeth form during development? As in many areas of comparative anatomy and embryology, the fundamental principles of dental ontogeny were outlined by nineteenth and early twentieth century researchers (e.g., C. Rose, O. Hertwig, A. von Brünn, W. Kükenthal, W. Leche, L. Bolk). More recently, *in vitro* and *in vivo* experiments with embryonic dental tissues have enhanced our understanding of the mechanisms and processes of odontogenesis. The literature in this field is substantial and technical, but there are useful reviews and texts for dental anthropologists, like ourselves, who are not developmental biologists (cf. Butler, 1956; Kraus and Jordan, 1965; Oöe, 1965; Gaunt and Miles, 1967; Townsend and Brown, 1981a; Ten Cate, 1994).

An embryo is made up of three basic germ layers, the ectoderm, mesoderm, and endoderm. Neural crest cells, a critical component of odontogenesis, may be considered a fourth primary embryonic tissue. All tissues and organ systems of the body can be traced ultimately to the primary germ layers and neural crest cells. Teeth are considered a skin organ along with feathers, scales, hair, and skin glands (Oster and Alberch, 1982). Like all skin organs, tooth development involves the action and interaction of the ectoderm and mesoderm. Enamel is of ectodermal origin while dentine, pulp, cementum, and periodontal fibers are of mesodermal origin. More specifically, enamel originates from a layer of epithelial cells lining the maxillary process and mandibular arch of the embryo. Dentine and pulp are derived from mesenchyme, a connective tissue with less regularly organized cells that underlies the epithelium. Oral mesenchyme derived from the neural crest is referred to as ectomesenchyme. Interactions between the epithelium and ectomesenchyme guide the entire process of odontogenesis, even prior to the appearance of the dental lamina, which is the first sign of 'teeth' in the embryo.

The mouth, or stomadeum, is evident early in embryonic development with its two main constituents, the maxillary process and the mandibular arch, derived from the first branchial arch (that goes back to ancient jawless fish). Tooth germs develop from the dental lamina, an ectodermal derivative from the surfaces of the maxillary process and mandibular arch. In humans, the horseshoe-shaped dental lamina appears during the second month *in utero* and exhibits ten sites of epithelial swelling in each jaw corresponding to the individual teeth of the deciduous dentition. Between the fifth and tenth months *in utero*, the free end of the lamina associated with each tooth germ develops successional lamina for permanent teeth positioned lingually to the growing deciduous tooth germ. This accounts for permanent incisors, canines, and premolars (which succeed the deciduous molars). Lacking deciduous precursors, development of the permanent molars involves a posterior extension of the lamina from the distal aspect of the deciduous second molar tooth germs. Initiation of permanent molar tooth germs extends over a five-year period, from four months *in utero* for first molars to the fourth or fifth postnatal year for third molars (Bhussry, 1976).

Oral biologists define six morphological stages of tooth growth prior to mineralization and eruption: (1) dental lamina; (2) bud stage; (3) cap stage; (4) early bell stage; (5) late bell stage; and (6) enamel and dentine matrix formation. Dental ontogeny is characterized by the physiological processes of initiation, proliferation, histodifferentiation, morphodifferentiation, apposition, calcification, and eruption. There is some correspondence between the physiological processes and morphological stages (e.g., initiation and formation of the lamina; apposition and enamel/dentine matrix formation), but proliferation and morphodifferentiation operate concurrently from the bud stage to the advanced bell stage. Histodifferentiation occurs after the bud stage of development and continues to the advanced bell stage (Bhussry, 1976). A detailed review of each morphological stage is not feasible in a limited space, and there are many books devoted exclusively to oral embryology, histology, and development (cf. Bhaskar, 1976; Ten Cate, 1994). Instead, we focus on those developmental events that influence the external details of the crowns and roots that are of special interest to dental morphologists and anthropologists.

A cross-section of a tooth germ is illustrated in Fig. 3.1. The three primary constituents of the tooth germ are the enamel organ, the dental papilla, and the enclosing follicle (dental sac). The enamel organ is separated from the dental papilla by a basement membrane. At this stage of development, the inner enamel epithelium is a single layer of undifferentiated cells, immediately superior to the basement membrane. Cells of the

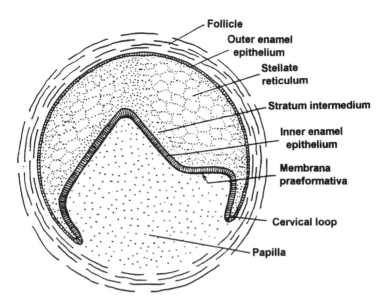

Follicle
Outer enamel epithelium
Stellate reticulum
Stratum intermedium
Inner enamel epithelium
Membrana praeformativa
Cervical loop
Papilla

Fig. 3.1   Cross-section of a tooth germ. After Butler (1956). Used with permission from the editors of *Biological Review*.

outer enamel epithelium form the lateral and superior walls of the enamel organ. Positioned between the inner and outer enamel epithelia are the *stratum intermedium* and *stellate reticulum*. The *stratum intermedium*, immediately adjacent to the inner enamel epithelium, interacts with this cell layer to produce enamel. Even though histologically distinct, they operate as a single functional unit during amelogenesis (Ten Cate, 1994). The cells of the *stellate reticulum* provide nutritive and/or mechanical functions, but their control function in crown formation is unclear. Along the basal edges of the enamel organ is the cervical loop where cells of the inner and outer enamel epithelia come in close contact. The dental papilla is highly vascularized and morphologically unstructured although cell concentrations are densest immediately below the inner enamel epithelium. The papilla is bounded by the basement membrane occlusally and the follicle basally. The follicle is formed partly from cells at the base of the dental papilla that flatten and migrate to enclose the whole of the tooth germ (Gaunt and Miles, 1967). In addition to the ectomesenchyme of the papilla, other types of mesoderm also contribute to the formation of the follicle (Tonge, 1971).

The inner enamel epithelium is particularly important because the cells from this layer proliferate through mitosis. For multi-cusped teeth, the rate of cellular proliferation in all directions (mesial, distal, lingual, and buccal) exceeds the rate at which the tooth germ increases in diameter. This results in a folding of the inner enamel epithelium with the primary folds corresponding to what become the major cusps of the crown. Artificial removal of the follicle at this stage of development results in a disk-shaped structure with little epithelial folding, indicating the important mechanical function of the enclosing follicle (Gaunt and Miles, 1967). In addition to rapid cell growth of the inner enamel epithelium, the dental papilla expands and exerts pressure on the enamel organ from a basal direction. The *stellate reticulum*, with its large water-absorbing cells, balances the pressures of growth from the papilla. While pressure from the papilla may stimulate the initial invagination of the inner enamel epithelium, the folding of this layer is primarily dictated by the proliferation of epithelial cells (Butler, 1956; Gaunt and Miles, 1967). The final folded configuration of the basement membrane is of special relevance as it is ultimately preserved as the dentinoenamel (amelodentinal) junction of the completed crown.

When cells of the inner enamel epithelium mature, they differentiate into specialized ameloblasts. Cells at the cusp tips are the first to differentiate, thus closing off avenues to further growth through mitosis. However, around the base of the crown, the last cells of the inner enamel epithelium retain their power to proliferate. This region encircling the base of the crown is the cingulum. In some mammals, the cingulum is visible as a ledge or shelf on the completed crown. Even when not visible, this cingular zone serves as a potential site for the origin of supernumerary cusps and styles(ids).

From the folded configuration of the inner enamel epithelium, two processes work in concert to produce enamel and dentine and the final form of the crown. Cells of the inner enamel epithelium differentiate into mature nondividing ameloblasts. Amelogenesis, the enamel-forming process, is initiated by ameloblasts located on the superior surface of the basement membrane. Starting at the cusp tips and moving down the slopes of the crown, ameloblasts secrete proteins (e.g., amelogenin, enamelin) as they migrate radially from the basement membrane toward the eventual crown surface. Enamel proteins form a soft fibrillar organic matrix that eventually mineralizes into a hard tissue. Also starting at the basement membrane, but growing inward rather than outward, odontoblasts produce pre-dentine during the phase referred to as apposition. Later, during the calcification phase, mineralization results in the hard dentine. Induced by adjacent cells of the inner enamel epithelium, odontoblasts differentiate earlier than ameloblasts so dentine becomes visible in the developing tooth prior to enamel. Cells of the papilla then induce the differentiation of ameloblasts and the ensuing process of amelogenesis. Amelogenesis and dentinogenesis are thus dependent on one another and are to a significant extent synchronous, albeit granting the temporal precedence of dentine formation over enamel formation. Together, they constitute the process of apposition (Ten Cate, 1994).

During amelogenesis, the enamel rod segments left in the wake of the migrating ameloblasts are calcified almost immediately during the first stage of mineralization. This occurs well before these cells reach the eventual crown surface. Deposition of inorganic salts in the enamel matrix begins first at the cusp tips and proceeds down the slopes of the cusps in a wave-like manner. The last areas to calcify are at the deepest invaginations separating the major cusps and the cervical third of the crown. By the end of the fourth and final phase of crown mineralization, 96% of the enamel is made up of inorganic hydroxyapatite crystals with the remaining 4% comprised of remnants of the original fibrillar protein matrix and water (Ten Cate, 1994).

The sequence of cusp calcification is regular and predictable, with only slight variations from the general theme. In human deciduous upper second molars, the paracone is the first cusp to calcify, followed in order by the protocone, metacone, and hypocone. In the deciduous lower second molars, the protoconid calcifies first, followed by the metaconid, hypoconid, entoconid, and hypoconulid (Kraus and Jordan, 1965; E. Turner, 1967). Two generalizations from Kraus and Jordan (1965) are particularly relevant to dental morphology. First, the mesiobuccal cusp is always the first to calcify, supporting the notion that the paracone and protoconid are homologues of the primitive reptilian cone, assuming, of course, that ontogeny recapitulates phylogeny in this instance. Second, calcification ordinarily proceeds from the mesial cusps to the distal cusps with the hypocone in the upper molars and the hypoconulid in the lower molars as the last major cusps to calcify.

Root formation begins immediately after the crown is developed. Hertwig's epithelial root sheath, an extension of the cervical loop, is a key element in this process. Subsequent to crown formation, this sheath turns inward toward the pulp immediately below the cervical enamel line. The only part of the papilla not enclosed is the primary apical foramen at its base. The sheath then expands in a basal direction to surround the dental papilla on its lateral borders. During the early stages of root formation, the growth of Hertwig's sheath is primarily vertical in orientation. For multi-cusped teeth, tongues developing from the sheath grow horizontally to form inter-radicular projections that divide the primary apical foramen into two or more secondary apical foramina. This temporary inward growth of the sheath is influenced by the lobes of the completed crown and the distribution of bundles of blood vessels that have differentiated within the papilla. The inter-radicular projections coalesce in a nonvascularized central area at the base of the root. There is some relationship between the number and placement of roots and the number of cusps on the crown, but this correlation is not perfect (Butler, 1956; Gaunt and Miles, 1967).

Mirroring many aspects of life, 'timing is everything' in the development of the dentition. From the initial thickening of oral epithelium that forms the dental lamina to the eruption of teeth, dental ontogeny follows a prescribed pathway with well-defined temporal constraints. Kronmiller *et al.* (1991) demonstrate that epidermal growth factor (EGF) mRNA is critical in inducing the formation of the dental lamina. Mandibles from day-9 mouse embryos treated with antisense oligodeoxynucleotides designed to block the action of EGF mRNA show total inhibition of odontogenesis (i.e., no dental lamina formed). When day-9 mandibles were treated with retinoids (the active agent of vitamin A), mitotic activity of the oral epithelium increased rather than decreased. This produced supernumerary loops of epithelium and epithelial swellings in the region of the diastema, an area that does not normally exhibit lamina at this stage of development (lamina initially restricted to presumptive incisor and molar regions). Hence, retinoids produce a combination of supernumerary, fused, and missing teeth. While supernumerary and fused teeth reflect the enhanced mitotic activity of the epithelium, why would some teeth be missing? Kronmiller *et al.* (1992) suggest that the enhanced growth caused by retinoids carries some tooth germs past that point in time when induction ordinarily takes place with the result being inhibition and agenesis. When fetal tooth germs are treated with retinoids at a later stage of development, the primary effect is on enamel histodifferentiation (Mellanby, 1941; Mellanby and Holloway, 1956). Glasstone (1952, 1979) explanted day-20 rabbit molar germs that had been halved in a tissue culture and found that both halves developed the morphology of the entire molar crown. When the procedure was repeated on day-22 molar germs, each half was no longer capable of developing into a whole tooth but only exhibited the appropriate morphology for the section of crown represented in the initial division. Developmental biologists working with teeth provide many such examples that illustrate the importance of timing and context in tooth formation (Kollar and Baird, 1971; Kollar, 1972; Kollar and Kerley, 1979). The general processes of odontogenesis are clearly guided by genic activity at specific times, in specific contexts, and in specific pathways.

Heterodonty and morphogenetic gradients

A homodont dentition is characterized by teeth of the same general size and form. Living reptiles, for example, exhibit rows of uniform teeth with single conical cusps that function primarily for grasping. Heterodont dentitions minimally involve teeth of different size, but more often are characterized by both size and morphological distinctions. From an evolutionary vantage, morphological differences translate into functional differences. Single-cusped teeth are effective at grasping and piercing while multi-cusped teeth are suited for shearing, crushing, and/or grinding. In mammals, the four types of teeth (incisors, canines, premolars, molars) are enhanced or reduced relative to specific dietary demands. An examination of herbivore and carnivore dentitions (cf. Peyer, 1968; Hillson, 1986) shows clearly the relationship between teeth and primary food items. Herbivores require more-or-less uniformly shaped teeth with complex occlusal surfaces that crush and grind vegetal matter. To maximize these surfaces, premolars are often enlarged and serve the same function as molars (i.e., they become molariform). Carnivores, by contrast, need highly differentiated teeth with pointed cusps and sharp crests for piercing (canines), slicing and dicing meat, a less abrasive material than plant cells and fibers. The premolars of a carnivore often have a single cusp (discounting molarized P4s) and their distal molars are reduced in size or lost entirely. More varied diets require dental compromises so the multi-purpose teeth of omnivores retain a generalized form. Molars have low rounded cusps and premolars are intermediate in form between canines and molars. Humans, along with bears, pigs, and other primates, have a dentition of this type.

Teeth, like many segmented and repeated structures in invertebrates and vertebrates, are metameres that show duplication with variation (Weiss, 1990). Variation in crown and root form is not random within the dentition but shows gradients where adjacent teeth are most similar to one another. These gradients are not continuous in the human dentition as there are distinct divisions in crown form at the boundaries of morphological classes (e.g., from upper canine to first premolar). In many species, however, class boundaries are not as distinct. One can find incisiform canines, caniniform premolars, and molariform premolars that serve as morphological bridges between different tooth classes. In the human dentition, lower canines are typically incisiform while lower first premolars are occasionally caniniform. Gradients across tooth classes are less evident in the maxillary dentition of humans because upper first premolars, with invariably well developed lingual cusps, are never caniniform. Still, there are unquestionable tooth classes and these exhibit gradients.

Butler (1937, 1939) conceptualized gradients in the dentition as morphogenetic fields. He presupposed that at an early stage of development, tooth germs distributed around the maxilla and mandible were equipotent; that is, they all possessed the same set of instructions that would allow them to develop the form of any class of tooth. Butler (1963) noted that while teeth develop generally in an anteroposterior direction, the determinants of tooth type have less to do with the timing of development than with the position of the tooth germs in the jaw. At some point in development, perhaps preceding the time at which the tooth germ is histologically recognizable, Butler inferred the action of a field substance or morphogen that issues instructions (figuratively speaking) for a tooth to become either an incisor, a canine, or a molar. At a secondary level, Butler proposed that 'pattern genes' operating on different tooth germs within a morphological class would determine how each member of the class responded to levels of a specific field. This sequence shows the hierarchically nested nature of the developing dentition with induction events determining, in order, tooth, class of tooth, type of tooth within a class, and primary and secondary cuspal morphology of individual teeth.

Kollar and Baird (1971) demonstrated that the instructions for tooth class are issued by the mesenchyme rather than the epithelium. In reciprocal exchanges of incisor and molar epithelia and mesenchyme, molar mesenchyme induces incisor epithelia to form molars while incisor mesenchyme induces molar epithelia to form incisors. When mouse incisor mesenchyme and molar epithelia were transplanted to the anterior eye chamber (interocular graft), the graft survived for five weeks and developed a well-formed incisor. Also using interocular grafts, Lumsden (1979) transplanted the presumptive molar region of a day-9 mouse embryo that showed no signs of an M1 enamel organ. This graft developed not only M1 but also M2. When this molar region was transplanted at 12 days, the entire set of molars developed.

Such embryological experiments, along with paleontological observations, led Osborn (1978) to develop a 'clone model' to explain the development of tooth classes and gradients. Osborn presupposed the existence of three primordia at an early stage of odontogenesis associated with specific classes of teeth: an anterior primordium (deciduous and permanent incisors), a canine primordium (deciduous and permanent canines), and a posterior primordium (deciduous molars and permanent premolars and molars). In contrast to the field model, these primordia are not assumed to be equipotent. The clone model dispenses with the notion that extrinsic field substances or morphogens generate gradients along the tooth row. Rather, growth of the tooth germ is wholly intrinsic, or controlled from within. Each primordium starts from a single stem precursor, or progenitor, with additional primordia cloning off from this stem precursor, and then off first order primordia of the stem precursor, and so on until the tooth formula is complete. From the molar stem precursor (last deciduous molar or first permanent molar), growth proceeds in both anterior and posterior directions. Gradients develop because the mesenchyme inducing the formation of additional teeth undergoes more mitotic divisions than the stem precursor. Thus, the stem precursor expresses the initial instructions of the primordial clone to the greatest degree while later teeth develop from primordia with an 'aging mesenchyme' of reduced tooth-forming potency (Butler, 1982). In support of his argument, Osborn noted that the teeth first reduced and/or lost in evolution are the teeth most anterior and most posterior to the stem precursor. For the molar clone, this would be the most anterior premolar and third molar. Butler (1982), however, observed that in many mammals, the largest molar is not M1 but M2 or even M3. If the mesenchyme at the expanding end of a clone is of reduced tooth-forming potency, Butler wondered how secondary primordia could show more enhanced development than the stem precursor of the same clone.

Butler (1982) acknowledges that the original formulation of his 'field theory' has problematic aspects. He deems it unlikely that a morphogenetic field could persist for the prolonged period between the initiation of the incisor and molar primordia early in embryonic development up to the initiation of permanent third molar primordia, often occurring several years later. Lumsden's (1979) ability to produce an entire molar tooth series from an interocular transplant of early embryonic molar tissue also challenges the notion of extrinsic morphogens acting on different members of a tooth class. Osborn (1978) points out additional problems with the field model, but as Butler (1982) notes, there are also inconsistencies in the clone model. Both models offer useful insights into the nature of gradients in the dentition, but some pieces of this intriguing puzzle continue to elude researchers. The final solution will likely include elements of both the field and clone models (Weiss, 1990).

Long before the debate over fields vs. clones, Dahlberg (1945a) adapted Butler's concept of morphogenetic fields to the human dentition. In so doing, he introduced additional terms and concepts that have been widely adopted in the study of human dental variation. In contrast to Butler, who defined three primary dental fields in mammals, Dahlberg viewed the human dentition in terms of four fields by separating premolars from the post-canine or molar field. These four fields, or 'tooth districts,' correspond to the four morphological classes of teeth: I, C, P, and M. Within each tooth district, there is a 'key' tooth that is its most stable member in terms of development and evolution. Excepting the lower central incisor, the key tooth in the upper and lower tooth districts is the most mesial member (UI1, UC, UP1, UM1; LI2, LC, LP1, LM1). The distal members of a district are 'variable' teeth and, within the molar district, variability increases from M1 to M2 and from M2 to M3. Stability vs. variability within tooth districts is measured by tooth size, numerical variations, and morphology. The key tooth in a district shows: (1) the lowest coefficient of variation in size within a population; (2) the lowest frequency of hypodontia and hyperodontia; (3) the highest frequency of crown and root trait expression; and (4) the least amount of fluctuating asymmetry (section 'Fluctuating asymmetry'). The variable teeth, by contrast, show more intragroup size variation, are more likely to be missing or show adjacent supernumerary teeth, and are more asymmetric (Townsend and Brown, 1981a). Morphologically, variable teeth have diminished trait frequencies and/or reduced trait expressions.

Was Dahlberg (1945a) justified in distinguishing premolars as a separate field or tooth district? Adherents of both the field (Butler, 1939, 1982) and clone (Osborn, 1978) models suggest, on developmental and phylogenetic grounds, that the premolars are an anterior extension of the molars and therefore place them within the molar field/clone. We interpret this to mean that as the 'field substance' or 'molar clone' diffuses or grades from the pole tooth (either dm2 or M1 in humans), the teeth least affected would be the most anterior (P1) and posterior (M3) members of the spreading field/clone. On this assumption and addressing strictly the human dentition, we can marshal evidence to either include or exclude premolars from the molar field. In the lower premolars, for example, LP2 exhibits more pronounced lingual cusps and a higher frequency of multiple lingual cusps than LP1. The effect is to make LP2 a more molariform tooth than LP1. Not uncommonly, LP1 fails to express a lingual cusp with a free apex and may appear caniniform. In terms of size, LP2 is also larger than LP1, especially in its buccolingual dimension (cf. Mizoguchi, 1988a; Kieser, 1991). In contrast to the lower premolars, UP1 is larger than UP2 in both diameters. However, neither of the upper premolars, with their typical 2-cusped morphology, is more molariform than the other. For root characteristics, Wood and Engleman (1988) show the ancestral hominid condition for upper premolar root number is two or even three roots. In recent humans, two- and three-rooted phenotypes are much more common on UP1 than UP2, again suggesting a distinct premolar field. Tomes' root of the lower premolars, a common but not invariant trait in early hominids

(Wood *et al.*, 1988), is also more frequent on LP1 than LP2. The pattern of hypodontia also supports the notion of a premolar field. Agenesis in the premolar district almost invariably involves UP2 and LP2, not the first premolars (Garn *et al.*, 1963). To summarize, the evidence is equivocal regarding a separate premolar field; in terms of crown form and size, lower premolars could be considered anteriorly graded members of the molar field/clone. Upper premolars, by contrast, show no such pattern. For our purposes, we treat UP1 and LP1 as key teeth for scoring root number. For crown traits, the decision on which member of the class is the key tooth is less obvious. It could well be UP1 and LP2 for the reasons noted.

Some authors have pointed out that morphogenetic fields describe but do not explain regularities in a system that exhibits gradients (Osborn, 1978; Weiss, 1990). Despite this caveat, the field concept is a useful methodological guide to the analysis of crown and root traits in families and populations. Carabelli's trait, for example, can be scored on UM1, UM2, and UM3. Which tooth's trait frequency is the best indicator of the genotypic variation underlying this feature and is thus preferred for genetic analysis and population characterizations? One could circumvent this question by including frequencies for all three molars, but this introduces redundancy into the equation as these are correlated variables (section 'Intertrait association'). Another strategy would be to use a composite frequency of M1 through M3 expressions, but this presents both practical (many subjects do not have UM3, due either to age or agenesis) and theoretical (developmental instability of UM3) difficulties. Moreover, Carabelli's trait is differentially expressed on the upper molars. Its primary focus is UM1 where it is most common and pronounced. From that point, frequency and expression become increasingly lower on UM2 and UM3. For these reasons, workers focus on UM1 as the tooth that best represents the genetic factors underlying Carabelli's trait expression. This rationale has been extended to many other crown and root traits (e.g., UI1 shoveling, LM1 cusp 6, cusp 7, deflecting wrinkle). In some instances, however, morphological features expressed on key teeth are so stable they are almost invariant (e.g., hypocone and hypoconulid of first molars). To study patterns of variation for these traits, it is necessary to turn to more variable teeth such as the second molars.

Butler (1963) says no tooth in a field, including key teeth such as M1, show all the patterns of that field to a maximal extent. This is evident in some morphological traits in the human dentition. The generalization that key teeth show the highest frequencies and most pronounced expressions of a specific trait within a field holds in many instances (e.g., hypocone,

Carabelli's trait, cusp 7, deflecting wrinkle, distal trigonid crest, two-rooted upper premolars, Tomes' root, 3RM1). In other cases, however, the key tooth has the highest trait frequency while the variable distal members of the field show more pronounced degrees of expression (e.g., cusp 5, cusp 6, protostylid). While traits generally exhibit unidirectional gradients within a tooth district for frequency and/or expression, one interesting exception involves 4 and 5-cusped lower molars, determined by the absence (4) or presence (5) of the hypoconulid. Five-cusped LM1 are found in high frequency (85–100%) in all human populations. The lower second molar shows a much higher frequency of 4-cusped forms, especially in Western Eurasians (chapter 5). For LM1 and LM2, the pattern of hypoconulid variation mirrors that of other crown traits; if LM1 is 4-cusped, LM2 will almost invariably be 4-cusped. What sets the hypoconulid apart from other traits is that it reappears on LM3 even when absent on LM2, so 5–4–5 lower molar cusp number sequences are common (cf. Greene, 1967; Grine, 1981) and even modal in some populations (Thomsen, 1955). This up-down-up sequence (5–4–5) for lower molar cusp number violates tenets of both the field and clone models. Upper molar cusp number, determined by the presence or absence of the hypocone, rarely if ever shows a comparable reversal at UM3. For the upper molars, one can observe 4–4–4, 4–4–3, 4–3–3, and even 3–3–3, but not 4–3–4.

Ontogeny and morphology

If ontogeny is the intermediary between genotype and phenotype, what ontogenetic processes do genes direct to produce morphological phenotypes of tooth crowns and roots? The five major processes appear to be: (1) folding of the inner enamel epithelium during proliferation; (2) ameloblast differentiation; (3) enamel/dentine matrix formation; (4) crown and root mineralization; and (5) growth of Hertwig's epithelial root sheath. Genes do not produce traits directly but only indirectly through their control of the rates, timing, and orientation of specific odontogenetic processes. Genes controlling mitotic activity are implicated in two processes, the folding of the inner enamel epithelium and the expansion of Hertwig's sheath. Differentiation of the ameloblasts, apposition, and mineralization do not involve mitosis. In these cases, genes may function to determine the timing of onset, rates, orientation, and cessation of these processes. Some insights into the developmental events that are regulated by genes are made possible by observations on the morphological surface of the dentine that is ordinarily obscured by the enamel cap.

Dentinoenamel surface

The basement membrane that separates the inner enamel epithelium and dental papilla is ultimately 'fossilized' following mineralization of the crown and root. This dentinoenamel (hereafter shortened to dentine) surface is the surviving remnant of this membrane, formed during the proliferative phase of tooth development when the inner enamel epithelium is folded. Ameloblasts covering this surface secrete enamel protein to form a soft matrix which, when mineralized, forms an overlying enamel cap that obscures its starting point on the dentine surface. If the folding of the inner enamel epithelium is a primary determinant of crown form (Butler, 1956, 1982; Kraus and Jordan, 1965), what is the relationship between the cusps, ridges and other morphological structures of the enamel surface and the morphology of the dentine surface?

Anthropologists and paleontologists, out of necessity, emphasize features that are observable on the accessible enamel crown surfaces. Few researchers have studied dentine surfaces and their relationship to corresponding enamel surfaces. Korenhof (1960, 1961), one of the pioneers in this research, focused on the dentine surfaces of upper molars in fossil and living primates and recent humans. He made impressions of the dentine surface from tooth caps where the roots had been destroyed so he could compare directly the external and internal surfaces of the enamel crown (Fig. 3.2). A comparison of enamel and dentine surfaces reveals two types of concordant expression and two types of discordant expression. Concordant expression (i.e., trait present or absent on dentine surface is correspondingly present or absent on enamel surface) suggests that a trait's manifestation does reflect folding of the inner enamel epithelium. Discordant surficial relationships are equally interesting. Some cuspules, ridges, and fissures apparent on the dentine surface are not expressed on the enamel surface. To borrow a phrase from Kraus (1963), they have been 'enameled over.' In some instances, this involves total concealment of the underlying morphology while, in others, a distinctive dentine phenotype for, say Carabelli's trait, is greatly muted in expression on the enamel surface. There are also instances where the enamel surface expresses an accessory cusp or ridge, but there is no indication of a comparable structure on the dentine surface. This appears to affect what Korenhof called 'edge tubercles' of the mesial (protoconule, mesial marginal tubercle, mesial paracone tubercle) and distal (cusp 5) marginal ridge complexes and many small ridges and crests on the occlusal surface. Kraus (1963) found the opposite situation in developing human deciduous second molars. He commonly found multiple tubercles along the mesial marginal ridge of the

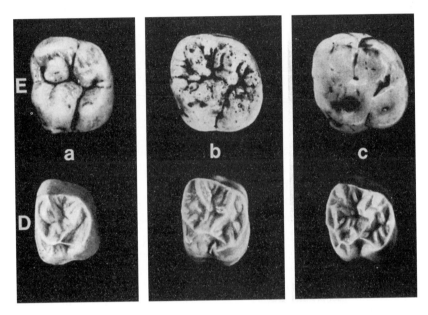

Fig. 3.2   Corresponding enamel (E) and dentine (D) surfaces in six human upper molars. Dentine surfaces exhibit sharper ridges and crests than the enamel surfaces. Many crests, ridges, cusps, and tubercles on the enamel crown have dentinal precursors. In d, for example, large mesial marginal tubercles expressed on the final crown are preceded by small crests on the dentine surface. For f, however, a distal marginal tubercle (cusp 5) on the final crown has no apparent dentinal precursor. Carabelli's trait, distinctly expressed on the dentine surfaces of all six molars, is more variable on the enamel surface; see, for example, the muted expressions of the trait on a and b despite large dentinal precursors. After Korenhof (1960) (lettering added). Used with permission from Uitgeversmaatschappij Neerlandia, Utrecht.

soft enamel crown, but these tubercles were rarely evident after mineraliz-ation; thus, he felt they were 'enameled over.' This situation might, however, reflect a problem with the cross-sectional sample used by Kraus. That is, the presence of multiple edge tubercles could be a signal of greater fetal disorganization. Fetuses with multiple tubercles may have died earlier than those lacking tubercles. Another possible explanation for the disparity between the findings of Korenhof and Kraus is that they were observing permanent and deciduous molars, respectively, and the relationship between the enamel and dentine surfaces of these teeth may differ. From Korenhof's plates, it appears unlikely that Carabelli's trait and the hypocone develop without a precursor on the dentine surface. For these two traits, being 'enameled over' introduces most, if not all, of the discordance between dentine–enamel surface contrasts.

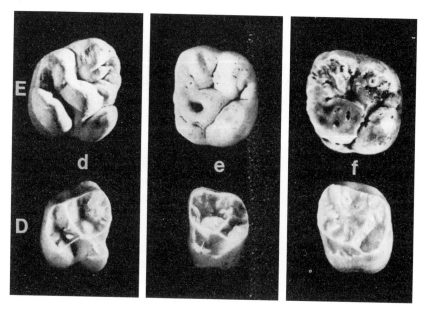

Fig. 3.2 (*cont.*)

Sakai *et al.* (1967) studied the dentine surface of the upper canines and arrived at conclusions parallel to those of Korenhof. As a rule, ridges of the dentine surface show sharper relief than those of the enamel surface. On both the labial and lingual surfaces of the upper canine, the five ridges evident are: mesial and distal marginal and accessory ridges and a median ridge. When trait frequencies and expression are compared between the two surfaces, marginal ridges are more common and pronounced on the dentine surface while accessory ridges are more common and pronounced on the enamel surface.

Major and secondary cusps

Cusps project to varying degrees from the occlusal surfaces of the crown so early workers felt they must grow faster than other parts of the crown. Butler (1956) suggested, to the contrary, that cusps constitute those areas of the crown where cells of the inner enamel epithelium first cease to proliferate. Rather, at the cusp tips, there is precocious maturation of cells of the inner enamel epithelium into functional secretory ameloblasts that makes further growth impossible. Cusps project not because they grow more rapidly in an occlusal direction but because they represent loci on the crown that stop growing first. Meanwhile, active proliferation of the inner enamel epithelium in a basal direction persists on the slopes of the cusps, into the fissures separating the cusps, and finally down the cervical third of the crown. This has the effect of elevating cusps above other aspects of the crown.

Of the crown traits listed in chapter 2, the hypocone and hypoconulid are the only major molar cusps that are frequently absent, especially so on the second and third molars but occasionally on the first molars as well. For this reason, they are treated as variable crown traits along with secondary cusps and other minor variants. Observations on the ontogeny of molar crowns suggest why this situation prevails. Both cusps are the last to form a soft enamel matrix and calcify; their subocclusal position relative to other major molar cusps also indicates late development. Logically, with developmental delays in timing, the hypocone and hypoconulid would be the first cusps reduced in size or even lost. As Dahlberg (1968:276) noted, 'timing variations reflecting different properties or strength of morphogenetic fields could very well be a part in the loss of certain elements of form such as the hypocone and hypoconulid.'

Although few dental morphologists score the size of major cusps other than the hypocone and hypoconulid, two additional primary cusps show dimensional reduction that could be predicted by the sequence and timing of cusp formation. In the upper molars, the metacone is the third cusp to calcify, and it also shows a tendency to size reduction. In the lower molars, hypoconulid absence is sometimes accompanied by reduction of the entoconid, the fourth and next to last cusp to form and calcify. Both the metacone and entoconid may be completely missing, although this is much less common than size reduction. Whether viewed in terms of developing fields or aging clones, timing events in odontogenesis appear to be critical in determining the occurrence, absence, simplification, and even hypertrophy of major cusps of the human molar crown.

Butler (1956:44) remarked, 'There is much evidence that secondary cusps arise as minor convexities of the inner enamel epithelium in the growth zone at the base of the primary cusp, and that the relief of the crown becomes more exaggerated as development proceeds.' Cusps that qualify as secondary include 6 and 7 of the lower molars and possibly cusp 5 and the protoconule of the upper molars. While variation in these traits likely reflects differential growth rates of the inner enamel epithelium during the proliferation stage, this generalization may not apply to all accessory cusps and tubercles, especially those of the mesial marginal ridge complex of the upper molars. In many instances, these tubercles lack precursors on the dentine surface so subsidiary genetic regulators must be operating during matrix formation.

Occlusal, marginal, and accessory ridges

It is not entirely clear how ridges are formed on a tooth crown but many are evident on the dentine surface, suggesting involvement of the inner enamel epithelium during proliferative growth. In Butler's (1956) view, ridges reflect lines of tension created by the major cusps pulling away from each other as the crown base expands. Ridges will remain elevated along with the cusps if proliferation is relatively slow on the slopes of adjacent and partly connected cusps. On the dentine surface of upper molars, this is particularly apparent in the oblique crest that connects the distal accessory ridge of the protocone to the median occlusal or distal accessory ridge of the metacone (Korenhof, 1960). Dentinal morphology suggests the mesial marginal ridge has a similar origin.

When shovel-shaped incisors exhibit attrition, it is evident that both enamel and dentine are involved in this trait's expression. American Indian, American white, and Japanese fetuses have incisors that show the full gamut of lingual marginal ridge expression in the soft enamel matrix. On this basis, Kraus and Jordan (1965:175) concluded that 'its presence in full-blown form in the precalcified tooth bud would appear to demonstrate the genetic control of mitotic activity in the cells of the inner enamel epithelium.' This may be true for other marginal ridges as well although accessory ridges might have a different origin (Sakai *et al.*, 1967).

Cingular traits

The human dentition exhibits relatively few morphological traits that originate from the cingulum – only *tuberculum dentale* of the upper anterior teeth, Carabelli's trait and the parastyle of the upper molars, and the protostylid of the lower molars. They are considered cingular in origin because of their location at the cervical third of the crown in the presumed, though not externally visible, *zona cingularis*. Of these traits, the ontogeny of Carabelli's trait is best known through the work of Kraus and Jordan (1965), although their observations on this feature might also be extended to other cingular traits.

Kraus and Jordan (1965) removed deciduous upper second molars and permanent first molars at varying stages of development in human fetuses, many of which were prepared and stained by C.G. Turner as a graduate student. On soft enamel crowns, they observed all manifestations of

Carabelli's trait expression, from absence and pits to large free-standing cusps. In terms of timing, they noted that all expressions of Carabelli's trait became evident on the crown just prior to initial calcification. On some molars, they found distinct invaginations on the lingual aspect of the protocone that they assumed would have developed into pit expressions on the final mineralized crown. In other instances, they observed distinct basal ridges around the protocone that were considered likely precursors of positive forms of cusp expression. The more pronounced basal ridges showed independent centers of calcification while smaller ridges calcified as an extension of protocone calcification. Molars that showed neither invaginations nor basal ridges were assumed to represent the absence of Carabelli's trait.

In chapter 5, we describe the population variation of Carabelli's trait. In this regard, Kraus and Jordan (1965:189) make the point that 'the differences, therefore, in the frequencies of various aspects of the trait among human populations must be thought of in terms of differential frequencies of genes regulating the speed and duration of mitotic activity within the inner enamel epithelium on the lingual aspect of the protocone.' Korenhof (1960) found that pit expressions on the external enamel surface corresponded to typical cingular shelves rather than invaginations on the dentine surface. While this supports the notion that pit expressions are definitely part of the Carabelli's complex (Reid *et al.*, 1992), it also hints at the operation of two different processes. If pits are not expressed on the dentine surface, then this form of Carabelli's trait expression is not due solely to the folding of the inner enamel epithelium. The invagination resulting in the pit phenotype occurs during amelogenesis so another set of regulations must be operating while the soft enamel matrix is being formed.

Hertwig's sheath

The expansion of Hertwig's sheath is a primary determinant of root number and form. The early convergence of inter-radicular tongues during sheath formation results in dividing the primary apical foramen into secondary apical foramen. When this occurs, two, three or more roots might form. When the inter-radicular tongues fail to converge at any point along the length of the root, the result is a single root with radicals marked only by developmental grooves. In the normally three-rooted upper molars, two roots might be joined while the third shows clear separation. The factors involved in the differential orientation of Hertwig's sheath are not well known.

Although Hertwig's sheath is primarily involved in root formation, it is

Fig. 3.3   Enamel pearl. Small droplets of enamel sometimes form in root bifurcations. In this example, the pearl is associated with a small enamel extension but makes no contact with the extension.

also implicated in the expression of enamel extensions. This sheath develops from the cervical loop of the enamel organ where cells of the outer and inner enamel epithelium come in close contact with no, or very little, intervening *stratum intermedium* and/or *stellate reticulum*. Ordinarily, the cells of this sheath do not mature into ameloblasts and produce enamel. However, the occasional presence of enamel pearls, small 'droplets' of enamel between the inter-radicular projections of the molar roots, shows the sheath retains enamel forming capabilities even after the initiation of root formation (Fig. 3.3). The basal tips of enamel extensions are usually oriented toward the furcation of the roots and, not uncommonly, extensions and pearls are found on the same tooth. The ontogenetic factors underlying enamel extensions remain obscure, but genes must modify the form of the cervical loop and/or prolong the timing of differentiation of the inner enamel epithelial cells in Hertwig's sheath.

Heterochrony

Gould (1977) defines heterochrony as 'evolution by change in the timing of development.' Complex morphological changes have often been attributed to either retarded or accelerated growth rates at certain points in development. Among dental researchers, Keene (1982, 1991) shows how different growth parameters can enhance our understanding of the relationship between odontogenesis and phenotypic trait expression. His concept of morphogenetic triangles involves three temporal markers: (1) time of initiation of proliferation; (2) time of initiation of differentiation;

and (3) time of crown completion. The onset of stage 1 is the point where cells of the inner enamel epithelium can first be distinguished from other epithelial cells of the enamel organ; during this proliferative stage, cells divide and expand in area, ultimately folding into the configuration represented by the dentinoenamel surface. Stage 2 marks the beginning of apposition where cells of the inner enamel epithelium differentiate into ameloblasts; during this stage, differentiation of unspecialized cells into functional secretory ameloblasts begins at the cusp tips and continues down the slopes of the cusps on the occlusal surface and down the lateral borders of the crown in a basal direction. The final stage, time of crown completion, is marked by the point where all cells of the inner enamel epithelium have ceased to proliferate and differentiate and the soft enamel matrix of the crown is completed. By this time, mineralization has already started at the tips of the cusps.

In a geometric representation, Keene (1982, 1991) plotted P (onset of proliferation), D (onset of differentiation), and C (time of crown completion) along a horizontal axis denoting elapsed time (Fig. 3.4). The vertical axis contains the single variable Z, final crown height (that can be traced vertically to C). His morphogenetic triangle is formed by drawing lines from P to Z and D to Z. Keene shows how perturbations in the onset of P and D can influence both tooth size and morphology. For example, the time from P to D sets the size and form of the dentinoenamel surface while the time from D to C determines the size and form of the enamel surface. Although tooth size is ordinarily measured only on the external enamel surfaces, there are two factors contributing to tooth size: the dimensions of the dentinoenamel surface and enamel thickness. Individual variation in the onset of P, D, and C could thus influence tooth size in a myriad of ways. Morphologically, a lower molar tooth germ with a long period of proliferation (P to D) has the greatest potential to exhibit its complement of five major cusps. Another molar tooth germ with an early onset of differentiation would be developmentally frozen at a particular stage, with a concomitant loss and/or reduction of major cusps. Given the sequence of formation and calcification, this usually involves the hypoconulid and sometimes the entoconid. The trend in crown size and cusp reduction in hominid evolution may well be due to the variability in developmental rates, particularly those related to the timing of calcification (Dahlberg, 1968, 1971b).

We have emphasized how genes, in a broad sense, regulate ontogenetic processes. Beyond the structural gene that codes for amelogenin, the strongly conserved protein of the enamel matrix, researchers have yet to

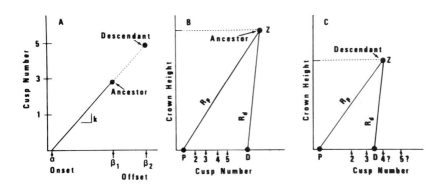

Fig. 3.4   Heterochronic models of tooth cusp number variation. (A) Single parameter model where delay in offset signal (*β*) results in prolonged growth period and an increase in cusp number from 3 to 5 (i.e., hypermorphosis). (B) and (C). Multiple parameter models shown through morphogenetic triangles (PDZ) where P: onset of proliferation, D: onset of differentiation, and Z: time to crown completion. In B (ancestor), five cusps appear early in ontogeny while tooth germ is in soft stage prior to differentiation. In C (descendant), growth rate (R_p) and cusp differentiation are retarded while onset of differentiation (D) is accelerated. These changes have the effect of reducing the number and size of cusps in human lower molars (e.g., loss of the hypoconulid and reduction of entoconid). After Keene (1991). Used with permission from Wiley-Liss, Inc.

decode the highly intricate genetic controls that guide odontogenesis from the formation of the lamina to the level of variable crown and root trait expression. Genes are clearly involved, and some are probably homeotic genes given the metameric structure of the dentition (Weiss 1990). Presently, we know little about the number or nature of these genes.

While the role of genes in dental development can be strongly inferred, extrinsic as well as intrinsic factors can influence the final form of the tooth crown and root. For example, within a small jaw, adjacent dental follicles may impinge upon one another because of space limitations. If no spatial adjustments are made (e.g., a buccal shift in the position of one follicle), the result is a crown that is compressed along its mesiodistal axis. If one follicle stands its ground, figuratively speaking, while its antimere shifts position in response to crowding, the result can be a compressed tooth on one side of the jaw and an antimere that is normal in form but misaligned in the arcade (Gaunt and Miles, 1967). This is one of the more obvious instances where an environmental factor (local crowding) disrupts a developmental program. More commonly, the effects of environmental factors on the crowns and roots are more subtle than a severely compressed tooth. The actions of extrinsic factors can be inferred from the lack of perfect correspondence in the teeth of monozygotic (MZ) twins who are genotypically identical (chapter 4) and also from discrepancies in trait expression between antimeres, or asymmetry, our next topic of discussion.

Fluctuating asymmetry

Most organisms exhibit paired or bilateral structures with corresponding right and left sides. Symmetry exists when one of these structures (e.g., wings) is metrically and morphologically identical on the two sides. While there is indeed a great deal of symmetry in nature, a closer examination of a structure's dimensional and morphological details often reveals left–right differences, or asymmetries, that range from subtle to pronounced. Van Valen (1962) defined three primary types of asymmetry. First, directional asymmetry involves traits that may or may not be normally symmetrical but when asymmetry is apparent, it shows a definite right-side or left-side bias (e.g., Broca's area on the left side of the human cerebral cortex). His second category, antisymmetry, pertains to structures that are normally asymmetric but show no consistent side bias (e.g., lateral hand dominance). Finally, fluctuating asymmetry characterizes traits that are normally symmetrical but exhibit occasional side differences with no left or right side preference. Waddington (1957) considered the minor and random deviations from symmetry (i.e., fluctuating asymmetry) to be accidents of development. The right and left sides of a body have the same set of genetic instructions so deviations from a blueprint for symmetry are caused by local background noise during development (i.e., 'developmental noise'). It is this type of asymmetry that is most common in the dentition.

Teeth in corresponding quadrants of the upper and lower jaws are normally symmetrical structures that exhibit mirror imagery. Specific antimeres, such as left UM1 and right UM1, are similar if not identical in size and form, often down to the smallest morphological details of cusps, tubercles, ridges, and fissure patterns (Fig. 3.5). This symmetry, however, is not always perfect. In some cases, a left tooth may exhibit a specific trait while its antimere does not (i.e., presence–absence asymmetry). Asymmetry can and does occur, at varying levels, for all crown and root traits, overall tooth size, cuspal dimensions, hypodontia, and hyperodontia.

Odontometric asymmetry has been associated with congenital abnormalities, genetic syndromes, and elevated levels of inbreeding, and is also thought to be a general indicator of stress caused by nutritional and/or disease factors. Morphological asymmetry, while not used as a comparative measure of population stress, is nonetheless relevant from genetic and

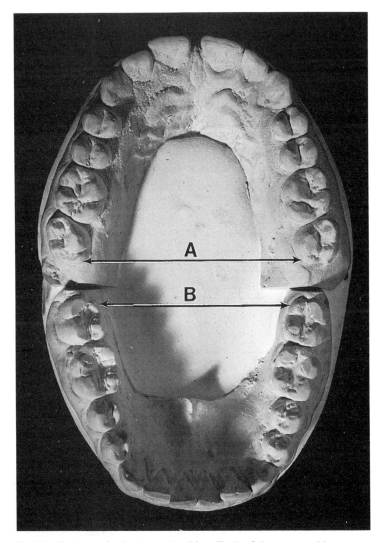

Fig. 3.5 Symmetry in the human dentition. Teeth of the upper and lower jaws, for the most part, exhibit antimeric symmetry and mirror imagery. Although most crown trait features are symmetrical in this dentition (e.g., the entoconid reduction noted by B), the hypocone exhibits asymmetry in size (noted by A).

methodological standpoints. A series of questions can be posed on dental morphological asymmetry. (1) How is asymmetry in crown and root traits quantified? (2) How asymmetric are morphological traits? (3) Is there any genetic basis to trait asymmetry? (4) Are asymmetries in different traits positively correlated? (5) Does a trait show different levels of asymmetry on different members of a tooth class? (6) If one antimere exhibits a trait and the other does not, which tooth should be scored? (7) How is asymmetry dealt with in estimating trait frequencies in populations?

More research has been carried out on odontometric asymmetry than morphological asymmetry. To quantify antimeric differences in tooth size, workers use variants of two primary methods. The first focuses on some measure of mean dimensional differences between right and left antimeres. If asymmetry is of the fluctuating type, the expected mean of L minus R or R minus L measurements will be zero. As this mean is not informative (except to demonstrate directional asymmetry), an alternative is to use absolute left–right differences that give a mean of zero only in cases of perfect symmetry. To make absolute antimeric differences comparable across morphological classes, with their teeth of varying dimensions, differences can be divided by average left and right tooth size (Harris and Nweeia, 1980; Kieser and Groeneveld, 1986). Garn and his collaborators (Garn and Bailey, 1977; Garn *et al.*, 1979; Smith *et al.*, 1982) prefer to quantify asymmetry by a measure of variation, the root mean square (RMS), which is the standard deviation of antimeric size differences. In addition to quantifying average side differences, a second method of quantifying tooth size asymmetry entails calculating intraclass or interclass correlation coefficients between left and right antimeres (cf. Bader, 1965; Siegel and Doyle, 1975a,b; Barden, 1980). For sample comparisons (e.g., male vs. female, MZ vs. DZ twins, group 1 vs. group 2, etc.), correlations are converted to z scores that can be assessed for significant differences in magnitude through t-tests or analysis-of-variance (Bailit *et al.*, 1970).

There is no consensus on a best method for quantifying asymmetry for either all-or-none or ranked traits. The most straightforward measures are percentage of symmetric cases (concordance) or percentage of asymmetric cases (discordance). If a trait is considered in terms of just presence (P) or absence (A), antimeres can exhibit PP, PA, AP, and AA phenotypes. Concordance is derived from PP + AA/total paired observations while percentage of asymmetric cases is computed by PA + AP/total paired observations. In cases where traits are scored on ranked scales, these same values are calculated with the added consideration that PP antimeres may exhibit asymmetry in degree of expression (Baume and Crawford, 1979, 1980). In such instances, it is useful to provide subsidiary information on degree of rank differences between sides (i.e., percentage of antimeres differing by 1 grade, 2 grades, etc.) (Scott, 1980; Saunders and Mayhall, 1982a). In addition to percentages of symmetry or asymmetry, some workers calculate correlation coefficients between left and right side trait expression. Instead of parametric correlations, workers use rank order (Kendall's *tau*, Spearman's *rho*), tetrachoric, or polychoric correlation coefficients.

Dental traits ordinarily exhibit a high degree of symmetry. Garn *et al.* (1966b) found that lower molar cusp number (LM1 and LM2) and groove pattern (LM1) showed an antimere concordance of about 95%. For cusp 5 (UM1), Harris and Bailit (1980) reported a side to side concordance of 84%. Dietz (1944) noted that only 2% of 732 American whites exhibited presence–absence asymmetry for Carabelli's trait (i.e., 98% concordance). Using four grades of trait presence, he added that only 7% of the individuals showed asymmetry in degree of expression, and these never exceeded a single grade difference. In ten world samples, Scott (1980) found a mean presence–absence asymmetry of 5.5% for Carabelli's trait although 20.6%, on average, exhibited some form of asymmetric expression. In most instances, concordance rates based on the presence or absence of morphological traits fall in the range of 85–95%. This range is lowered to somewhere between 50 and 80% when all forms of asymmetry are considered.

Saunders and Mayhall (1982a) critiqued concordance estimates based on their observation that the inclusion of absence–absence pairs as concordant members of a sample unduly lowers the actual percentage of asymmetry. They calculated the percentage of presence–absence asymmetry for Carabelli's trait in an Inuit sample with and without cases of bilateral absence and, not surprisingly, found that the elimination of the bilateral absence category elevated the frequency of asymmetry. The difference for UM1 is relatively modest (13.9% to 15.8%) but for UM2 and UM3, where the trait is less frequently expressed, the effect is dramatic (18.6% to 56.3% and 13.6% to 44.4%, respectively). They noted that since Carabelli's trait is less common on UM2 and UM3, there may be a relationship between total trait frequency and percentage of presence–absence asymmetry. That is, traits in low frequency may show more asymmetry than high frequency traits, but this is obscured by including the many absence–absence pairs one finds for low frequency traits. While they make an interesting point, either method of including or excluding bilateral absence pairs from percentage asymmetry calculations has its problems for standardizing between-trait and between-group comparisons. As Mizoguchi (1988b) noted, one can derive the same percentage asymmetry with a high frequency trait that has a low inter-side correlation or a low frequency trait that has a high inter-side correlation.

Correlation coefficients provide another avenue for assaying morphological asymmetry within and between groups. As a reference point, inter-side correlations for mesiodistal and buccolingual diameters mostly fall within the range of 0.80 to 0.95, not including the rarely measured and frequently asymmetric third molars. Inter-side correlations in morphology are slightly lower in magnitude and more variable among traits. To illustrate, upper and lower molar MD (mesiodistal) and BL (buccolingual) diameters of Pima Indians have mean antimeric correlations of 0.841 for males (s = 0.026) and 0.833 for females (s = 0.017); eight morphological traits observed on these same teeth have a mean Kendall's *tau* of 0.788 for males (s = 0.108) and 0.761 for females (s = 0.113) (Noss *et al.*, 1983b). In Mizoguchi's (1988b) analysis of metric and morphologic antimeric correlations in a Japanese sample, the mean inter-side correlation for 18 crown diameters is 0.864 for both males (s = 0.037) and females (s = 0.059). For morphological traits observed on key teeth (UI1, UP1, UM1, LP1, LM1), the mean tetrachoric correlation between antimeres is 0.766 (s = 0.170). For variable teeth (UI2, UP2, UM2, LP2, LM2), the mean correlation is 0.629 (s = 0.306). In an analysis of seven crown traits observed on 16 teeth, Baume and Crawford (1979) list antimeric correlations (Kendall's *taus*) for Caribs, Creoles, and four Mexican samples. For their large Carib sample, the mean *tau* value is 0.823 (s = 0.147) for key teeth and 0.768 (s = 0.274) for variable teeth. Calculating mean *tau* values across all four Mexican samples, the mean correlation is 0.792 (s = 0.143) for key teeth and 0.739 (s = 0.178) for variable teeth. The mean antimeric correlations calculated from these three data sets are remarkably close, especially given the disparate nature of the samples and the use of different sets of crown traits.

It appears that antimere correlations for morphological traits are more comparable across samples than measures of concordance or percentage asymmetry. In fact, the general position that low frequency traits show more bilateral asymmetry than high frequency traits, suggested by Ossenberg (1981) for nonmetric skeletal traits and Saunders and Mayhall (1982a) for nonmetric crown traits, receives some support from a consideration of antimeric correlations. An analysis of Mizoguchi's (1988b) data set that includes 14 traits observed on 26 teeth yields a Spearman's rank order correlation of 0.851 (P < 0.01) between total trait frequency and tetrachoric correlations between antimeres. How precise the relationship is between total trait frequency and degree of asymmetry is unknown at this time.

If fluctuating asymmetry in tooth size and morphology is a by-product of developmental noise, then left–right differences should have no genetic basis. Studies of antimere size differences in twins (Potter and Nance, 1976; Mizoguchi, 1987; Corruccini *et al.*, 1988) and families (Bailit *et al.*, 1970; Townsend and Brown, 1980) find no significant genetic variance component with heritability estimates approximating zero. Staley and Green (1974) and Mizoguchi (1989) reached the same conclusion in their analyses of morphological asymmetry in twins. Mizoguchi found no evidence for independent genetic effects for trait expression in the two sides of the dentition and notes that when asymmetry does covary between twins, it is more likely due to common environmental factors than to shared genes.

Many investigators have shown that additive genetic variance plays no discernible role in antimere size asymmetry, and this likely applies to morphology as well. If genes are involved in asymmetric expression, it is through such indirect effects as elevated levels of homozygosity, chromosomal abnormalities (e.g., Down syndrome), and developmental anomalies (e.g., cleft lip/palate). In such instances, an organism's ability to buffer environmental perturbations during development is challenged so individuals tend to exhibit more asymmetry than normal, i.e., show decreased developmental canalization (Lerner, 1954; Waddington, 1957; Bader, 1965; Barden, 1980). While there is no evidence to support direct genetic control of asymmetry, there is abundant experimental evidence that shows increased asymmetry can be induced by a variety of environmental stressors (e.g., noise, heat, cold) invoked during the period of dental development (Siegel and Doyle, 1975a,b; Siegel *et al.*, 1977; Siegel and Mooney, 1987).

Are asymmetries in the dentition correlated with one another? Some workers have concluded that asymmetries in crown size between different teeth are not correlated (cf. Townsend and Brown, 1980). Others fail to discern any relationship between morphological and odontometric asymmetries (Baume and Crawford, 1980; Noss *et al.*, 1983b; Mayhall and Saunders, 1986). Contrary to these results, Mizoguchi (1986) found correlated asymmetries in tooth size, especially among BL diameters of the posterior teeth. In a joint analysis of 10 nonmetric and 18 metric crown traits, Mizoguchi (1990) also detected interaction between size and morphological asymmetries. Given the nature of interactions in the dentition, correlated asymmetries are not a surprising finding. It seems unlikely that local developmental disturbances in one side of the jaw would always manifest themselves only in terms of single asymmetric dental variables. Although conjectural, the timing of events may be the key. Disruptive local conditions early in development could affect multiple variables simultaneously while later disruptive events would influence only single crown or cusp dimensions and morphologic trait expressions. Still, the fact that most researchers fail to reveal correlated asymmetries among tooth dimensions and between tooth size and morphological trait expression indicates that the relationships revealed by Mizoguchi (1986, 1990) are weak and difficult to detect under most circumstances.

In Dahlberg's (1945a) elaboration of the field model, the mesial members of a tooth district are its most stable elements while distal members are the most variable elements in terms of both evolution and development. Ordinarily, the enhanced variability of distal teeth is attributed to increased environmental inputs into their development. Key teeth, on the other hand, should better reflect the genetic program directing their ontogeny. If this interpretation of differences among members of the same morphological class is correct, then the variable teeth in a given tooth district should exhibit higher levels of asymmetry than the key teeth.

Barden (1980) found that the distal members of a tooth district do show elevated levels of dimensional asymmetry compared to the key teeth for individuals affected by Down syndrome. However, this pattern is not apparent in his control series where the mesial and distal elements show antimere correlations of similar magnitude. Several other researchers, however, report that mesial teeth exhibit less asymmetry than distal teeth, especially in the incisor and molar tooth districts (Garn and Bailey, 1977; Townsend and Brown, 1980, 1981a; Mizoguchi, 1986; Kieser and Groeneveld, 1986). Recalling our earlier discussion of the enigmatic 'premolar tooth district,' it is not surprising that first and second premolars violate predictions on how mesial and distal teeth should behave. In terms of average left–right differences or antimere correlations, first and second upper and lower premolars consistently show similar levels of fluctuating asymmetry.

As a rule, morphological traits do show the pattern of asymmetry predicted by the field model. When Carabelli's trait is expressed unilaterally or bilaterally, it is asymmetric 15.8% of time on UM1 but 56.3% on UM2 and 44.4% on UM3 (Saunders and Mayhall, 1982a). Trait expression on distal members of a field also show lower antimere correlations than the key teeth (Noss *et al.*, 1983b; Mayhall and Saunders, 1986; Mizoguchi, 1988b). Corruccini and Potter (1981) reported that the distal cusps of the lower molars (hypoconid, hypoconulid) showed more asymmetry than the protoconid, the most mesial cusp. Increased asymmetry of crown trait expression on distal teeth and even on distal cusps within individual teeth is likely due to environmental factors and developmental noise, although it is not possible to specify the source of this noise.

Methodologically, scoring bilateral structures that show some degree of fluctuating asymmetry appears to be a straightforward proposition. Workers should always score both antimeres for trait expression. However, after initial data collection, with observations on both sides of the dentition in hand, how should an individual's phenotype be quantified? Several options are apparent, but which can be justified on genetic and developmental grounds? To illustrate, assume an individual exhibits a grade 2 Carabelli's trait on their left UM1 and grade 0 on its antimere. The alternatives are to assign the individual: (1) both 0 and 2 phenotypes and analyze the teeth separately; (2) an average score of the two sides $[2 + 0/2 = 1]$; (3) the phenotype showing the lowest grade of expression [0]; or (4) the phenotype showing the highest grade of expression [2]. The unit of study in method 1 is teeth (tooth count) while in methods 2–4, it is the individual (individual count).

Which counting method gives a phenotypic value that bears the closest relationship to the individual's genotypic value? In this respect, method 1 is particularly problematic because it assumes an individual has genotypes corresponding to both trait presence and trait absence. Averaging sides would probably not seriously distort an analysis but this method is rarely used in morphological studies. Whether method 3 or 4 is adopted depends on the trait in question. Method 3 might be appropriate for the hypocone and hypoconulid, because interest is in the reduction and loss of major molar cusps. If a left LM2 is 5-cusped (hypoconulid present) and its antimere is 4-cusped (hypoconulid absent), emphasis could appropriately be placed on cusp reduction rather than cusp retention. In most instances, however, concern is more with the presence of structures rather than their absence. In these cases, the antimere exhibiting maximal trait expression is thought to best represent an individual's genotypic potential for that trait. The antimere that shows a lower degree of expression or even trait absence is thought to reflect the action of environmental factors that did not allow the tooth to express its full genetic potential for that trait. Of course, an alternative method that avoids the potentially thorny problem of asymmetry is to focus on just left side or right side trait expression, but this usually leads to marked data loss, especially in archaeologically-derived dental samples.

Although many types of morphological analyses can be accomplished efficiently using tooth, individual, or side count methods, with little loss of information, the most controversial aspect of scoring methodology involves the estimation of sample frequencies. Green *et al.* (1979) reviewed the options for estimating frequencies for bilateral nonmetric skeletal variants, focusing on total side, individual, and unilateral counting procedures. The total side method takes the total number of presence phenotypes observed on either the left or right sides and divides by the total number of observable sides. With well preserved remains, sample size will approximate 2n; with poor preservation, sample size falls between 1n and

2n. For the individual count method, sample size is always 1n. This method assigns each individual one phenotypic value whether or not the trait is observable on one or both sides. When only one side can be scored, the individual is assigned the phenotypic value of that side. In cases where both sides are scored, the single phenotypic value is usually the highest degree of trait expression. The side count, if rigidly followed, yields a sample size of 1n only in perfectly preserved remains.

Green *et al.* (1979) strongly endorsed the use of the total side count method for estimating the frequency of bilateral nonmetric traits. They recognize the major shortcoming of this method is the artificial increase in 'sample size' (n) that sometimes approaches 2n with well preserved materials. As sample size is a critical component of most statistical procedures, they cite a modification of the variance calculations for the oft-used Mean Measure of Divergence that takes inflated sample size into account. Ossenberg (1981) also argued for the use of total side counts but on different grounds. The individual count method, for example, gives equal weight to individuals who exhibit phenotypes of either presence–absence ($+/-$) or presence–presence ($++$); both are considered present ($+$). She feels that phenotypic equivalence is unjustified given the tenets of the threshold model of inheritance (chapter 4). That is, individuals who exhibit presence–absence are closer to the threshold for trait expression than individuals who exhibit bilateral presence so their underlying genotypic values differ as well.

Green *et al.* (1979) and Ossenberg (1981) make a good case for using the total side count method to estimate population frequencies for nonmetrical skeletal traits. Although these traits are superficially similar to nonmetric crown and root traits, there are some major underlying differences between nonmetric dental and skeletal variants. Most importantly, from the standpoint of counting procedure, is degree of asymmetry. The 16 nonmetric skeletal traits listed by Green *et al.* (1979) for a large composite California Indian series have a mean inter-side correlation of 0.423 (s = 0.150), about half that of most dental traits. Presence–absence asymmetry, including bilateral absence cases, averages 20.1% (s = 8.07). When bilateral absence is not included in percentage asymmetry, this frequency is elevated to 59.0% (s = 14.89). The prevalence of asymmetry in nonmetric skeletal traits may indeed necessitate the use of a total side count. However, the logic of their arguments for using this method in preference to individual counts may not extend to crown and root traits that show significantly less asymmetry.

Although dental researchers often use total side counts, the authors have traditionally preferred the individual counting method (cf. Turner, 1967;

Turner and Scott, 1977; Scott, 1980). Our reasoning is based partly on genetic and partly on statistical grounds. Ossenberg (1981) claims that individuals exhibiting grade 0 and grade 1 on the right and left sides are genotypically different from those who exhibit grade 1 on both sides. This is based on observations such as those presented by Falconer (1960) where two inbred strains of mice, with predominantly five or six lumbar vertebra, are crossed. In reciprocal crosses between the two strains, 15.5% and 25.0% of the offspring exhibit $5\frac{1}{2}$ vertebra (i.e., fifth vertebra sacralized on one side but not the other), so it appears there is some genetic basis to this intermediate asymmetric phenotype. This genetic intermediacy for asymmetric presence–absence phenotypes has not been demonstrated for crown traits. Green *et al.* (1979) discuss a procedure for adjusting the variance of Mean Measures of Divergence that takes into account the inflated samples sizes associated with the total side count method. They do not, however, provide similar adjustments for every other statistic and associated statistical distribution that is predicated on individuals serving as the basic unit of measurement. Because of this statistical consideration, along with our view that individuals are best characterized by a single phenotype, we use the individual count method in this volume to estimate population frequencies. For Carabelli's trait, it has been shown that trait frequencies in a single population estimated by tooth, individual, and unilateral side counts are almost identical (Scott, 1980; Bermúdez de Castro, 1989). For dental traits, but perhaps not for nonmetric skeletal traits, if there is a bias in any of the three methods, it is a modest bias.

Sex dimorphism

Dental researchers consistently find low levels of sexual dimorphism in human crown dimensions (cf. Moorrees, 1957; Garn *et al.*, 1964, 1966c; Mizoguchi, 1988a). Although male teeth are only 2–6% larger than female teeth, discriminant function analysis of tooth size can correctly classify the sexes 86% of the time (Garn *et al.*, 1977). Hypodontia and hyperodontia are also dimorphic. Females have a higher frequency of missing teeth and a lower frequency of supernumerary teeth than males (Brook, 1984). Do males and females show similar differences in the presence and expression of crown and root traits, which would require that data be analyzed and reported separately for the sexes?

Hundreds of samples have been tested for sex differences in crown and root trait expression. Not surprisingly, given the nature of sampling distributions, reports of significant sex differences for particular traits vary from one sample to another. To illustrate, many workers have found significant male–female differences in the expression of Carabelli's trait (Goose and Lee, 1971; Kaul and Prakash, 1981; Kieser and Preston, 1981; Townsend and Brown, 1981b; Mizoguchi, 1985; Scott *et al.*, 1983), but others find no sex difference in this trait (Garn *et al.*, 1966d; Turner, 1969; Scott, 1980; Townsend *et al.*, 1992). Mizoguchi (1985) provides an example of the irregularities one encounters when assessing sex differences in crown trait expression. He presents frequency distributions for 12 crown traits observed in males and females in two Japanese samples (n exceeds 100 for both males and females in 42 of 48 cases). In his tests for sex differences, six of 24 comparisons yield significant male–female differences at the 0.05 level. One sample shows a significant sex difference while the other does not in four of the six cases. Of the 12 traits, only Carabelli's trait shows a significant sex difference in both samples.

The only crown trait identified to date that appears to show a consistent sex dimorphism across diverse samples is the distal accessory ridge of the upper and lower canines (Scott, 1977a; Kaul and Prakash, 1981; Kieser and Preston, 1981; Scott *et al.*, 1983). Pima Indian male and female class frequency distributions for the lower canine distal accessory ridge are shown in Fig. 3.6. The chi-square value for this comparison is a robust 71.34 (4 d.f.; P < .0001). Sex dimorphism in the mesiodistal diameter of the lower canine is among the highest of all tooth dimensions in modern humans (Garn *et al.*, 1977). There is also a modest but significant correlation between the expression of the distal accessory ridge and the mesiodistal diameter of the lower canine (section 'Intertrait association'). For these reasons, Noss *et al.* (1983a) estimated the degree to which the dimorphism in tooth size influenced the dimorphism in the expression of the ridge. In mean grade of trait expression, the LC distal accessory ridge shows a male–female dimorphism of 75%. Controlling for the effects of crown size, the dimorphism is reduced to 47%. While tooth size plays some role in the sex difference shown by the canine distal accessory ridge, other factors are also at work. This trait often lacks a precursor on the dentine surface of upper canines (Sakai *et al.*, 1967) so the genes that control the expression of this trait involve both folding of the inner enamel epithelium and enamel matrix formation. The possibility that genes on the X chromosome influence the expression of the canine distal accessory ridges has not been explored.

When workers find significant male–female differences for a given crown trait, it is the males who ordinarily show the highest frequency and degree of expression. Harris (1980) contends that for shoveling of the upper central incisor, females exhibit more shoveling than males. In six composite

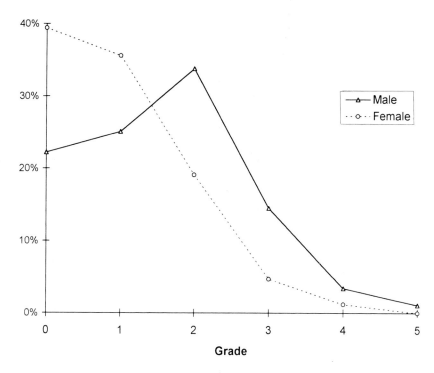

Fig. 3.6   Male and female class frequency distributions for the lower canine distal accessory ridge. This is the only dental morphological trait that shows a consistent, albeit moderate, sex dimorphism. In Pima Indians (379 males, 424 females), there is rightward shift in the class frequency distribution for males due to their higher trait frequency and more pronounced expressions.

geographic samples made up of 38 subsamples and more than 19,000 individuals, females show higher frequencies of shoveling in all but one instance. Subtracting the total female frequency from the total male frequency for shoveling, he found the following mean percentage differences: American blacks (+1.0); New World Indians (−1.5); Caucasians (−1.6); Asians (−2.5); Polynesians (−7.9); and Melanesians (−14.3). With the exception of Pacific Island groups, the differences in shoveling between males and females are minor and statistically nonsignificant in composite and individual samples. While Harris (1980) focused on data based on ranked scales, researchers who measure shoveling in terms of lingual fossa depth have found both a significant sex difference in favor of females (Rothhammer *et al.*, 1968) and no sex difference (Aas and Risnes, 1979a; Mizoguchi, 1985). Kirveskari and Alvesalo (1981) found greater lingual fossa depth in 13 Finnish 47,XYY males than in male and female relatives.

The differences were in the same direction for both UI1 and UI2 measurements, but they were statistically significant only for UI2. Overall, the case for a panhuman sexual dimorphism in shoveling is not well established. Even when differences are found, they are generally so slight that disentangling any underlying developmental mechanisms (e.g., X-linked dosage effect) would be a challenge.

Research indicates that genes on the sex chromosomes are involved in various facets of dental ontogeny. For example, the structural gene for amelogenin is located on the X and Y chromosomes (Lau *et al.*, 1989). Chromosomal abnormalities involving the sex chromosomes also influence crown and root morphology in a variety of ways. Individuals with Turner's syndrome (XO or 45,X) have smaller teeth than their relatives and other differences are found in crown and root morphology. Kirveskari and Alvesalo (1982) find XO individuals have reduced frequencies and expressions of shoveling, especially on UI2, the hypocone of UM1 and UM2, the hypoconulid of LM1 and LM2, and Carabelli's trait. The differences are not always statistically significant, but they are consistently in the direction of morphological simplification. Interestingly, in a sample of 87 Finnish 45,X females, Tomes' root and two-rooted lower premolars were far more common in individuals lacking an X chromosome (23–25%) than in relatives (2–3%) (Varrela, 1992). The simplification of the incisor and molar crowns stands in contrast to the increased complexity observed in lower premolar roots.

As the X chromosome exerts its primary influence on enamel while the Y chromosome promotes both enamel and dentine growth (Alvesalo and Varrela, 1991), crowns and roots may follow a different set of instructions from the sex chromosomes during development. In chapter 2, we did not include taurodontism on our trait list because of its rarity in modern human populations (it was common, however, in Neanderthals). This trait, which involves an enlarged and basally extended pulp chamber with a root furcation near the apex of the roots, shows the contrasting roles of extra X and Y chromosomes on development. Males who are 47,XYY exhibit relatively normal roots with a low frequency of taurodontism (Alvesalo and Varrela, 1991). Females with extra X chromosomes (XXX and XXXX) exhibit more taurodontism than is found in the general population. Although the sample of females with extra X chromosomes was only six, four of them showed hypo- or hypertaurodontism. Both cases of hypertaurodontism involved individuals with two extra X chromosomes (Varrela and Alvesalo, 1989).

Even though genes on the sex chromosomes are involved in dental development, crown and root traits still show little or no sex dimorphism at the phenotypic level. When differences are found, they are usually inconsistent among samples and low-order in magnitude. For this reason, crown and root traits share with autosomal genetic traits the advantage that male and female data can be pooled to estimate population frequencies. This advantage is particularly critical in the analysis of small skeletal samples where subdivision by sex often results in intolerably small samples. We feel that the methodological strategy of combining observations on males and females is justifiable in most cases although some workers do not agree with this assessment (cf. Harris, 1977). In chapters 5, 6, and 7, frequency data are based on combined male and female observations.

Intertrait association

Garn and his co-workers (Garn *et al.*, 1963, 1965, 1967, 1968a), among others, have demonstrated interrelationships among many kinds of dental variables, including tooth size, agenesis, eruption sequences, timing of mineralization, and so forth. Tooth dimensions, in particular, are strongly correlated among themselves. Third molar agenesis is related to other types of missing teeth, e.g., UI2, P2. Do morphological traits exhibit parallel interrelationships among themselves and between tooth dimensions and agenesis?

In population studies of tooth morphology, a working assumption is that crown and root traits are expressed independently of one another. Moreover, these traits are thought to show little or no interaction with crown size or agenesis. Actually, many authors have addressed these issues and found that the expression of some crown traits is associated with other crown traits and also with tooth dimensions and agenesis. These studies do not invalidate the working assumption of trait independence, but they do illustrate how aspects of the dentition, including morphology, operate in the context of an integrated system with overlapping developmental fields.

To assess interrelationships within the dentition, workers use a wide array of correlational methods. For odontometric traits, interclass correlation coefficients are commonly used, often in conjunction with multivariate statistics such as principal components or factor analysis. For morphological ('nonmetric') crown traits, correlation methods are more diverse. When these traits are treated as presence–absence variables, or dichotomized at a breakpoint other than absence, intertrait association is measured by contingency coefficients, *phi*, or tetrachoric correlations. As most crown traits have been classified on graded scales, rank order correlation coefficients (Kendall's *tau*, Spearman's *rho*) are also used to quantify relationships. Finally, because of certain advantages inherent in parametric statistics, others calculate interclass correlation coefficients between crown traits. As we will show, these different correlation methods give comparable results on the same data set, both in terms of assessing statistical significance and measuring degree of association.

Few attempts have been made to correlate dental morphological variables with non-dental variables (e.g., stature, fingerprint patterns, blood groups, etc.). For that reason, our focus is on trait interactions within the dentition. The first question centers on determining to what extent the expression of a given morphological variable is correlated with other morphological variables. Second, to what extent are morphological variables influenced by crown size? Finally, is morphological trait expression related to agenesis, in particular missing third molars?

Most studies aimed at measuring association between morphological traits or between morphological traits and tooth size focus on a limited suite of traits (cf. Garn *et al.*, 1966a,e; Keene 1968; Sofaer *et al.*, 1972b; Mizoguchi, 1978, 1985; Kirveskari and Alvesalo, 1982; Kieser and Becker, 1989; Reid *et al.*, 1991). For that reason, published and unpublished correlations among many morphological traits, derived from a large sample (n = 1251) of American Southwest Indians (Scott, 1973, 1977a,b,c, 1978, 1979), are presented as a baseline to compare the findings of other workers. Additionally, within- and between-field interactions among morphological traits and tooth size are assessed in a large sample of Pima Indians (n = 1528). Correlations from this analysis supplement, where possible, those derived from other American Southwest Indian groups (see next section below). In the section on size-morphology interaction, the primary tables come from the analysis of Pima Indian dentitions.

Interactions within and among crown traits

Members of the same morphological class, say first, second, and third upper molars, are similar in form and size (compared to other classes) and have the potential for expressing the same crown traits. Upper molar crown traits such as the hypocone, Carabelli's trait, and cusp 5, can be expressed at the same location on all three members of this class or tooth district. In other words, single traits can be observed on three separate teeth.

It has long been known that the expression of a particular trait on one member of a tooth district is not independent of trait expression on other members of the district. Such cases where a single trait is observed on two or three members of a tooth district are referred to here as within-field interactions. These particular relationships are expected given either the

Table 3.1. *Interclass correlation coefficients between two members of a tooth district for single morphological traits; referred to in text as within-field interactions.*

Trait (teeth)	Southwest Indians r	Pima Indians r
Shoveling (UI1–UI2)	0.469	0.585
Shoveling (LI1–LI2)	0.832	NA
Tuberculum dentale (UI1–UI2)	0.347	NA
Distal accessory ridge (UP1–UP2)	0.244	NA
Hypocone (UM1–UM2)	0.456	0.428
Carabelli's trait (UM1–UM2)	0.309	0.443
Cusp 5 (UM1–UM2)	0.411	0.593
Parastyle (UM1–UM2)	NA	0.230
Multiple lingual cusps (LP1–LP2)	0.175	0.320
Hypoconulid (LM1–LM2)	NA	0.098
Groove Pattern (LM1–LM2)	0.161	NA
Cusp 6 (LM1–LM2)	0.367	0.303
Cusp 7 (LM1–LM2)	0.227	0.317
Protostylid (LM1–LM2)	0.560	0.606

UI1/UI2: upper first/second incisor; UP1/UP2: upper first/second premolar; LP1/LP2: lower first/second premolar; UM1/UM2: upper first/second molar; LM1/LM2: lower first/second molar. NA: not available.

model of morphogenetic fields (Butler, 1939; Dahlberg, 1945a) or clones (Osborn, 1978). We do know, however, that these relationships are not perfect. The expression of the hypocone on UM1, for example, does not predict exactly the expression of the hypocone on UM2. If within-field correlations are not perfect ($r = 1.0$), what is the degree of relationship between single traits observed on different members of the same morphological class?

Interclass correlations are presented in Table 3.1 for 13 traits observed within tooth districts (shoveling counted as a single variable in upper and lower incisors). Given the large sample sizes for both Southwest Indians and Pima Indians, all correlations are highly significant ($P < .01$). The lowest correlation is for the hypoconulid (the presence or absence of which determines lower molar cusp number), but this value may reflect the low level of variation in trait expression on LM1 in Pima Indians. Groove pattern of the lower molars also shows a relatively low within-field correlation while shoveling of the lower incisors shows an exceptionally high correlation. Beyond these three traits, within-field correlations for other traits fall in the range of 0.25 to 0.60. For those traits observed in both Southwest Indians and Pima Indians, within-field correlations are similar in magnitude and order. Consistently high correlations are shown by the protostylid, shoveling, cusp 5, and the hypocone while Carabelli's trait, cusp 6, cusp 7, and *tuberculum dentale* have intermediate values. Accessory distal ridges of the upper premolars, multiple lingual cusps of the lower premolars, and the parastyle show the lowest within-field correlations.

In a study of Melanesian dentitions, Sofaer *et al.* (1972b) calculated within- and between-field correlations for five traits observed on ten teeth. Rather than using ranked scales, these authors scored traits as either nonaffected or affected. While their within-field correlations are not in exact agreement with those in Table 3.1, they fall in the primary range of correlation values noted above. They found high within-field correlations for shoveling (UI1–UI2: 0.67) and Carabelli's trait (UM1–UM2: 0.65) and lower values for the hypocone (UM2–UM3: 0.23), lower molar cusp number (LM1–LM2: 0.35), and lower molar groove pattern (LM1–LM2: 0.29). All correlations were significant at the 0.01 level. Axelsson and Kirveskari (1982), although primarily interested in trait associations between deciduous second molars and permanent first molars, found a low within-field correlation for lower molar cusp number (LM1–LM2: 0.144) in Icelandic school children that approximates the value found in Pima Indians.

Kirveskari and Alvesalo (1982) assessed shoveling by lingual fossa depth rather than grades of expression and found a within-field correlation (UI1–UI2) of 0.573 for this trait. This interclass correlation is almost identical to the values reported for Southwest Indians and Pima Indians and also that recorded by Sofaer *et al.* (1972b) for Melanesians. Mizoguchi's (1985) tetrachoric correlations for the shoveling field in Japanese samples have a mean of 0.759 that is only slightly higher than the values reported by other workers. These results suggest two things. First, method of scoring, whether by presence–absence, ranked scales, or direct measurement, does not significantly affect the outcome of a correlation analysis. Second, within-field correlations are probably not population specific but apply to a wide array of groups – in this instance, Europeans, American Indians, Melanesians, and East Asians.

When dental morphologists state that crown traits are expressed independently of other traits, they are not referring to within-field interactions. For example, cusp 6 expression on the lower first molar is not independent of cusp 6 expression on the lower second molar. For a basic descriptive characterization of a group, it is appropriate to report trait frequencies for all members of a tooth district. However, for assessing phenetic distance in population studies, it is recommended that each trait's frequency be based on one member of the morphological class.

Table 3.2. *Interclass correlation coefficients between members of two tooth districts for single morphological traits or between two different traits (in same or different tooth district).*

Trait (teeth)	Southwest Indians r	Pima Indians r
Shoveling (UI1–LI1)	0.514	NA
Shoveling (UI1–LI2)	0.460	NA
Shoveling (UI2–LI1)	0.309	NA
Shoveling (UI2–LI2)	0.295	NA
Tuberculum dentale (UI1–UC)	0.316	NA
Tuberculum dentale (UI2–UC)	0.318	NA
Distal accessory ridge (UC–LC)	0.341	NA
Hypocone (UM1)–Carabelli's trait (UM1)	0.334	0.303
Hypocone (UM1)–Carabelli's trait (UM2)	0.110	0.129
Hypocone (UM2)–Carabelli's trait (UM1)	0.352	0.247
Hypocone (UM2)–Carabelli's trait (UM2)	0.130	0.143
Carabelli's trait (UM1)–Protostylid (LM1)	0.286	0.328
Carabelli's trait (UM1)–Protostylid (LM2)	0.319	0.300
Carabelli's trait (UM2)–Protostylid (LM1)	0.216	0.313
Carabelli's trait (UM2)–Protostylid (LM2)	0.319	0.263

UI1/UI2: upper first/second incisor; UC: upper canine; LC: lower canine; UM1/UM2: upper first/second molar; LM1/LM2: lower first/second molar.

Within-field correlations for specific traits are generally acknowledged if not often calculated, but there are also several other types of associations between morphological traits. These associations involve: (1) single traits in opposing tooth districts (i.e., upper and lower jaws), (2) single traits on members of different tooth districts in the same jaw; (3) different traits on members of the same tooth district; and (4) different traits in opposing tooth districts.

The most consistent intertrait associations that do not involve within-field correlations for single traits are presented in Table 3.2 for Southwest Indians and Pima Indians. Most of these cases have been reported elsewhere (Scott, 1977a,b,c, 1978, 1979), but the method of analysis differs. In earlier studies, Kendall's *tau* values were calculated for six to ten individual samples. While the rank order correlations showed some sampling variance, the overall patterns of association were consistent. The values in Table 3.2 are, by contrast, interclass correlation coefficients computed for a combined Southwest Indian sample. The results of the two sets of analyses are effectively identical, suggesting that parametric (interclass correlations) and nonparametric statistics (Kendall's rank order correlations) are equally effective in detecting intertrait associations among ranked traits. The later study on Pima Indians involved observations on fewer traits so some correlations are noted as NA (not available). However, associations detected previously between Carabelli's trait and the proto-stylid (Scott, 1978) and Carabelli's trait and the hypocone (Scott, 1979) for individual Southwest Indian samples are supported by the correlations computed for Pima Indians.

Shoveling and the canine distal accessory ridge are two traits that show significant interjaw associations. For shoveling, all possible pairwise comparisons yield highly significant measures of association. Pairwise correlations are higher when they involve the stable upper central incisor rather than the more variable upper lateral incisor. While this would be predicted from the field model, it is unexpected that the lower central and lateral incisors show almost identical correlations to either upper incisor. The lower central incisor is more variable in terms of size and agenesis frequency than the lower lateral incisor so this tooth might be expected to show lower interjaw correlations. This does not, however, seem to be the case. The shoveling field, involving all upper and lower incisors, probably extends to lingual marginal ridge development on the canines of both jaws (Dahlberg, 1951), but shoveling was not scored on the canines for either Southwest Indians or Pima Indians. Additional traits not analyzed here that are likely to show significant interjaw correlations are upper and lower premolar odontomes and accessory ridges and upper and lower molar enamel extensions.

One trait that extends across two tooth districts is *tuberculum dentale* of the upper incisors and canines. The within-field (UI1–UI2) correlation for this variable closely approximates the two between-field correlations. In contrast to shoveling, the correlation involving the more variable lateral incisor does not differ from that for the central incisor.

Carabelli's trait figures prominently in the final two examples of intertrait association. This variable is correlated with the hypocone, a trait expressed on the same upper molar teeth as Carabelli's trait. Of the four correlations involving these two traits on the two upper molars, those involving Carabelli's trait of UM1 approximate 0.30 with hypocone expression on UM1 and UM2. There is a significant drop-off in correlation values (0.11–0.14) when hypocone expression is correlated with Carabelli's trait of UM2. This difference may, in part, be attributed to the reduced stability in Carabelli's trait expression on the more variable member of the upper molar district. However, this same pattern is not evident in the final intertrait association, that between Carabelli's trait of the upper molars and the protostylid of the lower molars. The four correlations in this instance are effectively identical, all approaching a level of 0.30.

The only intertrait association found in Southwest Indians and Pima Indians that has received much attention is that involving Carabelli's trait and the hypocone. Some early authors felt that large Carabelli's cusps in European dentitions compensated for reduced cusp number (i.e., hypocone loss) of the upper second and third molars. In other words, they were positing an inverse relationship between these traits. By contrast, Bolk (1915) found a higher frequency of Carabelli's trait on 4-cusped upper second molars than on tricuspal forms. Kotzschke and de Jonge (in Korenhof, 1960) suggested that reduced hypocones were associated with reduced expressions of Carabelli's trait. In a study of over 300 white male US Naval recruits, Keene (1965, 1968) found a relationship between diminished expressions of Carabelli's trait and hypocone reduction of both the upper first and second molars. We concur with these authors who conclude Carabelli's trait and the hypocone correlate in a positive rather than negative manner. Upper molars with large hypocones are more likely to have large Carabelli's cusps than those showing hypocone absence or reduction.

Principal components analysis (PCA) is a multivariate technique that reduces a complex set of interrelationships among many variables into a smaller number of components. The objective is to search for underlying structures, or components, which are related to significant proportions of the overall variance observed in a trait set. The analysis is designed so that each component is unrelated (orthogonal) to every other component. PCA is an exploratory tool that involves many options, including decisions on the number of components to extract and whether or not to orthogonally rotate components to maximize the loadings on each component.

Authors have employed PCA to assess the latent structures underlying tooth size variation within populations (cf. Potter *et al.*, 1968; Harris and Bailit, 1988). This analytical tool has been less frequently used to analyze the covariation among multiple tooth crown traits (cf. Mizoguchi, 1985).

PCA was conducted for 31 variables observed in the combined samples of Southwest Indians (excluding the Pima). Principal components were extracted from an unaltered correlation matrix (unities on the main diagonal) and then orthogonally rotated using Varimax criteria (Nie *et al.*, 1975). To determine the number of components to extract, we followed a conservative approach whereby all components were retained that contained more variance than the original variables, i.e., eigenvalues > 1.0 (Zegura, 1978).

Following the above criteria, 12 components, representing 62.4% of the total variance, were extracted from the correlation matrix for 31 variables. Those variables with the highest loadings on each of the 12 components are

Table 3.3. *Component loading scores from a principal components analysis of 14 morphological traits observed on 31 teeth; only highest loading scores are shown for 12 components with eigenvalues > 1.0 (for reference, trait with next highest score is shown in brackets).*

Trait (tooth)	Component	Component loading
Shoveling (UI1)	1	0.759
Shoveling (UI2)		0.596
Shoveling (LI1)		0.872
Shoveling (LI2)		0.839
[Winging (UI1)]		[0.180]
Carabelli's trait (UM1)	2	0.394
Carabelli's trait (UM2)		0.615
Protostylid (LM1)		0.674
Protostylid (LM2)		0.805
[Winging (UI1)]		[0.276]
Hypocone (UM1)	3	0.755
Hypocone (UM2)		0.760
Carabelli's trait (UM1)		0.550
[Protostylid (LM1)]		[0.188]
Tuberculum dentale (UI1)	4	0.744
Tuberculum dentale (UI2)		0.633
Tuberculum dentale (UC)		0.646
[Carabelli's trait (UM1)]		[0.235]
Mesial accessory ridge (UP1)	5	0.260
Distal accessory ridge (UP1)		0.594
Mesial accessory ridge (UP2)		0.661
Distal accessory ridge (UP2)		0.732
[Distal accessory ridge (UC)]		[0.256]
Cusp 5 (UM1)	6	0.785
Cusp 5 (UM2)		0.812
[Distal accessory ridge (UP1)]		[0.206]
Distal accessory ridge (UC)	7	0.692
Distal accessory ridge (LC)		0.824
[Groove pattern (LM2)]		[0.447]
Cusp 6 (LM1)	8	0.766
Cusp 6 (LM2)		0.774
[Deflecting wrinkle (LM1)]		[0.190]
Lingual cusp number (LP1)	9	0.741
Lingual cusp number (LP2)		0.578
[Groove pattern (LM2)]		[0.335]
Cusp 7 (LM1)	10	0.741
Cusp 7 (LM2)		0.708
[Distal accessory ridge (UP1)]		[0.263]
Groove pattern (LM1)	11	0.883
Groove pattern (LM2)		0.318
[Lingual cusp number (LP2)]		[0.207]
Deflecting wrinkle (LM1)	12	0.755
Groove pattern (LM2)		0.342
[Cusp 7 (LM2)]		[0.268]

UI1/UI2: upper first/second incisor; LI1/LI2: lower first/second incisor; UM1/UM2: upper first/second molar; LM1/LM2: lower first/second molar; UC: upper canine; LC: lower canine; UP1/UP2: upper first/second premolar; LP1/LP2: lower first/second premolar.

presented in Table 3.3. Variables with the next highest component loadings are shown in brackets to illustrate the magnitude of difference from traits characterizing each component.

With few exceptions, the interpretation of component loadings is straightforward, basically corroborating the pattern of intertrait associations already noted. That is, the first component represents the shoveling field of the upper and lower incisors. Carabelli's trait and the protostylid, reflecting both within- and between-field correlations, load on the second component. Carabelli's trait of UM1 but not UM2 loads on the third component with the hypocone of both upper molars. The fourth component involves lingual cingular ridges and tubercles (*tuberculum dentale*) that manifest themselves in a field gradient across two tooth districts, upper incisors and canines. The fifth component is represented by accessory occlusal ridges of the upper premolars, although the mesial accessory ridge of UP1 loads on this component to a lesser degree than the other three accessory premolar ridges. The final seven components reflect primarily individual fields where the same trait is expressed on two teeth within a single tooth district. That is, cusp 5 loads on component 6, cusp 6 on component 8, lower premolar lingual cusp number on component 9, cusp 7 on component 10, and groove pattern on component 11. An interjaw field is indicated for the distal accessory ridges of the upper and lower canines that load on component 7. The deflecting wrinkle was observed only on LM1 so it loads largely by itself on component 12, although groove pattern of LM2 loads weakly on this final component.

Both univariate and multivariate analyses suggest that tooth crown traits are largely independent of one another. The majority of components from the PCA reflect within-field interactions. In only two instances do two different traits load on single components. This result is in sharp contrast to a PCA of tooth size where 28 (or more) variables load on only a few major components.

Interactions between crown morphology and tooth size

Dental researchers ordinarily study either crown morphology or tooth size, but rarely consider both types of variables simultaneously. For that reason, the intricacies of size-morphology interaction in the dentition are incompletely understood. In fact, does tooth size have anything to do with crown morphology? Before addressing that question, we first take up the issue of interrelationships among crown dimensions.

Many workers have calculated interclass correlation coefficients among mesiodistal (MD) and/or buccolingual (BL) diameters of upper and lower tooth crowns. These correlations are consistently positive and significant within and between morphological classes, but they vary in magnitude from approximately 0.20 to 0.80 with most in the range of 0.30 to 0.60 (cf. Harris and Bailit, 1988). Is there patterning in this variation among correlation coefficients? As a correlation matrix for 28 variables (7 upper teeth and 7 lower teeth, minus M3, measured in two-dimensions) includes 378 correlations, the search for patterning requires a statistical design that reduces a complex system of variation into a smaller number of factors, each representing a greater percentage of the total variance than any single variable. Toward this end, factor or principal components analyses are the most commonly used designs.

In a familial study of tooth size in over 600 Pima Indians, Potter *et al.* (1968) employed factor analysis to extract latent factors from a correlation matrix involving 56 variables (left and right antimeres considered separately). Their first three factors accounted for 51.5% of the total variance in the system. For the rotated factor solution, the first factor centered on MD and BL measurements for posterior teeth, with factor loadings somewhat higher for molars than premolars. The second factor was for BL diameters of the anterior teeth. The highest loadings on the third factor were for MD diameters of the anterior teeth.

Harris and Bailit (1988) performed a principal components analysis of tooth size in a large sample of Solomon Islanders (n > 2000). In their unrotated solution, all 28 metric variables loaded on the first component so this was interpreted as an overall tooth size factor. The second component involved BL diameters for most teeth while the third and fourth components represented both MD and BL dimensions of the anterior teeth and molars, respectively. With Varimax rotation, their PCA solution was slightly different. Their first four components, accounting for 62% of the total variance were: (1) BL dimensions of the anterior teeth; (2) MD and BL dimensions of the posterior teeth; (3) MD dimensions of the anterior teeth; and (4) MD and BL dimensions of the premolars.

The findings of Potter *et al.* (1968) and Harris and Bailit (1988) are largely congruent, especially given the disparate nature of the samples studied. For the rotated solutions, both sets of researchers found that MD and BL diameters of the anterior teeth loaded on separate components. The only difference is that in the Pima Indian sample, both premolars and molars load on a common factor while they load on two separate components in Melanesians. These results are supported by the univariate correlations of Garn *et al.* (1968a) who found lower correlations between MD and BL diameters of the anterior teeth than for posterior teeth in Ohio whites. Although Mizoguchi (1981) only measured MD diameters, he found that canonical variates first distinguish general tooth size and then separate anterior from posterior teeth. On this basis, he proposed that anterior and posterior teeth, not morphological classes, were the primary functional units in the human dentition. This proposition receives support from the hominid fossil record where pronounced size differences are evident between the anterior teeth as a group and the posterior teeth as a group, especially in robust Australopithecines.

Our goal is not to assess the many studies that have dealt with interrelationships among crown dimensions. Workers sometimes find different patterns of component loadings in their PCA of tooth size, but there is general agreement that relatively few components underlie tooth size variability. What would happen if crown morphology was analyzed concurrently with tooth size? To address this issue, we performed a principal components analysis that included 28 metric variables and 24 morphological variables in Pima Indians. All morphological observations were made by one of the authors (GRS) while all crown diameters were measured by an odontometrician (R.H.Y. Potter).

A components analysis of 52 metric and morphologic variables results in 14 components with eigenvalues greater than 1.0. These components account for almost two-thirds (65.3%) of the total variance in the matrix. The key variables that load on each of these 14 components are summarized below:

(1) MD and BL dimensions of posterior teeth. All premolar and molar dimensions have loadings between 0.60 and 0.78 on the first component. The next highest loadings are for MD diameters of the upper (.52) and lower (.50) canines. Total contribution to variance: 23.3%.

(2) BL diameters of anterior teeth. Loadings for incisors and canines on this component range from 0.64 to 0.83. The next highest loading is 0.32 for the BL diameter of LP1. Total contribution to variance: 5.9%.

(3) MD diameters of the anterior teeth and distal accessory ridge (LC). Loadings for the incisors vary from 0.60 to 0.77 while the canines show slightly lower values (0.43 for UC and 0.55 for LC). For MD but not BL diameters, the canines fall about equally in component 1 and component 3, not an unexpected finding given their position between the incisor and premolar-molar fields. The distal accessory ridge of the lower canine has a loading of 0.30 on this component. Total contribution to variance: 4.9%.

(4) Carabelli's trait (UM1, UM2) and protostylid (LM1, LM2). This

component represents both the within- and between-field interactions noted earlier. The loadings for Carabelli's trait (0.58, 0.61) are slightly lower than those for the protostylid (0.77, 0.80). Total contribution to variance: 4.1%.

(5) Cusp 6 (LM1, LM2) and hypoconulid (LM1). This is primarily for the cusp 6 field but the hypoconulid of LM1 has a high negative loading on this component. This is an expected finding. By definition, as cusp 6 increases in rank, the hypoconulid is reduced in rank. Total contribution to variance: 3.5%.

(6) Shoveling (UI1–UI2) and winging (UI1). Shoveling on both incisors has a loading of 0.80 on this component while that for winging is 0.51. Winging is not strongly correlated with shoveling but as an isolated trait, with no field member, it loads with the variable with which it is most strongly correlated. Total contribution to variance: 3.4%.

(7) Hypoconulid (LM2), cusp 6 (LM2) and MD diameter (LM1, LM2). Loadings for the hypoconulid and cusp 6 are 0.70 and 0.46, respectively. This is only the second component where tooth size loads with morphological variables. Not surprisingly, the dimensions that load on this component are MD diameters of the lower first and second molars (0.39, 0.50). Total contribution to variance: 3.2%.

(8) Cusp 5 (UM1, UM2). This component is exclusively for the cusp 5 field of the upper molars; this variable on UM1 and UM2 has loadings of 0.85 and 0.87, respectively. Total contribution to variance: 2.8%.

(9) Carabelli's trait (UM1) and the hypocone (UM1, UM2). As noted in the discussion of univariate correlations, Carabelli's trait of UM1 but not UM2 shows relatively high correlations with hypocone expression of both upper molars. While this component is primarily the hypocone field (0.78, 0.79 loadings for UM1, UM2), Carabelli's trait of UM1 stands out from all other metric and morphologic traits with a loading of 0.38. Total contribution to variance: 2.7%.

(10) Parastyle (UM1, UM2). This component is for the parastyle field with loadings of 0.57 and 0.77 for UM1 and UM2, respectively. Total contribution to variance: 2.5%.

(11) Cusp 7 (LM1, LM2). This component, representing the cusp 7 field, has extremely low loadings on all variables except cusp 7 where they are 0.77 and 0.75 for LM1 and LM2, respectively. Total contribution to variance: 2.4%.

(12) Lingual cusp number (LP1, LP2). This component again represents a single morphological field – that for multiple lingual cusps of LP1 and LP2 which have loadings of 0.79 and 0.75, respectively. Total contribution to variance: 2.3%.

(13) Tuberculum dentale (UC). As lingual tubercles on UI1 and UI2 were not scored in the Pima study, the expression of this trait on the canine loads alone on this component. Total contribution to variance: 2.2%.

(14) Deflecting wrinkle (LM1). This trait was scored only on LM1 so it has no field counterpart; as it is uncorrelated with all tooth dimensions and other morphological variables, it loads on a single component. Total contribution to variance: 2.0%.

The PCA of metric and morphologic traits in the Pima Indian dentition shows a high level of integration among tooth dimensions and mostly independent factors for morphological traits. The first three components are dominated by metric variables while subsequent components reflect within-field interactions of single morphological traits. The Carabelli's trait-protostylid and Carabelli's trait (UM1)-hypocone components are exceptions, but these components were expected given their relatively high univariate correlations.

While the PCA suggests that tooth size and crown morphology are largely unrelated, other evidence shows these two types of variables are not entirely independent. To illustrate, 672 metric-morphologic correlations were calculated and only 69 of these (10.3%) were negative. Moreover, while some traits are clearly not affected by size, others show consistent, albeit moderate, correlations with particular tooth dimensions.

Most workers who assess size-morphology interactions limit their analysis to the specific tooth a trait is observed on. For example, shoveling is considered relative to the dimensions of the upper incisors, while Carabelli's trait is considered in the context of upper molar dimensions. We have noted, however, that all MD and BL diameters of the upper and lower teeth are correlated and the pattern of correlations among dimensions can be reduced to three or four major components involving, in particular, contrasts between MD and BL diameters and anterior and posterior teeth.

In Table 3.4, we present not only those correlations between the expression of a given trait and the tooth the trait was observed on but also the mean correlations for each trait with MD and BL dimensions of all anterior and posterior teeth (upper and lower teeth combined). This set of correlations shows, first and foremost, that there is not a strong relationship between crown size and morphological trait expression. There is

Table 3.4. *Interclass correlation coefficients between nonmetric tooth crown traits and mesiodistal (MD) and buccolingual (BL) diameters of: (a) tooth the trait was observed on; (b) all anterior teeth (mean correlation); and (c) all posterior teeth (mean correlation).*

Trait (tooth)	Tooth trait was observed on:		Anterior teeth		Posterior teeth	
	MD	BL	MD	BL	MD	BL
Winging (UI1)	0.049	−0.093	0.048	−0.007	0.033	0.033
Shoveling (UI1)	0.281	0.090	0.210	0.048	0.169	0.113
Shoveling (UI2)	0.136	0.107	0.101	0.065	0.121	0.073
Tuberculum dentale (UC)	0.155	0.299	0.099	0.203	0.090	0.109
Carabelli's trait (UM1)	0.187	0.184	0.097	0.052	0.145	0.136
Carabelli's trait (UM2)	0.146	0.114	0.056	0.072	0.108	0.112
Hypocone (UM1)	0.138	0.064	0.006	0.051	0.064	0.053
Hypocone (UM2)	0.102	0.052	0.008	0.070	0.058	0.051
Cusp 5 (UM1)	0.148	0.008	0.068	0.057	0.105	0.039
Cusp 5 (UM2)	0.084	0.009	0.034	0.026	0.057	−0.006
Parastyle (UM1)	0.077	0.129	0.095	0.019	0.116	0.087
Parastyle (UM2)	−0.013	0.002	0.028	0.038	0.045	0.048
Distal accessory ridge (LC)	0.301	0.145	0.232	0.148	0.174	0.148
Lingual cusp number (LP1)	0.011	0.073	0.031	0.050	0.017	0.013
Lingual cusp number (LP2)	0.033	0.126	−0.005	0.056	0.028	0.032
Deflecting wrinkle (LM1)	0.023	0.017	0.025	−0.031	−0.001	0.013
Hypoconulid (LM1)	0.174	0.013	−0.025	−0.002	0.063	−0.013
Hypoconulid (LM2)	0.311	0.148	0.100	0.111	0.158	0.092
Cusp 6 (LM1)	0.132	0.156	0.136	0.081	0.107	0.125
Cusp 6 (LM2)	0.177	0.123	0.050	0.040	0.095	0.057
Cusp 7 (LM1)	0.030	0.037	0.001	0.006	0.033	0.021
Cusp 7 (LM2)	0.023	0.034	0.035	0.039	0.027	0.040
Protostylid (LM1)	0.172	0.166	0.124	0.061	0.167	0.136
Protostylid (LM2)	0.184	0.206	0.116	0.063	0.150	0.136

UI1/UI2: upper first/second incisor; UC: upper canine; UM1/UM2: upper first/second molar; LC: lower canine; LP1/LP2: lower first/second premolar; LM1/LM2: lower first/second molar.

noteworthy variation among traits as to the degree of size interaction. Traits that consistently fail to show a correlation with tooth size are winging, parastyle (UM2), lower premolar lingual cusp number, the deflecting wrinkle, and cusp 7. Some traits show a weak relationship to size but only on the teeth where they are expressed. Examples of this include the hypocone, cusp 5 (UM1), the parastyle (UM1), the hypoconulid (LM1), and cusp 6 (LM2). The remaining traits that show patterned relationships with overall tooth size are shoveling (UI1), *tuberculum dentale* (UC),

Carabelli's trait, the hypoconulid (LM2), cusp 6 (LM1), and the proto-stylid.

A significant association between degree of shoveling expression and upper incisor dimensions has been reported by other researchers. Suzuki and Sakai (1966) noted that the MD diameter of the incisor crown in Japanese populations is distinctly larger in teeth showing pronounced shoveling expressions. Mizoguchi (1978, 1985) also found a consistent relationship between shoveling expression and MD diameters of the upper incisors in several Japanese samples. For Skolt Lapps and Finns, Kirveskari and Alvesalo (1979) reported high correlations between depth of the lingual fossa and MD diameters of both UI1 and UI2. In a Melanesian sample, Lombardi (1975) found that incisors with shoveling had significantly larger MD dimensions than non-shoveled forms.

With few exceptions, the authors who report a relationship between tooth size and shoveling expression concentrate solely on incisor dimensions. Mizoguchi (1985) calculated correlations between shoveling and MD diameters of 12 permanent teeth (M2 and M3 not included) and found that shoveling is positively correlated with all MD diameters, not just those of UI1 and UI2. Moreover, in his PCA of 10 crown traits and 12 MD diameters, shoveling had high loadings on the first component that was otherwise dominated by the 12 MD diameters.

In the Pima Indian dentition, the pattern of correlations is similar to that shown in Mizoguchi's (1985) analysis. In the Pima, however, shoveling was analyzed in the context of both MD and BL dimensions. The pattern of correlations is very distinctive. Shoveling is consistently correlated with MD diameters of all anterior teeth but has very low correlations with BL anterior tooth dimensions. Shoveling also shows moderate correlations with posterior tooth dimensions, but the distinction between MD and BL diameters, while still evident, is not as pronounced. The pattern of correlations between shoveling and all tooth dimensions parallels closely the loadings on the principal component for MD diameters of the anterior teeth. It appears that shoveling and the component for MD dimensions of the anterior teeth share some developmental factors in common, although these remain unknown at this time.

In contrast to shoveling, the relationship between *tuberculum dentale* of the upper canine and tooth size has not been explored. In the Pima sample, this cingular trait is distinctly associated with the BL diameter of the upper canine while the correlation with the MD diameter is only half as great. This finding comes as no surprise as any cingular addition to a tooth would tend to enhance its BL dimension. Unexpectedly, this canine trait consistently correlates with the BL dimensions of all anterior teeth, not just the tooth that it directly affects. This trait also shows low but consistently positive correlations with posterior tooth dimensions but there is not a distinct difference between MD and BL correlations. Although the pattern of correlations is not as clear as that shown by shoveling, expression of *tuberculum dentale* on the upper canine appears to be related to the principal component for BL dimensions of the anterior teeth.

The pattern of size interaction for the distal accessory ridge of the lower canine mirrors almost exactly that for shoveling. That is, it is most highly correlated with the MD diameter of the tooth on which it is expressed (LC), but it also shows consistently high correlations with all MD diameters of the anterior teeth. For both the lower canine and combined anterior teeth, the MD correlations are about twice as high as the BL correlations. Again, like shoveling, this trait shows lower though consistent correlations with posterior tooth dimensions with only a minor difference between MD and BL correlations. The overall pattern of DAR-size correlations suggests this trait is also linked to factors underlying the principal component for MD diameters of the anterior teeth.

Not only are Carabelli's trait and the protostylid correlated with each other, they also show the same type and level of association with tooth size. Their highest correlations are with those dimensions of the teeth expressing the trait, with effectively no difference between MD and BL correlations. Moreover, both traits show higher correlations with posterior tooth dimensions than with anterior tooth dimensions. In contrast to shoveling, *tuberculum dentale*, and the canine distal accessory ridge, which correlate differentially with either MD or BL dimensions, Carabelli's trait and the protostylid correlate almost equally with both dimensions. In our PCA of morphologic and metric variables, the first principal component was for both MD and BL dimensions of the posterior teeth so Carabelli's trait and the protostylid may be influenced by some of the same factors involved in the development of overall posterior tooth dimensions.

Garn *et al.* (1966e) found no relationship between tooth size and the expression of Carabelli's trait. By contrast, Korenhof (1960), Keene (1968), and Lombardi (1975) found that upper first molars with Carabelli's trait were significantly larger than those showing trait absence. Keene also observed that UM2 adjacent to UM1 that express Carabelli's trait are larger than UM2 adjacent to UM1 that lack the trait. Reid *et al.* (1991, 1992) found a linear relationship between molar crown area and degree of expression of Carabelli's trait when measured on an eight grade scale. Not only were teeth expressing Carabelli's trait larger, but teeth with more pronounced expressions were larger than those showing slight or moderate expressions. The same pattern was evident for all three upper molars, although it was most distinctive for UM1. Kieser and Becker (1989) calculated rank order correlations between Carabelli's trait and MD and BL diameters of UM1 in Lengua Indians, South African blacks, and South African whites and concluded there was no significant relationship between tooth size and Carabelli's trait expression. However, their small samples, broken down by males and females, adversely affect tests of the null hypothesis ($rho = 0$). Taking sampling error into account, the overall pattern and near equality of correlations with both MD and BL diameters correspond to our findings and those of other researchers. Mizoguchi (1985) presents interclass correlations between both Carabelli's trait and the protostylid and MD diameters of 12 upper and lower teeth. Although there is some variation in these correlations between males and females and between two samples (JP and EZ series), they are usually positive for both Carabelli's trait and the protostylid and higher for posterior teeth than anterior teeth.

The hypoconulid and cusp 6 also show some relationship to tooth size. In the Pima sample, the hypoconulid is invariably present on LM1 but does show variation in degree of expression. More pronounced hypoconulid expressions correlate positively with the MD but not BL diameter of LM1. Interactions beyond LM1 are not noteworthy. For the hypoconulid of LM2, which shows more variation in terms of presence–absence and degree of expression, there is a strong relationship with the MD diameter of LM2 while the correlation with the BL diameter is about half as great. Moreover, there is a low level of interaction between the hypoconulid of LM2 and both MD and BL dimensions of the anterior and posterior teeth. This probably reflects the influence of a general size factor. Cusp 6 of both lower molars shows a low but consistent interaction with tooth dimensions that involves MD and BL diameters to about the same degree.

Dahlberg (1961) found that 4-cusped lower molars in Melanesians showed significantly reduced MD diameters. There was no corresponding difference in BL diameters. Our findings support this observation. In this instance, there is direct involvement between trait presence and size as the hypoconulid sets the distal boundary in lower molars for mesiodistal measurements. The loss of this cusp would ordinarily lower the MD diameter. What is less obvious is the fact that hypoconulid expression on LM2 also correlates, albeit modestly, with other tooth dimensions. A relationship to the principal component for MD and BL dimensions of the posterior teeth is, however, not as distinctive as that shown for Carabelli's trait and the protostylid.

If we set up a 4-point scale characterizing the level of size–morphology interactions in the dentition, using the terms strong, moderate, weak, and absent, most tooth crown traits would fall in the weak or absent categories. No morphological trait shows a strong relationship to tooth size, at least in terms of MD and BL diameters. The only traits that show a moderate level of interaction are UI1 shoveling, UC *tuberculum dentale*, and LC distal accessory ridge. What is most interesting is not their correlations of *c.* 0.30 with the specific teeth they are expressed on but their patterns of correlation across MD and BL diameters of all upper and lower teeth. Shoveling and the LC distal accessory ridge are somehow related to the principal component for MD diameters of the anterior teeth. Both load more strongly on this component than BL diameters of all teeth and most MD diameters of the posterior teeth. In contrast to these traits, *tuberculum dentale* of the upper canine shares some common basis with the principal component for BL diameters of the anterior teeth. As there is a relationship between cingular ridges and tubercles on all three upper anterior teeth, one can understand how the presence of such traits would enhance BL diameters. However, the lower anterior teeth rarely exhibit visible cingular traits and their BL diameters also correlate with this upper canine trait. Finally, Carabelli's trait and the protostylid show a consistent but admittedly weak relationship with tooth size. As their correlations with MD and BL diameters of the posterior teeth are nearly equal and exceed the correlations for anterior teeth, these traits are apparently related to the principal component for MD and BL dimensions of the posterior teeth.

It is difficult to discern in many cases whether the presence of a particular trait contributes directly to larger crown diameters or if larger crown size enhances trait expression. For example, do the lingual marginal ridges that characterize shoveling contribute to the MD dimensions of the anterior teeth or are larger teeth more likely to express lingual marginal ridges? While this question is interesting from a developmental and phylogenetic standpoint (e.g., Neanderthals had large anterior teeth and pronounced shoveling), it should be emphasized that it has little bearing on the pattern of geographic variation shown by shoveling. Coefficients of determination (r^2) show that tooth size, at most, contributes only about 10% of the overall variance in shoveling expression. This percentage is much less for other traits. Crown size and morphology interact weakly, if at all, so it appears that the factors involved in the development of morphologic traits are largely independent of tooth size.

Interactions between crown morphology and agenesis

Garn *et al.* (1963) have shown a relationship between third molar agenesis and congenitally missing teeth in other tooth classes. Individuals missing one or more third molars are more likely to be missing upper lateral incisors, lower central incisors, and second premolars. The model developed by Brook (1984) also indicates a relationship between tooth size and agenesis. Individuals with smaller teeth are more likely to be missing third molars, which may account in part for the higher frequency of third molar agenesis in females. While there is some interaction between tooth size and agenesis, are crown traits also associated with congenitally missing teeth?

In a study of American white Naval recruits, Keene (1965) found a significant relationship between hypocone reduction (tricuspal forms) of the upper first and second molars and third molar agenesis. In a subsequent article, he also reported that individuals lacking Carabelli's cusp or showing only variable pits, grooves, and depressions were twice as likely to show upper third molar agenesis than individuals who exhibit some form of cusp expression (Keene, 1968). Also finding a relationship between the expression of the hypocone and Carabelli's trait, Keene (1968:1025) concludes 'It would appear that Carabelli's trait and the hypocone are somewhat parallel in their variability and the expression of each is dependent to some extent upon those factors that are involved in the structural and numerical reduction observed in the maxillary molars.' Garn *et al.* (1966e) did not find a relationship between Carabelli's trait and third molar agenesis, but their study was based on lower third molars rather than upper third molars.

Parallel to the work of Keene (1965, 1968) on the hypocone and upper third molar agenesis, Davies (1968) found a significant relationship between hypoconulid reduction on the lower first molar and third molar agenesis. Individuals with a full complement of teeth had a 4-cusped LM1 frequency of 8% while those missing one or more third molars had an incidence of 12.5% and, when the cusp was present, showed fewer well developed hypoconulids. Individuals missing one or more teeth other than third molars (but may also lack M3s) had the highest incidence of 4-cusped LM1 at 25.3% and showed even more hypoconulid size reduction than the two samples with and without third molar agenesis. Contradictory results were obtained by Anderson *et al.* (1977) who found no relationship between hypoconulid loss on the lower first molar and third molar agenesis.

There may be a weak association between agenesis and the expression of some crown traits. The hypocone and hypoconulid are the best candidates for such a relationship. They are the last major molar cusps to differentiate and calcify and are the most likely to be reduced in size or completely lost. In many essentials, these cusps are the variable within-tooth analogue to the variable within-field third molars. However, these relationships may vary between populations. Eskimo-Aleuts, for example, have one of the world's highest frequencies of upper and lower third molar agenesis (cf. Pedersen, 1949). They also have one of the highest frequencies of 3-cusped upper second molars, an expected finding if there is indeed a correlation between hypocone loss and agenesis at the individual level. However, they also have an extremely high frequency of 5-cusped lower second molars (chapter 5). The hypocone and hypoconulid, despite showing certain parallels in ontogeny and phylogeny, are not correlated at either the individual or population level. We have not yet reached the point where generalizations can be made on their relationship to third molar agenesis.

Summary

The dentition is the product of a highly integrated and strongly canalized developmental system (Saunders and Mayhall, 1982b). Barring significant environmental insults and chromosomal abnormalities, the genetic control mechanisms guiding dental ontogeny from the formation of the lamina to tooth eruption and root formation operate in a precise and orderly fashion to produce teeth, classes of teeth, and structural details of the crowns and roots. There is some latitude in these control mechanisms or the subject matter of this book, dental variation, would be moot.

From a phylogenetic standpoint, some components of the teeth are so important they have stood the test of evolutionary time. For example, the complete amino acid sequence for amelogenin, the principal protein in the soft enamel matrix, has been deciphered for cows, pigs, mice, and humans and, in all four groups, the sequences are nearly identical (Ten Cate, 1994). As an evolutionarily conservative component of teeth, amelogenin is a poor candidate for the study of human variation. Although not as conservative as amelogenin, general hominoid tooth crown form has remained relatively stable for more than 30 million years. The genetic control mechanisms guiding odontogenesis, however, have some latitude to produce differences in the surficial morphology of the crown and the number and form of roots, as well as variations in crown size and tooth number. Regarding morphological variants, these are produced principally through three odontogenic processes: folding of the inner enamel epithelium, enamel matrix formation, and mineralization. The genes underlying these processes are most likely regulatory rather than structural genes. The information they convey to cells involves the timing of specific developmental events (start and stop signals) and also the rates of change (e.g., mitotic rates; secretory rates of ameloblasts) between the initiation and termination of a particular process. The presence of fluctuating asymmetry for surficial morphological traits shows that genetic signals can sometimes be partly obscured by environmental factors. The extent to which external factors impinge upon the development of specific crown and root traits is addressed in chapter 4.

Dental morphological traits do exhibit some fluctuating asymmetry, but the level of asymmetry for these traits and the dentition as a whole is relatively low. Ordinarily, crown and root traits are expressed bilaterally. When one tooth exhibits a diminished expression of a trait compared to its antimere, our operational assumption is that the tooth showing the greatest expression is the best reflection of the genetic factors underlying the development of that trait. For this reason, we prefer the individual count method over the total tooth count or unilateral count methods, which artificially enhance or minimize sample size, respectively.

In human populations, males and females consistently differ in tooth size. Quite simply, males have larger teeth than females. A dimorphism also exists for missing and supernumerary teeth. For tooth crown and root traits, if there are male–female differences in trait frequency and expression, they are subtle. With the exception of the canine distal accessory ridge, no crown or root trait has been shown to differ significantly between males and females across a diverse array of samples. Hence, we conclude that most morphological traits lack sexual dimorphism. Combining data on males and females is justified in most cases.

The mesiodistal and buccolingual crown diameters of any given tooth are significantly correlated with the dimensions of every other tooth (in both jaws). As these correlations are not equal, multivariate analysis suggests that relatively few common factors are involved in controlling particular dimensions and/or suites of teeth (e.g., anterior vs. posterior; MD vs. BL). Agenesis in one tooth district also shows a significant relationship to agenesis in other tooth districts, as well as to crown size. Different morphological traits sometimes show interrelationships (e.g., Carabelli's trait-hypocone) but the highest correlations are found between the same trait on different members of a tooth district (i.e., within-field interactions). For this reason, we feel the safest course of action in population characterizations is to avoid redundancy by focusing on how a trait is expressed on only one member of a tooth district. In most instances, the mesial member of a field, the 'key' tooth, is preferred because it is the most stable in terms of development (e.g., least amount of asymmetry, maximal trait development) and evolution (e.g., shows the most conservative features). When traits are almost invariant on the key tooth (e.g., hypocone, hypoconulid), it is necessary to shift attention to an adjacent field member (e.g., M2). Third molars, with their prolonged period of postnatal development, are the most likely candidates for significant environmental modification so their use is not encouraged for population characterizations of crown and root trait morphology. The only exception would be the frequency of third molar loss/reduction that shows interesting patterns of within- and between-group variation.

4 Genetics of morphological trait expression

Introduction

McKusick's (1990) massive compendium *Mendelian Inheritance in Man* catalogs more than 12,000 autosomal dominant, autosomal recessive, and X-linked phenotypes, only 41 of which are dental variables. Fully one-third of these are developmental defects of the enamel and dentine, including several varieties of *amelogenesis imperfecta*. After such defects, seven phenotypes with allegedly simple modes of inheritance are variants of hypodontia (missing teeth). The remaining dental traits run the gamut from fused teeth and dental ankylosis to syndromes involving dental abnormalities. With few exceptions, these dental phenotypes are rare, reported in only one or a few families.

For commonly occurring crown and root traits, McKusick notes only six with presumed simple modes of inheritance. Listed as autosomal dominants are Carabelli's anomaly, upper lateral incisor interruption grooves (*dens invaginatus*), premolar odontomes (*dens evaginatus*), upper central incisor winging, and shovel-shaped incisors. Taurodontism is listed as an autosomal recessive. Judiciously, McKusick marks none of these entries with an asterisk, his method for signifying phenotypes with well established modes of inheritance.

Despite the seeming paucity of simple genetic markers in the dentition, it is almost axiomatic that tooth development in general and dental morphology in particular are under strong hereditary control. Gabriel (1948:7) asserts 'there can be no doubt of a genetic pattern for tooth morphology extending to even minute detail of root formation.' Kraus and Furr (1953:554) add 'It often has been stated in the past, and constantly reiterated at the present time, that the entire dentition, in its gross morphology, is governed strictly by the action of genes.' From an anthropological vantage, Moorrees (1962:101) notes that 'The use of the dentition as a criterion of race stems from the recognition that form and size of the teeth, for the most part, are genetically determined.' In a review of the genetics of dental development, Garn (1977:82) concludes 'It is true that by all conventional tests, dental development is to a larger extent under

131

genetic control than is known for most other calcified tissues, including even the round bones of the hand and foot.' Comments to this effect are common in the dental literature, but deducing the nature of this genetic control has not been straightforward for any facet of normal human dental variation, including tooth size, morphology, and number. This point is made evident in many review articles on dental genetics (Tobias, 1955; Kraus, 1957; Krogman, 1960; Witkop, 1960; Osborne, 1963, 1967; Dixon and Stewart, 1976; Biggerstaff, 1979; Nakata, 1985; Townsend *et al.*, 1994).

Patterned racial variation in crown and root traits demonstrated during the first half of the twentieth century, in conjunction with limited twin studies (e.g., Korkhaus, 1930), led early workers to conclude these traits had some genetic basis. Few, however, explored modes of inheritance through family or twin studies. By 1950, it was still not clear if inheritance mechanisms were simple or complex. As dental morphology encompasses a suite of traits that are expressed in terms of presence or absence, some early workers thought this phenotypic dichotomy might be due to the action of dominant or recessive genes at a single locus.

The quest for simple modes of inheritance

The rediscovery of Mendelism in 1900 stimulated a revolution in animal and plant genetics, but this had little immediate impact on physical anthropology. The genetics of normal human variation, as opposed to rare 'inborn errors of metabolism' (Garrod, 1909), was in its infancy during the first half of the twentieth century, led by the discovery and analysis of red blood cell antigens (ABO, MN, P, and Rh) that were shown to obey Mendel's laws for simple autosomal inheritance. For the most part, physical anthropologists were occupied with metrical and morphological observations on fossils, primates, human skeletons, and living humans. The assumption was that the anatomical traits used in comparative studies had some hereditary basis but few attempted to demonstrate this through a genetic analysis of families and twins.

The modern synthetic theory of evolution formulated in the 1930s and 1940s (Dobzhansky, 1937; Mayr, 1942; Simpson, 1944) had a greater impact on genetic thinking in physical anthropology than the rise of Mendelian genetics four decades earlier. As the population models developed by S. Wright, R.A. Fisher, and J.B.S Haldane focused on gene frequencies and the forces that bring about genetic change through time and space, anthropologists started asking genetic questions about the traits long used for making racial comparisons. In a landmark Cold Spring

Harbor volume *The Evolution of Man*, Lasker (1950) contributed a paper 'Genetic analysis of racial traits of the teeth.' He noted how genetic studies of teeth had lagged behind racial studies and reviewed a wide array of morphological traits of the tooth crowns and roots that he felt were amenable to genetic analysis. As a testimony to Lasker's prescience, he did not focus just on shovel-shaped incisors, Carabelli's cusp, and lower molar groove pattern and cusp number, as did most of his contemporaries, but listed many variables that have only recently become commonplace in dental morphological analyses (e.g., cusp 6, cusp 7, enamel extensions, supernumerary roots, fused roots).

Coincident with the publication of Lasker's paper was W.C. Boyd's (1950) monograph *Genetics and the Races of Man*. Unlike previous workers who classified human races based on skin color, anthroposcopic features, measurements, and indices, Boyd used gene frequencies from the ABO, MN, and Rh blood group systems to categorize human variation. Although his racial classification differed little from earlier ones based on traditional phenotypic methods, Boyd's work stimulated physical anthropologists to think more in terms of gene frequencies. Of course, to reduce phenotype frequencies to gene frequencies, a trait had to have a simple mode of inheritance, and, to demonstrate this, it was necessary to examine phenotypic trait expressions in families. A number of studies over the next 25 years were aimed at answering just this question for dental morphological variables. If they did have simple modes on inheritance, they could be analyzed using the processual models developed by population geneticists for selection, gene flow, and genetic drift.

The first crown trait assessed for mode of inheritance through pedigree analysis was Carabelli's trait. Using the proband method (more often employed with rare traits), Kraus (1951) assessed the transmission characteristics of Carabelli's trait in one large and seven small Mexican and Indian families in Arizona. The genetic model he tested assumed the following phenotype–genotype correspondence: trait absence = cc; pits or grooves and small tubercles = Cc; and pronounced tubercles = CC. His hypothesis of autosomal codominant inheritance could not be rejected in eight pedigrees that included 10 mating types and 19 observable offspring. He conceded that other genetic factors might have some influence on the three postulated genotypes but was unable to test this proposition. Kraus did not have the most critical mating type (absence × absence) for assessing dominant or codominant inheritance. Absence × absence matings should not produce presence phenotypes in offspring where the mode of inheritance is either autosomal dominant or codominant (putting aside the issue of incomplete penetrance).

Following Kraus' analysis of Carabelli's trait, Tsuji (1958) assessed the inheritance of this feature in family samples drawn from a Japanese population. With population frequencies of 39.6% for males and 31.2% for females, he first concluded the inheritance pattern of Carabelli's trait was autosomal rather than X-linked. Although the author classified the trait on a four grade scale, the 28 pedigrees he illustrates are noted only for trait presence or absence. Tsuji, unlike Kraus, did have the critical absence × absence mating types (12 in all) and, in one such family, all the offspring expressed the trait, violating the expectations of a simple autosomal dominant model. Despite this exception, Tsuji (1958:22) averred 'It is very probable that the presence of the cusp is controlled by a simple Mendelian dominant gene, but the penetrance of the gene is incomplete, as is clearly shown by an exceptional family (No. 445) in which the children with the cusps are born from parents without any trace of the cusp.'

Stimulated by the family study of Kraus (1951) and the potential for estimating gene frequencies in prehistoric populations, Turner (1967, 1969, 1991) tested an autosomal codominant model using population data. His general procedure was to assign genotypes to three forms of phenotypic expression, generally absence, intermediate, and pronounced, with absence and pronounced expressions corresponding to the two homozygous genotypes and intermediate expressions corresponding to heterozygotes. As with any two allele, single locus, codominant system (e.g., MN blood groups), he estimated gene frequencies by the gene counting method. These gene frequencies (p and q) were used to estimate the expected number of genotypes which, when compared to the observed phenotypic distribution, could be tested for goodness of fit to Hardy-Weinberg equilibrium. In one paper, Turner (1967) found a simple autosomal codominant model fit the observed distributions for Carabelli's trait, shovel-shaped incisors, and the protostylid. Following the same methodology, DeVoto and his colleagues tested, with mixed success, a codominant model of inheritance for shovel-shaped incisors and Carabelli's trait in South American populations (DeVoto and Perrotto, 1971; DeVoto *et al.*, 1968). They found the model fit their population data on deciduous incisor shoveling but deviated significantly from expectations for permanent incisor shoveling. For Carabelli's trait, they went so far as to calculate 'gene frequencies' for samples from different schools in northwest Argentina.

Portin and Alvesalo (1974), using a single-generation method of genetic analysis, compared the variation in shovel-shaped incisors between two types of siblings: those from affected individuals and those from the general population. Their observed and expected values differed significantly for both simple autosomal dominant and recessive models of inheritance, but the difference was nonsignificant for a model of intermediate autosomal inheritance. The authors concluded cautiously that 'the hypothesis of one locus with more than two alleles involved and the polygene hypothesis would be equally compatible and cannot be ruled out' (Portin and Alvesalo, 1974:62). Despite this qualification, these authors are often credited with proposing a simple mode of inheritance for shovel-shaped incisors, and it appears in McKusick's (1990) volume under autosomal dominant traits.

In one of the few genetic analyses that does not involve Carabelli's trait or shovel-shaped incisors, Escobar and his co-workers (1976) analyzed the segregation pattern of bilateral rotation of the upper central incisors (i.e., winging) in 37 Guatemalan families. They ruled out X-linked inheritance because of a nonsignificant ratio of affected males to females in matings involving normal females married to affected males. Using Morton's (1959) method of segregation analysis, they tested the hypothesis that the segregation frequency did not differ significantly from 0.50 (i.e., autosomal dominant). Their overall segregation frequency of 0.42 did not differ significantly from the expected value so they concluded the trait was inherited as an autosomal dominant. However, 12 normal × normal matings resulted in 22 affected and 25 normal offspring, an unexpected finding for a trait thought to be inherited as a simple autosomal dominant. To circumvent this complication, they invoked the concept of incomplete penetrance. In their summary, they 'suggest a dominant trait with 84% penetrance and variable gene expression' (Escobar *et al.*, 1976:114).

The simple modes of inheritance for certain tooth crown traits suggested by pedigree analysis and population models have been critiqued on a number of grounds. Goodman (1965) reduced the eight pedigrees illustrated by Kraus (1951) to mating type and offspring proportions and concluded the data set was too small to clearly establish a mode of inheritance for this variable. He also tested population data for Carabelli's trait provided by Kraus (1959) for goodness of fit to Hardy-Weinberg equilibrium and found significant deviations from expected values for three different samples. Goose and Lee (1971) examined Carabelli's trait in English families and also tested an autosomal codominant model of inheritance. The number of offspring from the six possible mating types was small (n = 32), but they did find exceptions to the model, including two offspring with presence phenotypes (intermediate expression, or Cc) from absence × absence matings and also one absence (cc) × tubercle (CC) mating yielding an absence offspring. On these grounds, the authors rejected a simple dominant–recessive model and proposed the inheritance pattern was more likely multifactorial. In a subsequent study of Chinese immigrant families in England, Lee and Goose (1972) presented more convincing evidence that Carabelli's trait was not inherited in a simple manner. In this family sample, there were a large number of absence × absence mating types and these produced 22 offspring exhibiting trait absence and 30 offspring with trait presence (all intermediate expression).

By the early 1970s, dental researchers were starting to have serious doubts that dental morphological traits had simple modes of inheritance. The few family studies that had been conducted often had telling exceptions to expected segregation ratios, and the concept of incomplete penetrance was becoming a less convincing means of explaining away these exceptions. Sofaer (1970) provided a general critique of assessing modes of inheritance using population rather than family data. The main problem, however, was the long acknowledged observation that most dental morphological traits were not simply present or absent but exhibited variable expressivity of presence phenotypes. How could a single dominant or codominant allele be translated into a phenotype that could take the form of a small furrow up to a large free-standing cusp? One would have to invoke modifier genes at other loci or a large component of environmental variance to account for what is often a substantial range in expression.

Quasicontinuous variation

The notion that discontinuous qualitative phenotypes are associated with qualitative, or simple, modes of inheritance was widespread throughout much of biology, medicine, and anthropology during the first half of the twentieth century. Expressed in presence or absence form, both rare and common traits were subjected to pedigree analysis (or inspection) to determine if inheritance was autosomal or X-linked and dominant or recessive. Autosomal recessive inheritance was sometimes proposed on the basis of a single family if one or a few affected offspring were produced by consanguineous parents normal for the trait. In larger family samples, exceptions to a proposed model of inheritance were often explained by incomplete penetrance.

In 1960, J.H. Edwards wrote a brief paper 'The Simulation of Mendelism.' Edwards, primarily targeting the rarely occurring traits of interest to medical geneticists, made the point that presence–absence traits sometimes mimic the segregation patterns of simple Mendelian inheritance where, in reality, inheritance is complex. The empirical foundation for this

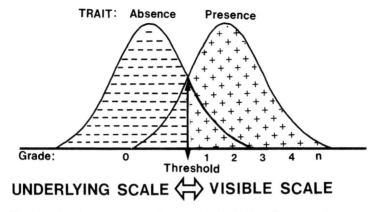

TRAIT: Absence   Presence

Grade: 0   1 2 3 4 n
Threshold

UNDERLYING SCALE ⟺ VISIBLE SCALE

Fig. 4.1   Quasicontinuous variation and threshold effect. Two overlapping normal distributions illustrate the continuous genetic basis of quasicontinuous traits. A threshold separates a visible scale from an underlying scale. When an individual has a genotype to the right of the threshold, they present a visible phenotype that can be scored as slight, moderate, or pronounced, depending on distance from the threshold (see grades of expression). Individuals with genotypes below the threshold fail to exhibit any visible trait manifestation (trait absence), but there is also genotypic variability underlying the absence phenotype depending on genotypic position relative to the threshold. For convenience, the threshold is shown at the point where the two distributions overlap (technically, this threshold could occur at any point along either of the two distributions with corresponding changes to the visible and underlying scales).

observation was laid by Sewall Wright (1934) and Hans Grüneberg (1952) who had conducted large-scale breeding experiments on guinea pigs and mice, respectively. In their analyses of skeletal and dental anomalies, Wright and Grüneberg found that many discontinuous phenotypes refused to obey Mendel's laws for simple dominant–recessive inheritance. They both concluded the mode of inheritance for such qualitative traits was polygenic, or complex. Wright (1968) coined the phrase 'threshold dichotomies' to distinguish such traits from 'point dichotomies' that do have simple modes of inheritance. Grüneberg (1952) referred to such variables as 'quasicontinuous' to distinguish them from discontinuous traits that have simple modes of inheritance and continuous traits that are ordinarily assumed to have complex modes of inheritance.

To envision the relationship between phenotypic and genotypic variation for quasicontinuous variables, it is useful to depict two normal distributions with one threshold (Fig. 4.1). The distribution has both visible and underlying scales. The visible scale is represented by phenotypic values that are directly observable; an individual is scored as either 0 (absent) or 1 (present) for a particular trait. While the visible scale is discontinuous, the underlying distribution, determined by both genetic and environmental factors, is continuous. In one respect, the two scales meet at the threshold, which marks a point of discontinuity on an otherwise continuous distribution. An important consideration is the fact that there is genotypic variation underlying the absence phenotype. Individuals who fail to express a trait may be close to or far removed from the threshold genotypically, but they are identical in phenotypic value. While Grüneberg (1952) noted some traits are phenotypically '0' or '+', with no variation in expression among presence phenotypes, most quasicontinous variables exhibit variation in expression. Individuals just above the threshold manifest the trait to a slight degree while those far above the threshold exhibit pronounced trait expression. Any number of intermediate categories of expression are possible between these extremes.

Grüneberg (1952) outlined several attributes that distinguish quasicontinuous variants from traits with simple modes of inheritance. For simplicity, these are reduced to two major categories. First, a continuous distribution may occupy any position relative to a physiological threshold. If the threshold bisects the distribution at its midpoint, 50% of the population is above the threshold (trait presence) and 50% below the threshold (trait absence). If the distribution shifts to the left of the threshold, trait frequency goes down in a population. If the shift is to the right, trait frequency increases. One implication is that there should be a positive correlation between trait incidence and expressivity. Populations with higher frequencies of a trait should have more individuals exhibiting pronounced trait expressions than populations with lower trait frequencies. The second set of distinguishing criteria is based on the assumption that the multiple genes responsible for the expression of quasicontinuous variables are additive and sensitive to environmental effects. As with continuous variables, quasicontinuous traits can be influenced by prenatal and postnatal effects, e.g., maternal age and/or parity, maternal physiology, sex, and growth differences between the two sides of the body (asymmetry).

The concept of quasicontinuous inheritance was not widely appreciated in anthropology until the 1960s. To that point, its application was limited to geneticists working with experimental animals (Robertson and Lerner, 1949; Dempster and Lerner, 1950; Grüneberg, 1952, 1963) and rare disease entities in humans (Carter, 1969; Edwards, 1969). The introduction of this concept to anthropology is attributable largely to R.J. and A.C. Berry (Berry and Berry, 1967; Berry, 1968) who were systematizing the analysis of nonmetric cranial traits in human skeletal material. To justify the use of nonmetric traits in the assessment of biological affinity, they had to address the genetical basis of these variables. As few nonmetric cranial variants are directly observable in living humans, precluding analysis through conventional family studies, they relied on analogies to animal models. As the earlier work of Wright and Grüneberg focused on nonmetric variants in guinea pigs and mice, Berry and Berry (1967) drew on the quasicontinous or threshold model to explain the underlying genetic basis of analogous traits manifested in human skeletons.

Sofaer (1970) was among the first dental researchers to claim that nonmetric tooth crown traits were quasicontinuous variables. This conclusion was not based on formal genetic analysis but was inferred from animal models and the observation that nonmetric crown traits exhibited variable expressions of trait presence. Scott (1971, 1972, 1973, 1974), Bailit and his colleagues (Bailit *et al.*, 1974, 1975), and Harris (1977) examined crown traits in populations and families to determine whether or not such traits conformed to the distinguishing characteristics of quasicontinuous variation as set forth by Grüneberg (1952).

Scott (1973) analyzed the segregation patterns of 20 crown traits in American white families and found that many approximated the expectations of autosomal dominant or recessive inheritance. However, in each mating type analysis, there were ordinarily a few offspring that violated the tenets of simple autosomal inheritance. While incomplete penetrance might have explained some exceptions, one pattern was evident that did not conform to the expectations of simple Mendelian models. That is, traits in high frequency (e.g., Carabelli's trait) simulated a pattern of dominant inheritance while low frequency traits (e.g., cusp 7) conformed more closely to recessive inheritance models. When all traits were combined, there was a highly significant difference between high and low frequency traits in the proportions of presence and absence offspring from presence × presence, presence × absence, and absence × absence mating types. The ratio of presence to absence offspring from presence × absence mating types for nine low frequency traits (range: 4–27%) was 0.6 while the corresponding ratio for 11 high frequency traits (range: 53–95%) was 2.2. Such a contrast would not be expected if trait expression was dictated by simple Mendelian inheritance, but such a difference would be predicted by the threshold model. For low frequency traits, the genotypes of affected parents are close to the physiological threshold while nonaffected parents could have genotypes just below or far removed from the threshold. The overall effect is to produce many more nonaffected than affected offspring in presence × absence matings. Conversely, for high frequency traits, affected parents could be close to or far above the threshold while nonaffected

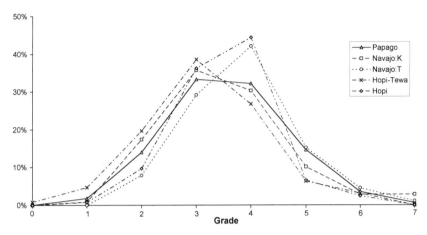

Fig. 4.2   Class frequency distributions for upper central incisor shoveling in American Southwest Indians. What happens when a quasicontinuous trait becomes a continuous trait? While shoveling is polymorphic in most world populations, this trait approaches or attains a frequency of 100% in many American Indian populations. In other words, the entire distribution of shoveling expression in these groups has shifted to the right of the threshold, with all individuals exhibiting phenotypes along the visible scale. When scored on an eight grade ranked scale (0: absence; 1: slight ... 7: pronounced), the shoveling distribution in American Indians is effectively normal, supporting the idea that the underlying genetic basis of such morphological traits is actually continuous, like other polygenic traits with complex modes of inheritance.

parents would, by necessity, have genotypes closer to the threshold. Many more affected offspring would thus result from these presence × absence unions.

One of the criteria for quasicontinuous inheritance is a correlation between penetrance and expressivity. Grüneberg (1952:108) noted that 'As the continuous distribution crosses the critical level, the first few abnormals will generally only be slightly affected; as the distribution is further shifted in the same direction, an increase in the percentage of abnormals will go together with the appearance of more severely affected individuals.' To assess this proposition, Scott (1973) observed class frequency distributions for shoveling (UI1), *tuberculum dentale* (UC), and Carabelli's trait in 18 samples from six major geographic races (European, Asiatic Indian, African, Polynesian, Melanesian, North American Indian). He found significant correlations between incidence and degree of expression for shoveling (r = 0.94) and *tuberculum dentale* (r = 0.77), although the correlation for Carabelli's trait was low and nonsignificant (r = −0.08). Shoveling is noteworthy in this regard as it varied from 44% in Asiatic Indians to 100% in American Southwest Indians. In some American Indian populations, shoveling loses its status as a nonmetric variant as it is present in all individuals. Relative to the threshold model, there is no longer an underlying scale – the entire distribution is visible. If measured on a fine enough scale (eight ordinal classes or direct measurement, but not Hrdlička's four grade scheme), shoveling in American Indians closely approximates a normal distribution (Fig. 4.2). This situation could be predicted from the threshold model, although such a possibility was not entertained by medical geneticists studying rare traits barely manifest on the visible scale. Despite the close relationship between the predicted and observed class frequency distributions of shoveling, Carabelli's trait does not show a significant correlation between incidence and expressivity. This may indicate other factors, such as major genes or modifier genes, play a role in the expression of this trait.

One of the most ambitious genetic studies on the quasicontinuous nature of tooth crown morphology was conducted by Harris (1977) who examined 11 traits on 21 teeth in 315 Solomon Island families (333 parents and 661 offspring). With this large sample at his disposal, Harris tested nine predictions from the model of quasicontinuous variation. As Grüneberg noted, quasicontinuous variants should be subject to the same kinds of environmental influences that moderate the development of continuous traits. Harris, however, found that a common measure of environmental stress in the dentition, namely fluctuating asymmetry, was not influenced by maternal age, parity, or sex. The criteria of the quasicontinuous model that are indicated by a majority of the traits center on the primary assumption of a continuous distribution occupying any position relative to a physiological threshold. For example, Harris found the probability of a younger sibling being affected increased with the number of affected siblings already born. Moreover, offspring with more pronounced trait expressions had a higher frequency of affected relatives. Neither result would be expected for traits with simple modes of inheritance. Harris also found an association between total trait frequency and segregation ratios. For 16 of 21 traits, high frequency traits simulated the ratios expected for Mendelian dominance while low frequency traits simulated recessive inheritance ratios. While Harris concluded tooth crown traits were quasicontinuous variants with polygenic modes of inheritance, he qualified this by noting that not all traits were necessarily inherited in the same manner. Some traits may have a large environmental component of variance while others are influenced primarily by genetic factors. Many traits may be quasicontinous and polygenic, but this does not preclude the action of major genes, modifiers, dominance, and epistasis influencing the expression of other traits.

Complex segregation analysis

When traits do not conform to the expected segregation patterns of simple dominant or recessive models of inheritance, most authors conclude inheritance is polygenic or multifactorial. For most quantitative variables, polygenic inheritance implies that trait development is regulated by the action of genes at many loci, each with a small and additive effect, in concert with environmental effects of varying magnitudes. Until recently, it was not feasible to ask whether genes had differential effects within a polygenic milieu.

In his model of quasicontinous variation, Grüneberg (1952) noted there was no sharp distinction between major and minor genes that were only limiting cases on a continuous scale. He used the terms 'good genes' for traits with distinctly simple modes of inheritance, 'poor genes' for traits influenced by a major gene but affected by modifiers, and 'quasicontinuous characters' where all genes directing trait development were of roughly comparable importance. While 'good genes' for tooth crown traits have not been consistently identified in family studies, the possibility remains that the development of some traits is moderated by 'poor genes.'

By the early 1970s, N. Morton, R.C. Elston and their co-workers had developed statistical methods to test for major genes within a polygenic system. In essence, this involved a search for Grüneberg's so-called 'poor genes.' This method for teasing out the presence of major genes for traits that do not conform to simple modes of inheritance is called complex segregation analysis.

In 1980, Kolakowski *et al.* employed the complex segregation model of Morton and MacLean (1974) to analyze the inheritance of Carabelli's trait in 1152 individuals from 358 Solomon Island families. The model they used allowed traits to be trichotomized, so they set up the categories of 0 (absence), 1 (intermediate expression), and 2 (pronounced expression). From their family data, they estimated seven parameters, four of which were used to test the following hypotheses: (1) no major locus; (2) no sibling environmental resemblance; and (3) neither major locus nor sibling environmental resemblance, polygenic heritability being sufficient. All three hypotheses were rejected. They concluded most of the variation in Carabelli's trait could be accounted for by a major locus, although the effects of a shared sibling environment were also significant (*c.* 19% of variance in liability for the trait). The estimated H value (polygenic heritability) was 0.018, signifying only a small fraction of the variance of this trait could be explained by the action of additive genes at many loci.

Nichol (1989, 1990) applied complex segregation analysis to 19 tooth crown traits using a sample of Pima Indian families (166 parents, 434 sibs) from the A.A. Dahlberg collection (Dahlberg *et al.*, 1982). In contrast to Kolakowski *et al.* (1980), Nichol employed the method of Morton *et al.* (1971) that only allowed traits to be dichotomized. As most of Nichol's 19 variables were scored on graded scales following the ASU standards (Turner *et al.*, 1991), he used different breakpoints (e.g., 0/1–5, 0–1/2–5, 0–2/3–5, etc.) to test genetic hypotheses.

Nichol (1989, 1990) focused on two primary models of inheritance: (1) a generalized two allele, single locus model; and (2) the model of quasicontinuous inheritance, or polychotomized normal distribution of liability. Of the three primary parameters that can be estimated for the first model, Nichol assumed penetrance (t) to be complete and the frequency of sporadic or nonheritable cases (z) to be zero, thus placing emphasis on G, the allele for affection. The degree of dominance (d) for G was tested at three levels where d was set equal to 1 (dominant), 1/2 (codominant), and 0 (recessive). The model for the polychotomized normal distribution of liability involved the estimation of two parameters, the standardized deviate for the threshold of affection (Z) and transmissibility (T).

Using a varying number of breakpoints for each trait, dependent on how much variation was evident in trait expression, Nichol performed a total of 76 individual tests. Of this total, about half (39) conformed to a model of polygenic inheritance while the remainder (37) best fit the expectations of the two-allele, single locus model. For traits that conformed to the two allele model, dominant inheritance was indicated in more runs (26/37 = 70%) than recessive inheritance. Traits that best fit a dominant mode of inheritance included shoveling, double-shoveling, Carabelli's trait, the hypoconulid, cusp 7, and multiple lingual cusps of the lower premolars. The only crown traits that fit a model of recessive inheritance were the hypocone and transverse ridge. Traits conforming to the polygenic model included winging, canine distal accessory ridge, lower molar groove pattern, and the deflecting wrinkle. Despite finding a large number of traits that fit the two-allele, single locus model, Nichol (1989:56) noted 'In 30 of the 39 cases when the polygenic model is favored, the trait frequency is below 20% and above 80%, whereas this is true in only 11 of the 37 cases when the single-locus model is the best fit.' If there are major genes affecting the expression of some crown traits, trait incidence should not be a factor in the rejection or acceptance of a particular genetic model.

The model of codominant autosomal inheritance for Carabelli's trait proposed by Kraus (1951) gave workers some hope that crown trait phenotype frequencies could ultimately be reduced to gene frequencies. The ramifications of complex segregation analysis with major gene effects, as opposed to single gene effects, are not as clear. What are the practical implications for population studies of tooth morphology? Can the frequencies of these 'major genes' be estimated? This question was not addressed by Kolakowski *et al.* (1980) or Nichol (1989). Their work does suggest, however, that relatively few genes and loci are involved in crown trait development. Even if the inheritance of dental traits is not simple in the traditional Mendelian sense, neither does it appear to be as complex as the genetic mechanisms underlying most polygenic traits (e.g., stature, IQ) involving dozens or even hundreds of loci. At the present time, there is no terminology or method to distinguish 'simple' polygenic traits from 'complex' polygenic traits.

Family resemblance and heritability

To this point, we have reviewed familial studies of tooth crown traits that focused on simple modes of inheritance, the threshold model of quasicontinuous variation, and the actions of major genes within the bounds of a polygenic system. Many family studies, however, do not address any of these issues, all of which revolve around how a trait is inherited. An alternative methodology, adopted largely from quantitative genetics, emphasizes phenotypic correlations among family members and assumes polygenic inheritance at the outset.

In discussing traits with complex modes of inheritance, workers often introduce the general formula: phenotype = genotype + environment. For anthropologists interested in population history, phenotypic traits that are largely the result of environmental influences should not be used to estimate population affinities given the assumption that phenotypic similarity reflects primarily genetic similarity. Much confusion has been generated by studies where traits strongly influenced by environmental factors were used indiscriminately with traits controlled largely by genetic factors. Thus, it is of considerable importance for anthropologists to understand the genetic and environmental contributions to the phenotypic variation they observe, whether fingerprint patterns and ridge counts, anthropometric and osteometric measurements, or tooth crown size and morphology. Lacking other means to address this question, the property of heritability has been applied widely to polygenic traits of interest to anthropologists.

Heritability is one of the most useful yet beguiling concepts adopted from quantitative genetics by anthropologists and dental researchers. To animal and plant geneticists, heritability is a powerful tool to set up breeding programs given that one of its definitions is the 'correspondence between phenotypic values and breeding values' (Falconer, 1960:165). In this sense, traits with low heritabilities are difficult to alter no matter what phenotypes are manifested in selected male and female mating pairs. By contrast traits with high heritabilities are readily amenable to manipulation through directed breeding schedules. For human phenotypic variation, there is no interest in breeding values. Instead, anthropologists estimate heritability as a guide to the genetic and environmental components of phenotypic variance. As heritability is calculated through estimates of familial resemblance, quantitative geneticists who work with domestic or laboratory animals can set up elaborate breeding experiments to tease out general and specific genetic and environmental components of variance (Mather, 1949; Kempthorne, 1957; Falconer, 1960; Wright, 1968, 1969). Humans are less easily manipulated. To estimate heritability for tooth crown traits, workers have to focus on 'natural experiments' in the general population – twins and families.

Twin studies

The twin method has enjoyed great popularity in human genetics research during the twentieth century. Its use is principally to determine the relative contributions of genes and environment to anatomical, physiological, and/or behavioral traits whose modes of inheritance are unknown or assumed to be complex. The primary assumption of the twin method is that monozygotic (MZ), or identical, twins share 100% of their genes in common so any measurable differences between such twins are environmental in origin. It is acknowledged, however, that similarities between MZ twins are partly environmental in origin, given their shared intrauterine environment and common postnatal environment (unless reared apart). Thus, there is a genetic basis to the phenotypic similarities observed between identical twins, but there may be an environmental contribution as well. For that reason, most twin studies are not limited to MZ twins but include contrasts with dizygotic (DZ), or fraternal, twins who share the same genetic coefficient of relationship as ordinary siblings (0.50) but are subject to many of the same shared environmental effects as MZ twins.

Because crown traits are qualitative in the sense they can be scored as present or absent, many workers study MZ and DZ twins using concordance analysis, the conventional genetic method associated with discontinuous variables (Saheki, 1958; Aoyagi, 1967; Biggerstaff, 1970, 1973; Berry, 1978; Scott and Potter, 1984; Kaul *et al.*, 1985; Skrinjaric *et al.*, 1985; Zubov and Nikityuk, 1978; Townsend *et al.*, 1988, 1992). The most frequently asked question in these analyses is whether concordance rates differ significantly between the two twin types. The premise is that if genes are responsible for the expression of a trait, concordance rates should be significantly higher in MZ twins than DZ twins. Other authors ask whether or not the concordance frequencies for both twin types differ significantly from the frequencies expected by chance, given that a certain amount of concordance is expected for variables that occur commonly in a population. For example, a trait with a population frequency of 70% (p; where p = presence and q = absence) would be concordant in randomly matched pairs 49% (p^2) of the time for presence–presence and 9% (q^2) of the time for absence–absence based on chance alone. The expected rate of discordance given this population frequency is 42% (2pq). Thus, a concordance rate for a trait is meaningful only in the context of that trait's frequency in the general population.

In addressing methods of twin analysis for qualitative traits, Neel and Schull (1954:272) state that 'If, under a given environment, both members of a twin-pair develop the same phenotype, the twins are said to be 'concordant'; if they have differing phenotypes, they are said to be 'discordant'.' Although this seems straightforward, two aspects of dental trait expression – variable expressivity and antimeric asymmetry – complicate the question as to what constitutes the same phenotype. Consider, for example, a single twin pair, A and B, with trait expression scored on both the left and right antimeres, l and r. If a hypothetical trait is scored in this twin pair as Al = 2, Ar = 1, Bl = 1, Br = 0, is there concordance or not? If the trait is dichotomized as present or absent and the individual count method is used, this pair is concordant (++) as both twins exhibit the trait. With the same dichotomy but following the tooth count method and making only homolateral (Al–Bl; Ar–Br) comparisons, the left antimeres are concordant (++) and the right antimeres are discordant (+/−). Heterolateral (Al–Br; Ar–Bl) comparisons also result in one concordant and one discordant pair. If the requirement is that all four antimeres in a twin pair have to exhibit the trait to exactly the same degree, this pair is discordant. Another complication arises when members of a twin pair both fail to express a given trait, i.e. Al = 0, Ar = 0, Bl = 0, Br = 0. Is this pair concordant? Some workers say yes, the twins have the same phenotype, albeit absent–absent. Other workers, influenced by the proband method of medical genetics (more appropriate for the analysis of rare traits), exclude this twin pair from concordance rate calculations (cf. Berry, 1978; Kaul *et al.*, 1985).

For Carabelli's trait, five different methods have been used to estimate concordance rates in twins. Table 4.1 summarizes the results of these

Table 4.1. *Concordance rates for Carabelli's trait expression in monozygotic (MZ) and dizygotic (DZ) twins. Table is divided into sections to distinguish the different methods used to calculate concordance (N=number of twin pairs; C=concordance rate; H=heritability estimate). At the bottom of the Table are expected concordance rates for traits that occur commonly in a population (e.g., 40–90%).*

Concordance method	MZ twins		DZ twins			Reference
	N	C	N	C	H	
00 and ++ [tooth count]	421	0.881	119	0.647	0.663	Saheki, 1958
	222	0.901	106	0.868	0.250	Aoyagi, 1967
	54	0.944	85	0.694	0.817	Zubov and Nikityuk, 1978
	28	0.964	38	0.605	0.909	Skrinjaric et al., 1985
00 and ++ [individual count]	75	0.844	56	0.768	0.328	Scott and Potter, 1984
++ only [tooth count]	326	0.764	316	0.624	0.372	Berry, 1978
	80	0.429	120	0.455	-0.048	Kaul et al., 1985
00, 11, ...77	99	0.565	93	0.483	0.159	Biggerstaff, 1973 (males)
	85	0.470	95	0.431	0.069	Biggerstaff, 1973 (females)
	80	0.550	78	0.244	0.405	Townsend et al., 1992
00, 11, and 22	80	0.788	78	0.538	0.541	Townsend et al., 1992
	61	0.951	55	0.501	0.902	Townsend and Martin, 1992

	Frequency	(00 & ++)	(++ only)
Expected concordance rates for commonly occurring variables:	0.4	0.52	0.16
	0.5	0.50	0.25
	0.6	0.52	0.36
	0.7	0.58	0.49
	0.8	0.68	0.64
	0.9	0.82	0.81

studies and provides 'heritability' (H) estimates derived through Holzinger's (1929) formula: $H = C_{MZ} - C_{DZ}/1 - C_{DZ}$, where C equals concordance rates for the two twin types. Also shown are concordance rates expected by chance for traits with a population frequency ranging from 40 to 90%, the effective world range of Carabelli's trait frequencies (Scott, 1980; Mizoguchi, 1993).

Across all studies, concordance rates for Carabelli's trait vary from 0.470 to 0.964 for MZ twins and from 0.244 to 0.868 for DZ twins. In all but one instance, concordance rates are higher in MZ twins than DZ twins though the difference is slight in several other cases. Estimates of H vary from -0.048 to 0.909 (mean = 0.447, s = 0.323). As expected, studies that include both 00 and ++ as concordant report higher concordance rates than those that count only ++ as concordant. Moreover, when the method employed requires exact correspondence on the eight-grade Dahlberg scale, concordance rates for MZ twins fall to about .50. Interestingly, the only workers who set this requirement arrived at similar results for MZ twins but were divergent on rates for DZ twins (Biggerstaff, 1973; Townsend *et al.*, 1992). When Townsend *et al.* (1988, 1992) required correspondence for three rather than eight grades of expression (absent, intermediate, and pronounced), their concordance rates were elevated to the range of studies that used the presence–absence dichotomy.

Three of the studies noted above limit concordance rate estimations to Carabelli's trait (Biggerstaf, 1973; Skrinjaric *et al.*, 1985; Townsend *et al.*, 1992). Other workers included a wide range of nonmetric dental traits in their analyses. Without addressing details for specific traits, the general findings include: (1) concordance rates, with few exceptions, are higher for MZ twins than DZ twins; (2) concordance rates for both twin types are usually higher than those of randomly matched singletons from the population; and (3) there is a wide range of variation between traits in concordance rates. From these studies, we can conclude most tooth crown traits have some genetic basis although the degree of genetic determination is variable among traits.

Concordance analysis has its limitations. Any estimate of heritability derived through contrasts in MZ–DZ concordance rates should be considered approximations. However, as tooth crown traits are often in high frequency (i.e., 20–80%) and show a wide range of expression, it is possible to go beyond concordance analysis in twin studies. Taking advantage of the variable expressivity of crown traits, several workers have applied other statistical methods to estimate crown trait heritability.

Mizoguchi (1977) analyzed crown morphology in a large Japanese twin series (191 MZ and 75 like-sexed DZ pairs). To measure the level of association within MZ and DZ twin pairs, he employed tetrachoric correlation coefficients, given the assumption that crown traits are discrete in expression but have underlying continuous distributions (i.e., they are quasicontinuous). For 14 traits, he found tetrachoric correlations ranging from 0.47 to 0.95 for MZ twins, with a mean of 0.717. For DZ twins, correlations ranged from 0.07 to 0.92, with a mean of 0.501. His control series of unrelated matched pairs had a mean correlation of −0.069. Only one trait showed higher correlations for DZ twins than MZ twins. Heritability estimates, derived from the formula $h^2 = 2(r_{MZ} - r_{DZ})$, where $r =$ tetrachoric correlation, ranged from −0.47 to 1.63, with a mean of 0.456. The majority of estimates (11/18) fell between 0.40 and 0.80.

Prior to the 1970s, twin analysis typically involved the calculation of intraclass correlation coefficients for MZ and DZ twins that were used to estimate heritability in the formula: $H = (r_{MZ} - r_{DZ})/(1 - r_{DZ})$, where r signifies intraclass correlation coefficients. Heritability estimates for a wide array of biological and behavioral traits were derived through this method, but these calculations did not take into account potential sources of error variance. For example, MZ and DZ twins should not differ significantly in either means or variances for a given trait since any difference might indicate inherent differences in the twinning process or greater environmental covariance in MZ twins. To overcome some of the difficulties associated with traditional twin analysis, Christian and his colleagues (Christian *et al.*, 1974, 1975; Christian and Norton, 1977; Kang *et al.*, 1978; Christian, 1979) developed methods, based on analysis of variance, to test assumptions about MZ and DZ twin means and variances. They estimated heritability using four variance components: within and among mean squares for MZ and DZ twins, abbreviated to WMS_{MZ}, AMS_{MZ}, WMS_{DZ}, and AMS_{DZ}. Total mean square, or WMS plus AMS, is denoted as TMS for either twin type.

Scott and Potter (1984) assessed within and among pair mean squares for 10 tooth crown traits observed on 14 teeth in an American white twin sample (79 MZ pairs, 59 DZ pairs, and 50 unrelated pairs acting as a control series – CS). Each trait had to meet five assumptions before passing to the final stage of analysis that involved estimates of genetic variance and heritability. The five assumptions, stated in the form of null hypotheses were: (1) $\mu_{MZ} = \mu_{DZ}$; (2) $\sigma^2_{MZ} = \sigma^2_{DZ}$; (3) $AMS_{DZ}/WMS_{DZ} > 1.0$; (4) $AMS_{CS}/WMS_{CS} = 1.0$; (5) $WMS_{CS}/WMS_{DZ} > 1.0$. Only seven of 14 traits met all five assumptions. Heritability estimates for these seven traits, based on the formula $h^2 = (WMS_{DZ} - WMS_{MZ})/[(TMS_{MZ} + TMS_{DZ})/4]$ ranged from 0.19 to 0.40, with a mean of 0.34.

Scott and Potter (1984) did not report heritability estimates for seven traits that violated one or more assumptions of the analysis of variance model. When heritability is calculated for these traits using the formula for within, among, and total mean squares, the h^2 values range from 0.29 to 0.96, with a mean of 0.59. This illustrates how heritability values, which seem reasonable, can conceal underlying problems associated with the · expected variance components within and among twin pairs. When heritability is calculated directly through MZ and DZ correlation coefficients, it is not possible to detect these problems.

Townsend and his co-workers (Townsend *et al.*, 1992; Townsend and Martin, 1992) have provided another new direction in the analysis of dental morphologic trait expression in twins. For a large sample of Australian white twins (n > 100 for both MZ and DZ twin pairs), they classified Carabelli's trait expression following Dahlberg's eight grade scale on the upper first molars of all twins that had both left and right antimeres. Several methods were applied to this data set, including concordance rate analysis, analysis of variance, polychoric correlation coefficients, and path diagrams. They addressed a wide range of questions, including correspondence between test/re-test data, degree of antimere asymmetry, sex differences between twin types and a control series in means and variances, the magnitude of homolateral and heterolateral correlations within each twin type, sex differences between male and female MZ and DZ pairs, model fitting via path analysis, and heritability estimates.

Townsend *et al.* (1992) and Townsend and Martin (1992) found no difference in means or variances of Carabelli's trait between the two twin types and the control series. No sex differences were evident, either in terms of the class frequency distributions of male and female twins or between homolateral and heterolateral correlations for male and female MZ and DZ twins. They also found an interesting similarity between polychoric correlations for test/re-test data (0.94 left, 0.96 right), antimere asymmetry (*c.* 0.90), and homolateral and heterolateral correlations between MZ twin pairs (0.81 for males; 0.88 for females). The average polychoric correlations for DZ twins were 0.24 for males, 0.32 for females, and 0.48 for unlike-sex pairs. With the exception of the correlation for unlike-sex DZ twins, the DZ correlations were less than half the correlations found for MZ twins, indicating possible dominance effects. However, when they employed model fitting through path analysis, these dominance effects were not apparent. In conclusion, Townsend and Martin (1992:408) state 'The favored model for explanation of the variation observed in the Carabelli trait within and between the twins is one incorporating additive genetic effects, together with both a general environmental component and an environmental effect specific to each side.' They do not discount the possibility of either dominance or epistasis but note that other types of relatives should be examined before these effects can be taken into account. The heritability of Carabelli's trait was estimated at 90% through path analysis. More conservative heritability values were derived for upper left (0.37) and right (0.51) first molars, within and among mean square components, when assessed by the method of Kang *et al.* (1978).

Family studies

Family studies of tooth crown traits have been directed primarily at ascertaining modes of inheritance. Most attempts to estimate the heritability of crown traits have been based on twins rather than families. Fewer studies have focused on parent–offspring correlations that can be used in the formula $h^2 = 2r_{OP}$ to estimate heritability (Falconer, 1960; Wright, 1969). Assuming large sample size, when correlations approximate 0.50, heritability approaches 1.0, signifying a large proportion of a trait's phenotypic variance is attributable to additive genetic effects. To estimate heritability, sibling correlations are somewhat less useful than parent–offspring correlations because of the complications introduced by dominance and shared environmental effects (Falconer, 1960; Wright, 1969).

Sofaer *et al.* (1972b) used familial correlations to test the tenet of the Butler–Dahlberg field model that the key tooth in a tooth district is genetically more stable than the distal and more variable members of a district. Making all possible pairwise comparisons in a sample of 229 Solomon Island family members, they formed 117 parent–offspring and 146 sibling pairs. Observations were made on shoveling (UI1, UI2), Carabelli's trait (UM1, UM2), the hypocone (UM2, UM3), the hypoconulid (LM2, LM3), and groove pattern (LM2, LM3). Their approach involved the use of intrafamilial correlations to determine if earlier developing teeth (UI1 vs. UI2, M1 vs. M2, or M2 vs. M3) were genetically more stable than later developing teeth. When sibling correlations were compared to parent–offspring correlations, six of ten comparisons showed distinct differences although only two were statistically significant. Despite the lack of close correspondence in parent–offspring and sibling correlations, all first degree relatives were pooled to derive a single correlation coefficient for comparisons between early and late developing teeth. For all five variables, the correlation for the earlier developing tooth was larger than that of the late developing tooth. The differences, however, were minor (0.01 to 0.18) and nonsignificant. While the authors acknowledge a lack of statistical significance, they still concluded that later developing

Table 4.2. *Intrafamilial correlation coefficients for upper central incisor shoveling in Japanese, Chilean, and Pima Indian families.*

Relationship	Japanese families[1]		Chilean families[2]		Pima Indian families[3]	
	n*	r	n*	r	n*	r
Father–son	27	0.410	22	0.141	97	0.401
Father–daughter	39	0.422	13	0.694	87	0.450
Mother–son	42	0.428	28	0.416	120	0.329
Mother–daughter	52	0.248	31	0.249	146	0.350
Brothers	22	0.529	20	0.427	146*	0.447
Sisters	47	0.381	42	0.178	145*	0.320
Brother–sister	94	0.389	65	0.397	194	0.362
Mean parent–offspring correlation		0.377		0.339		0.383
Mean sibling correlation		0.433		0.329		0.376

1: Hanihara *et al.*, 1975; 2: Blanco & Chakraborty, 1977; 3: Scott (unpublished data).
*number of sibships (n = 419 for brothers and 413 for sisters).

teeth of a given class showed a larger environmental component of variation than the earlier developing teeth for crown trait expression. This corresponds to what might be predicted from the field model and has been cited many times to support the position that trait expression on the key tooth in a tooth district (or at least M2 vs. M3) is under stronger genetic control than expressions on the more variable distal teeth. While Sofaer and his colleagues may be correct, their own data do not support this hypothesis. For siblings only, correlations for three of five traits are higher for the more distal tooth. Only by pooling sibling and parent–offspring pairs do they find correlations that match their expectations and, even then, fail to reject any single null hypothesis. For the most part, their correlations between first degree relatives are low.

Hanihara *et al.* (1975) and Blanco and Chakraborty (1977) performed intrafamilial analyses of shovel-shaped incisors across all types of first degree relatives in Japanese and Chilean samples, respectively. For both studies, shoveling was measured in terms of lingual fossa depth. The familial correlations reported from these two studies are shown in Table 4.2 along with correlations on Pima Indian families (Scott, unpublished data) where shoveling was scored on an eight grade scale.

While ethnicity, sample size, and method of measurement vary across the three studies, the overall results are consistent. Correlations for the seven types of first degree relatives range between 0.14 to 0.69 (Blanco and Chakraborty, 1977), 0.25 to 0.53 (Hanihara *et al.*, 1975), and 0.32 to 0.45 (Scott unpublished data). The mean parent–offspring correlations are 0.38,

0.34, and 0.38 while mean sibling correlations are 0.43, 0.33, and 0.38. The average heritability based on parent–offspring correlations is 0.75 for the Japanese sample, 0.68 for the Chilean sample, and 0.76 for the Pima Indian sample. Given the congruence of results between these studies, it appears that the heritability of shoveling falls around 0.75, at least in Asian and Asian-derived groups who have high frequencies of this trait.

For Carabelli's trait, the results from intrafamilial correlation analysis are not as consistent as those reported for shoveling. Alvesalo *et al.* (1975) matched up 177 sibling pairs from a sample of 294 Finns and found a low and nonsignificant Kendall's *tau* value of 0.02. This value is much lower than the sibling correlation of 0.42 for Carabelli's trait in Melanesians (Sofaer *et al.*, 1972b) and 0.57 for American whites (Scott, 1973). Based on their low sibling correlation, Alvesalo *et al.* (1975:195) concluded 'These results suggest that the variation in the expression of the character is not due to genetic factors.'

Harris and Bailit (1980) assessed cusp 5 (metaconule) in a large sample of Melanesian families. Using Spearman's *rho*, they calculated six in-trafamilial correlations for trait expression on all three upper molars. For UM1, correlations ranged from 0.12 for sibling–sibling to 0.32 for sister–sister comparisons with a mean intrafamilial correlation of 0.22. For UM2, the mean intrafamilial correlation for this trait was only 0.06. Interestingly, the mean correlation of 0.20 for cusp 5 of the upper third molar was almost as high as the mean for the upper first molar. For pooled relatives, heritability was estimated at 0.65 for UM1, 0.15 for UM2, and 0.98 for UM3. The high estimate for UM3 is associated with a standard error three times greater than for cusp 5 on UM1. Harris and Bailit noted a general tendency for females (i.e., sister–sister, mother–daughter) to have slightly higher heritabilities than males, and explained this as a possible indication of X chromosome involvement in cusp 5 expression.

Townsend *et al.* (1990) observed cusp 6 in sibling pairs among the Yuendumu, an Australian aboriginal group. In this sample, the frequency of cusp 6 was more common in siblings of affected individuals than in individuals from the general population (81.8% vs. 69.8% for LM1), a finding in accord with the threshold model of inheritance. They also found a significant *phi* coefficient of 0.30 between siblings in cusp 6 expression on LM1. Heritability was not estimated from this correlation, but they note 'the heritability does not seem to be high' (Townsend *et al.*, 1990:272).

Heritability and crown traits

Heritability, despite its many applications and successes in animal and plant breeding, has generated certain misunderstandings in the anthropological and popular literature. What do authors imply when they claim the heritability of a trait is low when they do not estimate heritability (cf. Biggerstaff, 1973; Alvesalo *et al.*, 1975)? This value is not a measure of 'degree of genetic determination.' Say, for example, the development of trait X is controlled by genes at three loci, A, B, and C. Assume further that all individuals in a population are homozygous at these three loci (i.e., AAbbCC). In such a population, any variation in trait expression is entirely environmental in origin – genetic variance and heritability both equal zero. An heritability of zero does not vitiate the fact that the development of the trait is controlled by genes – it is only the within group variation in trait expression that is determined by environmental factors. In other words, heritability is the degree of genetic contribution to observed variation between individuals in a population, not to the trait as it develops in an individual.

Heritability is a population specific value. It can be influenced by gene frequency differences between groups and by population size. In small isolated populations, increased levels of homozygosity brought about by genetic drift and inbreeding lower trait heritabilities. Positive assortative mating also lowers heritabilities although this may not be a problem with crown traits that are not highly visible mating signals, although incisor winging could be exceptional in this regard. Populations in varied environmental settings tend to have traits with lower heritabilities than populations living under more uniform environmental conditions (Falconer, 1960). Heritability is also time-specific within populations. As an environment can change through time, a population with stable gene frequencies may exhibit different heritabilities in different generations. Heritability is thus limited to individuals within a population at the time data are collected, not to all members of a population over a number of generations. Measurement techniques also impact heritability estimates. Scoring qualitative traits involves some level of intraobserver error and the effect of such error is to lower rather than raise heritability estimates.

Given that heritability is a population specific value, it is telling that three sets of researchers studying twin series from three different populations arrived at similar heritability estimates for Carabelli's trait. Mizoguchi (1977), Scott and Potter (1984) and Townsend *et al.* (1992) calculated heritabilities of 0.46, 0.38, and 0.44 (mean of left and right UM1) for Carabelli's trait in Japanese, American white, and Australian white twins, respectively. Similar results were also reported for shoveling where Hanihara *et al.* (1975), Blanco and Chakraborty (1977), and Scott found heritabilities ranging from 0.68 to 0.76 based on intrafamilial correlations.

Granting that twin studies and family studies do not necessarily yield comparable heritability estimates (although for many traits they do), let us assume that the estimates of heritability for Carabelli's trait (c. 0.40) and shoveling (c. 0.70) are adequate approximations. If so, does this difference in heritability values make sense?

Shoveling is the quintessential quasicontinuous trait. The correlation between total shoveling incidence and degree of expression is very high across populations ($r > 0.90$). In populations where shoveling attains a frequency of 100%, it exhibits a normal distribution like that of conventional continuous quantitative traits (e.g., stature). In short, it behaves like a character that has a strong additive component of genetic variance. Carabelli's trait, by contrast, does not show a correlation between trait incidence and expressivity. Its segregation pattern in families follows or at least approximates what is expected for a simple dominant or codominant trait (Kraus, 1951; Tsuji, 1958; Scott, 1973, Nichol, 1990). Although the inheritance of this trait is not 'simple' in the traditional sense, complex segregation analysis suggests the action of genes at a major locus (Kolakowski *et al.*, 1980; Nichol, 1989, 1990). The lower heritability values reported for Carabelli's trait, compared to shoveling, may signify that about half of this trait's variation is attributable to additive genes at many loci, while a significant amount of variance is due to the actions of a major dominant gene. Since Carabelli's trait shows some antimere asymmetry and MZ twin discordance, environmental factors influence trait expression to some extent but this is secondary to genetic factors.

The inheritance of Carabelli's trait may have parallels to several widely acknowledged simple genetic markers. By way of example, the ability to taste phenylthiocarbamide (PTC) is controlled by two alleles (Tt) at a single locus, with T dominant to t. Both homozygote dominants (TT) and heterozygotes (Tt) result in 'taster' phenotypes. When paper infiltrated with PTC is placed on a subject's tongue, they either do or do not experience a taste reaction. Subjects are scored simply as tasters or nontasters. On the surface, this tasting ability appears to be an unambiguous case of simple autosomal dominant inheritance. No one questions that genes are involved in determining an individual's tasting phenotype. However, when PTC is administered in serial dilutions, there is a range of variation in the ability to taste PTC at different strengths. Some individuals taste PTC in a highly diluted form while others have no taste reaction until the concentration of PTC reaches a high level. Part of this variation may be due to dosage effect, with TT genotypes being more sensitive to lower dilutions than Tt genotypes. Much of the variation, however, has been attributed to environmental factors (e.g., age, diet, smoking). With environmental factors underlying phenotypic variance, the heritability of

PTC would not be 1.0 even though it is inherited as a simple Mendelian trait. The evidence for major gene involvement in the development of certain, though by no means all, crown traits suggests that their distinction from traits like PTC may be quantitative rather than qualitative in nature.

Falconer (1960:167) lists heritabilities for a number of phenotypic traits of domestic and laboratory animals. He observes that 'On the whole, the characters with the lowest heritabilities are those most closely connected with reproductive fitness, while the characters with the highest heritabilities are those that might be judged on biological grounds to be the least important as determinants of natural fitness.' For example, egg production, litter size, ovary size, and conception rate have low heritabilities. Characters with higher heritabilities include tail length, abdominal bristle number, coat color, and thickness of back fat. The pattern is clear. The latter traits are less critical to fitness and therefore have more freedom to vary in their genetic backgrounds.

On Falconer's scale of heritability, where would dental characters fall? This, it seems, would depend wholly on the character. In this day of easily masticated foods, humans often take their teeth for granted, even buying artificial replacements as the need arises. Other animals, outside of a few pampered pets, do not have this luxury. The dentition is and was essential for survival in tooth-bearing animals and their ancestors, including our own. This is why the normative dental blueprint that guides ontogeny from formation of the dental lamina to tooth eruption must be directed by genes that are homozygous at many loci. Given the importance of timing in odontogenesis, there is little room for error – if developmental events are not closely coordinated by genes acting at specific times and places, results could range from no tooth formation (i.e., anodontia, a very rare occurrence in humans) to aberrations in tooth number and form. This is certainly true for the highly conserved gene for the enamel protein amelogenin that varies little in its primary amino acid sequence among mammals. Granting occasional missing and supernumerary teeth, dental formula is also strongly conserved. The 2–1–2–3 formula of humans can be traced back to the Oligocene, over 30 million years ago. Excluding changes in the form and size of the canines in hominid evolution, the basic cuspal morphology of human teeth also shows great evolutionary conservatism. In other words, the genes that act to produce (1) teeth, (2) enamel, (3) number of teeth, and (4) morphology of the primary cusps are not free to vary like the many polymorphisms we see in blood groups, serum proteins, hemoglobin structure, etc. With limited underlying genotypic variation, if these 'characters' were analyzed in families, they would have low heritabilities. This is not because genes are not involved in the development of these primary 'dental characters' but quite the opposite. The genes that are involved are critical to survival and reproduction (ask an edentulous moose). For this reason, many stages of dental development are resistant to genetic change. Mutations that affect tooth development (e.g., genes for *amelogenesis imperfecta*) introduce potential odontogenetic polymorphisms, but, in a natural environmental setting, such mutations are subject to strong selective pressures. The old adage 'if it ain't broke, don't fix it' seems to have wide application in dental ontogeny and evolution.

Butler (1982) observed that minor variations of the tooth crowns and roots (e.g., crown size, supernumerary cusps, root number variants) are the characters most amenable to evolutionary change. In contrast to features of the normative dental blueprint noted above, such characters have more freedom to vary within and between populations. This variation is attributable to greater underlying genetic diversity. A mutation that affects the form of, say, the protostylid, has little affect on an individual's fitness. Such a polymorphism could be maintained (and has been) not only within a species but even within genera, families, and higher taxonomic levels.

If Falconer's (1960) generalization on fitness and heritability is correct, one would expect crown traits to have high heritabilities. Most of the traits Mizoguchi (1978) analyzed have heritabilities between 0.40 and 0.80 and the two traits for which there is some consensus – Carabelli's trait and shoveling – fall within this range. These values are similar to those for characters listed by Falconer as 'least essential' to fitness., e.g., abdominal bristle number in Drosophila (0.5), tail length at 6 weeks in mice (0.6), body length in pigs (0.5), etc. The heritability of tooth size also falls in this range (*c.* 0.6–0.8). More degrees of freedom (i.e., greater heterozygosity) are available for tooth size and presence–absence attributes of tooth crowns and roots than for the more fundamental aspects of the dentition. It is just these dental elements that show variation within and among early fossil hominids and recent human populations.

Crown morphology and the diagnosis of twin zygosity

Incidental to their trait classification system for the lower first premolar, Kraus and Furr (1953) noted the prospects for using dental morphology to diagnose twin zygosity. Given the limited array of genetic markers available at that time, they suggested crown morphology might provide the best suite of traits for determining zygosity. With advances in serological genetics, immunogenetics, and molecular genetics, we can see, in retrospect, they overstated the case for dental morphology. However, several studies have shown that a careful assessment of crown form and relief is a useful and inexpensive adjunct to other methods of zygosity determination. Although Kraus and Furr (1953) did not report the results of any zygosity diagnosis in their study, Kraus and his co-workers (1959) did use the battery of traits defined for lower first premolars, along with Carabelli's trait, to diagnose the zygosity of four triplet sets. Of the four sets, one was monovular (100% concordance among all three individuals), one was triovular (concordances of 0.50, 0.40, and 0.36 among the three individuals), while the two remaining sets included a monovular pair and a single fraternal individual (i.e., an MZ twin pair and DZ singleton). These were the same results suggested by fingerprint analysis. While the findings from this small scale study were not definitive, they demonstrated the potential for further research along these lines.

Lundström (1963) used crown morphology to assess zygosity in 70 twin pairs from Michigan and 54 pairs from New York. By observing differences in cusp number, fissure patterns, crown form, and lingual variations of the anterior teeth, Lundström diagnosed correctly 66 of 72 MZ pairs (91.7%) and 51 of 52 DZ pairs (98.1%). His overall diagnostic precision was 94.4%. Subsequently, Wood and Green (1969) diagnosed the zygosity of 32 like-sexed twin pairs using seven morphologic traits of the lower second premolar defined by Ludwig (1957). With the requirement that all seven traits had to be scored the same for a diagnosis of monozygosity, these authors correctly assigned zygosity about 85% of the time when they focused on homolateral comparisons. However, when heterolateral comparisons were made, this lowered the incidence of correctly diagnosing MZ status, although the diagnosis of DZ twins remained at about the same level. From this, they felt the procedure promised hope in zygosity assignment but only when trait comparisons were made on homolateral teeth

With a larger array of crown traits, Townsend *et al.* (1988) made a blind diagnosis of 120 Australian white twin pairs. Two observers assessed anterior crown form (size, shape, labial and lingual surface features), posterior crown form (occlusal patterns, cusp number), and Carabelli's trait. Twins were scored as concordant or discordant for major trait categories; allowing for minor variations in UI2, UM2, and LP2, twin pairs scored as discordant for two or more criteria were diagnosed as DZ. Following a meticulous methodology, the authors misclassified only 3 of 120 twin pairs yielding an overall diagnostic accuracy of 97.5%. Interestingly, no MZ twins were diagnosed as DZ; all three misclassifications were DZ twins diagnosed as MZ. In accord with the field concept, most MZ discordances were observed in the morphology of distal members of a field (i.e., UI2, UM2, LP2). The authors feel this may account, in part, for the lower level of zygosity diagnosis achieved by Wood and Green (1969) who focused exclusively on morphological traits of LP2. Despite a high level of success in zygosity determination, Townsend *et al.* (1988:11) conclude 'Our findings would suggest more caution as MZ twin pairs usually show some differences in dental morphology. Nevertheless, these differences are often rather subtle and would only be distinguished by an experienced eye.'

Diagnosing twin zygosity on the basis of dental morphology, though interesting and useful in some contexts, provides only general information on the genetics of tooth crown development. No specific knowledge is gained on the genetics of single dental traits as zygosity diagnosis involves the simultaneous assessment of multiple variables. The hidden assumption of these studies is that all traits and features are equally useful and equally genetic in nature, but, as Townsend *et al.* (1988) note, some teeth and some traits are influenced by environmental factors more than others. However, the fact that MZ twin dentitions are so overwhelmingly similar that their zygosity can usually be identified from the subtleties of crown relief indicates that genes are principally responsible for these similarities.

Environmental effects on crown morphology

In 1912, Franz Boas showed that European immigrants to the United States differed significantly in cephalic index and other anthropometric variables from their parental populations. This work demonstrated that changes could be induced in complex biological attributes in a single generation through a change in environmental conditions. This ability to physically adjust to a changing environmental milieu is referred to as plasticity. Using laboratory animals, one can ascertain through controlled experiments how a wide variety of environmental variables, especially nutritional factors (vitamins, trace elements), influence the development of specific biological variables. With human populations, such experimentation is not usually feasible, although there are other ways to assess plasticity indirectly, through 'natural' experiments, migrant populations, and generational analysis.

The secular trend toward increasing stature in western European countries over the past few centuries is a classic illustration of human plasticity. Stature, a typical quantitative trait with a relatively high heritability (*c.* 0.80), is clearly alterable by environmental factors, most of which relate to health and nutrition. Tooth size, with heritability values similar to those for stature, also exhibits plasticity as evidenced by generational differences between fathers vs. sons and mothers vs. daughters (Garn *et al.*, 1968b; Lavelle, 1972) and secular differences over short periods of time (Ebeling *et al.*, 1973). In controlled experiments on rats and mice, workers have shown how different nutritional regimens provided to pregnant or lactating mothers influence tooth size in their offspring (Searle, 1954; Paynter and Grainger, 1956) and such maternal influences may affect human tooth size as well (Bailit and Sung, 1968).

While tooth size may be influenced by environmental factors (Perzigian, 1984), no worker has demonstrated this level of plasticity in discrete morphological traits of human tooth crowns or roots. Many 'natural experiments' in human history have been conducted in the sense of populations moving great distances to settle in new and radically different environmental settings from those of the founding populations. European colonists who settled the disparate geographic areas of the Americas, South Africa, and Australia have modern descendant populations with very similar tooth morphology, all of European character and quite distinct from the indigenous populations of those continents. Lasker (1945) made this same point when he compared dental characteristics between Chinese populations and Chinese immigrants born in the United States. When the Norse colonized Iceland (AD 830) and Greenland (AD 986), this involved a move from a moderate maritime environment to a more demanding subarctic setting. In adapting to these new conditions, the Norse in the North Atlantic diverged significantly from their parental Scandinavian populations in palatine and mandibular tori and other features of the craniofacial complex (Hooton, 1918; Halffman *et al.*, 1992; Scott *et al.*, 1992), but their crown and root morphology did not deviate from that of the ancestral population (Scott and Alexandersen 1992).

What kinds of environmental factors might influence tooth morphology? Mean annual temperature, humidity, ultraviolet radiation, altitude, and other environmental stressors studied by physiological anthropologists have never been demonstrated to have any influence on the development of crown or root traits. Episodic undernutrition and infectious diseases disrupt crown formation and leave a permanent mark on the teeth in the form of bands and pits referred to as linear enamel hypoplasia. While such insults affect gross morphology and ultrastructure of the enamel and dentine, no one has shown that the expression of specific crown traits is influenced by these events. Experimental studies on laboratory rats have assessed the role of trace elements (e.g. boron, fluorine, molybdenum) in tooth formation, and these do have some affect on the gross morphology of molar teeth; however, this affect is limited primarily to cusp height and the width and depth of fissures and grooves on the crown surface (Paynter and Grainger, 1956, 1962; Kruger, 1962, 1966; Grainger *et al.*, 1966).

Møller (1967) reviewed a number of studies that involved 'natural experiments' among human populations with varying levels of fluoride in their drinking water. Human molars, like rat molars, are influenced by fluoride levels. In fluoride areas, teeth tend to be larger with lower, rounder cusps and shallower grooves and fissures. There is no indication, however, that the tooth crown and root traits described in chapter 2 are influenced by trace elements. In a rare effort to relate a trace element to variation in a human crown trait, Cox *et al.* (1961) found no relationship between fluoride levels and the expression of Carabelli's trait. Future studies should assess this possibility in more detail.

Current views on the genetics of dental morphology

The prevailing view on the genetics of tooth crown morphology during the first half of the twentieth century was that such traits were under strong hereditary control. This position was based largely on their patterned geographic variation and long-term phylogenetic conservatism as indicated by dental remains in the hominid and hominoid fossil record. Genetic analysis played only a minor role in forging this view.

Prior to 1950, a few twin studies focusing on dental morphology had been conducted, and crown traits had often been observed to show a 'familial disposition' (noted as early as 1842 by von Carabelli). When Kraus and other workers initiated more systematic studies on the genetics of dental morphology, these traits proved difficult subjects for analysis. Even though Kraus (1951) reported that Carabelli's trait was inherited by two codominant alleles at a single locus (a point referenced to this day in McKusicks's catalog of Mendelian inheritance), he expressed reservations about this model at the outset. Anyone who has observed Carabelli's trait in a large number of subjects knows that phenotypic expression is not broken down into discrete categories but grades imperceptibly from slight expressions to large free-standing cusps. Even those workers who argued initially for a simple mode of inheritance had to invoke modifying genes at other loci or environmental factors to account for this level of variable expressivity.

When later family studies cast doubts on finding simple modes of inheritance for dental morphological traits, genetic research shifted in several directions. The threshold model for quasicontinous variants proved to be a useful way of looking at crown traits. Variable expressivity, a major impediment to single locus models, was not only explained but predicted given the tenets of the threshold model. Some researchers then asked whether or not major loci could be detected within the bounds of a more complex genetic system controlling trait expression. Others shifted their attention to the question of heritability with some efforts aimed at sorting out genetic and environmental components of variance (Nichol, 1990).

Most lines of evidence (e.g., zygosity diagnosis in twins, familial correlations, population variation) indicate that genes are a major controlling factor in crown and root trait development. Trait expression is, however, affected to some extent by environmental factors. The dentition, as a system, is characterized by bilaterality and mirror imagery between the two sides of the jaw in crown size and morphological trait expression. However, antimeres sometimes exhibit different expressions of a trait. As an individual has only one genotype operating on both sides of the dentition, most of this fluctuating asymmetry is attributable to environmental factors. Regarding twins, it remains a truism that MZ twins share all their genes in common and differences between such twins are also ascribable to environmental factors. Although MZ twin pairs show numerous similarities in general crown form and details of morphological minutiae, both of which have been used with a high level of success in zygosity determination, MZ twins do not show perfect concordance for discrete trait expression (Fig. 4.3). If genes were the only factor in morphological trait expression, one would expect perfect concordance in MZ twins. While proximate explanations are elusive, it may prove significant that the correlations between antimeres are similar to the correlations between MZ twins (Scott and Potter, 1984; Townsend and Martin, 1992). Possibly, the environmental factors responsible for antimere asymmetry are similar to those that preclude complete concordance of trait expression in MZ twins.

With our emphasis on crown trait expression in permanent teeth, we have not noted one additional complicating factor in genetic analysis. That is, some individuals express crown traits on their deciduous teeth but fail to express the trait on their permanent teeth. The reverse situation also occurs but less commonly. To illustrate, Townsend and Brown (1981b) compared degree of Carabelli's trait expression on dm2 and UM1 and found that the trait was expressed concordantly (absent–absent or present–present) on both teeth about 80% of the time. However, when there was discordance in trait expression between the permanent and deciduous molars, 90% of the cases involved presence on the deciduous molar and absence on the permanent molar. In other words, some individuals who have genes for Carabelli's trait expression are scored as absent for the Carabelli's trait phenotype when only permanent molars are analyzed in twins and families.

Although never fully explored, another factor complicating the genetic analysis of crown traits is level of observation. Morphological crown traits are by convention and convenience observed on the enamel surfaces of the teeth. In those circumstances where the enamel is removed from the crown,

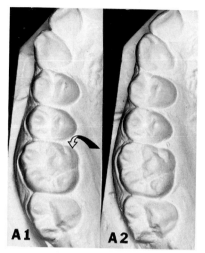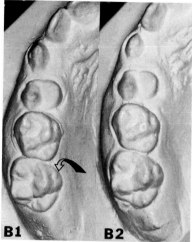

Fig. 4.3 Carabelli's trait expression in monozygotic twins. Shown are the upper right teeth of two pairs of monozygotic twins. Twins B1 and B2 exhibit identical expressions of Carabelli's trait. The A twins, however, are slightly discordant with A2 showing more pronounced trait expression than A1, indicating the influence of environmental factors. Even though Carabelli's trait is not perfectly concordant in the A twins, note that other ridges and cusps of both twin pairs are exceptionally similar, befitting their status as 'identical' twins. (After Scott, 1991a. Used with permission from Academic Press.)

either by accident or design, the surface of the dentine is directly observable (cf. Korenhof, 1960, 1961). As for trait expression, three situations can pertain to the enamel and dentinal surfaces: (1) a trait visible on the enamel surface is also visible on the dentinal surface; (2) a trait visible on the enamel surface is not visible on the dentinal surface; and (3) a trait visible on the dentinal surface is not visible on the enamel surface. The latter situation poses the greatest problem to assessing the genetic basis of tooth crown traits. If an individual expresses a feature on the dentinal surface that is not visible on the enamel, the trait is scored as absent. The enamel, in such an instance, may hide an 'underlying' phenotype. This may account for some of the antimeric asymmetry evident on the external crown surfaces and also discordance in trait expressions between members of MZ twin pairs. Lacking superficial and underlying surfaces, these qualifications and comments are not applicable to root traits.

Ramifications of genetic studies to population studies

No single crown or root trait has been shown to have a simple mode of inheritance. Complex segregation analysis indicates some traits are influenced by the actions of a major gene, but no method has been developed to tease these genes out of the background noise introduced by other genetic and environmental factors. At present, dental trait frequencies cannot be reduced to gene frequencies.

For those traits that conform to the threshold model of inheritance, Falconer (1960) suggests the best parameter to characterize a population is total trait frequency. This value, which marks the physiological threshold, characterizes the entire continuous distribution of a trait. To enhance comparability between workers, this places a premium on establishing a standard for trait expression just at the point of the threshold. Theoretically, any level of expression on the visible scale could serve to characterize a continuous genotypic distribution but, as before, workers have to agree on this level to arrive at comparable results.

In their recent opus on human genetic variation, Cavalli-Sforza *et al.* (1994:317) acknowledge that 'dental data on northern Asia, southeast Asia, and the Americas are generally in excellent agreement with those from single genes.' The dental data they refer to are crown and root trait frequencies (cf. Turner, 1983a, 1984, 1985a, 1985b, 1986a). As human geneticists who deal with 'simple' genetic markers, they seem perplexed by this concordance of results. Their expressed concerns on genetic grounds are the unknown levels of heritability for crown and root traits and the number of independent genes reflected by phenotypic data of this nature. As noted earlier, heritability values have been estimated for many traits and are relatively well established for a few. For traits expressed on the stable teeth within a tooth district, heritability generally assumes a middle range value (0.40–0.80). At this time, we do not know how many independent genes are involved in the development of any specific trait. On developmental grounds, we assume that specific dental phenotypes, while polygenic, are more comparable to skin pigmentation (few genes) than to compound quantitative traits like stature (many genes).

While environmental factors influence trait expression to some extent, they have not been shown to significantly affect population trait frequencies. Admittedly, some crown and root traits seem not as genetically stable as others, making them less useful for taxonomic purposes. This seems especially true for traits expressed within the same field (e.g. unstable shoveling at UI2). While there is still no definitive list of key traits that are the most useful in population comparisons, the traits we focus on in the next chapter have been determined, through extensive experience with dental variation across time and space, to be the most stable traits from a genetic and evolutionary standpoint.

5 Geographic variation in tooth crown and root morphology

Introduction

Early in the twentieth century, world-renowned paleontologist William King Gregory (1922) expressed the view that tooth crown morphology varied hardly at all among the major races of humankind. Traits he noted as exceptions to this 'generalization' included shovel-shaped incisors, distinguishing Asians from non-Asians, Carabelli's cusp, distinguishing Europeans from non-Europeans, and molar cusp numbers and patterns, for which Europeans showed more cusp reduction (i.e., hypocone and hypoconulid loss) and groove pattern simplification (i.e., from Y to +) than Asians and Africans.

In 1922, when Gregory published his treatise *The Origin and Evolution of the Human Dentition*, his belief about limited human dental variation was fundamentally correct as far as was then known. Few crown traits had been defined, root variants were largely unknown, and, as Sir Arthur Keith (1931) noted, much of the world remained in a state of 'dental darkness.' Prior to 1920, Europeans were the primary focus of dental researchers while groups representing other major geographic races had barely been examined, if at all.

Campbell's (1925) monograph on the dentition of Australian Aborigines was a major cornerstone for subsequent characterizations of non-European dental morphological variation as well as other dental anthropology topics. Shaw (1931), stimulated by Campbell's effort, observed crown and root traits in the Bantu dentition, along with dental pathologies and jaw characteristics. Hrdlička (1920) had earlier described shovel-shaped incisors in Native Americans, followed by multiple morphological characterizations of this group by Nelson (1938), Goldstein (1948), Dahlberg (1951, 1963a), and others, including P.O. Pedersen (1949) and C.F.A. Moorrees (1957) who wrote invaluable monographs on the dentitions of linguistically related East Greenland Eskimos and Aleuts, respectively.

Despite an upsurge in interest in dental morphological variation during the middle of the twentieth century, it was still not possible, even as late as

1970, to characterize more than a few crown traits on a worldwide scale. Shovel-shaped incisors, Carabelli's trait, and lower molar cusp number and pattern were exceptional in this regard, but some major geographic areas were still poorly known even for these widely studied traits, especially Africa, India, and Central and far Northeast Asia. To adequately characterize dental variation around the world and through recent times: (1) more crown and root polymorphisms had to be isolated, defined, and classified to supplement the limited trait set of earlier workers; (2) more populations had to be characterized for nonmetric dental trait variation, especially in geographic areas where few samples had been previously studied; and (3) more attention needed to be given to archaeologically-derived dentitions.

The crown and root traits reviewed in chapter 2 represent the cumulative efforts of many workers to expand the number of characters available for analysis. The primary architects of this list are A.A. Dahlberg and C.G. Turner II, with contributions, in chronological order, from C.S. Tomes, W.K. Gregory, L. Bolk, A. Hrdlička, M. Hellman, F. Weidenreich, E.K. Tratman, P.O. Pedersen, B.S. Kraus, V. Alexandersen, K. Hanihara, D.H. Morris, G.R. Scott, E.F. Harris, H.L. Bailit, and C.R. Nichol. Each of these workers defined one or more crown or root traits which, taken together, make up a set of more than 30 largely independent variables.

In Lasker's (1950) article on genetic aspects of nonmetric crown and root traits, he detailed the promising uses for these traits in anthropological studies, but attempted no synthesis of dental variation because the comparative baseline was still too narrow. Since that date, hundreds of articles on dental morphological variation have been published covering a wide range of local and regional populations. Contributions to this literature come in all shapes and sizes, from massive monographs and dissertations to abstracts and one page annotations. Many population studies include observations on one or a few traits in small samples, while others provide observations on many traits in multiple samples. Studies of small and large scope both provide valuable building blocks for a worldwide synthesis, some just provide more blocks than others. Contributors to this database are also a diverse lot, though most are associated with some branch of anthropology, dentistry, or genetics. While dental morphology is a secondary or tertiary interest of some researchers with specialties in other areas (e.g., orthodontia, genetics, archaeology), a few workers have devoted most of their professional energies to dental morphological studies. The combined efforts of workers with both primary and secondary interests in dental morphology make the following synthesis possible.

To assay geographic variation, the first step is to depict the frequency variation for 23 crown and root traits, as presently known, among the major subdivisions of humankind. Trait frequencies are presented in some combination of graphs, tables, and summary text, depending on the availability of comparative data. Our rationale for the geographic subdivisions and sources of information used is presented below. This chapter is primarily descriptive. In chapters 6 and 7, the historical insights that can be obtained from the temporal and spatial patterns of dental morphological data are shown through biological distance analyses at regional and global levels.

Geographic subdivisions of humankind

In 1776, J. F. Blumenbach recognized that all humans belonged to a single species. On the basis of anthroposcopic observations (i.e., skin color, hair color and form, and the form of the nose, cheeks, lips, face, and chin), he divided the species into five principal varieties: Caucasian, Mongolian, Ethiopian, American, and Malay. From his description of their geographic range, the Caucasian variety included European, Asiatic Indian, North African, and Middle Eastern populations. The Mongolian variety encompassed most Asians as well as Finnish and Lappish populations in northern Europe and Eskimos from Alaska to Greenland. His Ethiopian variety included all African populations except groups today defined as Afro-Asiatic speakers who occupy the northern third of the continent. All Native Americans, excepting Eskimos, were included in his American variety. Blumenbach added that 'Americans came very near to the Mongolians, which adds fresh weight to the very probable opinion that the Americans came from northern Asia' (in Count, 1950:38). His fifth and final division, the Malay variety, included mainland and insular Southeast Asians along with Pacific Island populations. While Lapps and Finns have since been shown to fall within the Caucasian sphere, morphological and genetic studies over the past 200 years are in general accord with Blumenbach's five major varieties.

In 1817, the noted French paleontologist G. Cuvier simplified Blumenbach's scheme when he defined three rather than five major human races: 'Caucasian or white, Mongolian or yellow, and Ethiopian or Negro' (in Count, 1950:44). While recognizing some of the distinct features of Malays and Americans, he was reluctant to place them on the same taxonomic level as his three major varieties. Following the classificatory efforts of early natural historians like Blumenbach and Cuvier, later nineteenth and early twentieth century physical anthropologists divided humans in many different ways, depending on whether they were lumpers (3 races) or splitters (up to 63 races). In contrast to many of his nineteenth-century contemporaries, C. Darwin believed the variety of human racial classifications 'shows that they (i.e., races) graduate into each other, and that it is hardly possible to discover clear distinctive characters between them' (in Count, 1950:138). This opinion has been echoed by many twentieth century researchers, especially those who emphasize the importance of clinal variation (cf. Montagu, 1964). While workers in anthropology and allied fields have abandoned racial classification as an end in itself, it is still necessary to adhere to some organizing principle to characterize biological variation above the population or country level. Even critics of the race concept cannot dispute the need for labels based on geography, biology, and other relevant data.

Subdividing humanity to describe between-group variation can be accomplished in several ways. In presenting large bodies of data, geneticists prefer to list countries and/or populations based strictly on geographic divisions (Mourant, 1954; Mourant *et al.*, 1976; Roychoudhury and Nei, 1988). In so doing, indigenous groups and recent colonizing populations, despite significant genetic differences, are listed under the same heading (e.g., North America: Eskimo, American Indians, American whites, American blacks, etc.). In their recent tome on population genetics and history, Cavalli-Sforza *et al.* (1994) discuss variation in five substantive chapters corresponding to five major geographic regions (Europe, Africa, Asia, Australia and the Pacific, North and South America). Although his classification has geographical overtones, the anthropologist S.M. Garn (1971) used additional lines of evidence to define nine major human races: Europeans, Asiatic Indians, Africans, Asians, Australians, Melanesians, Polynesians, Micronesians, and American Indians. Recognizing Garn, but using other lines of classificatory evidence, we propose five major human subdivisions, each containing further regional subdivisions, based on geography, language, bioarchaeology, natural and cultural history, and dental variation. We intentionally do not include genetic information because one of our aims is to independently compare the taxonomic findings of dental anthropology with those of genetics.

Our subdivisions are not called races as we have no interest in entering into the debate on whether or not human races exist. We are cognizant of the complexities of human population history – the human penchant to migrate, invade, colonize, and trade, and the potential for such movements to stimulate the free exchange of genes between groups. As we will show, the fine details of a tooth crown or root can aid in reconstructing prehistoric movements, as well as demonstrating the great differences that develop between groups given extended periods of isolation.

Our framework depends, to some extent, on research in historical linguistics. Many anthropologists argue that language and biology are independent phenomena; one cannot predict biology from language nor language from biology. This is certainly true at the individual level (i.e., any human can learn any language) and also the group level when social pressures instigate the adoption of a new language (e.g., émigrés to the United States learning English). Moreover, it is clear that some exceptional groups show biological relationships in one direction and linguistic relationships in another. For example, the Turkic-speaking Azerbaijani are more closely allied biologically to Middle Easterners and Europeans than to other Turkic groups such as the Kazakhs of Central Asia. However, discounting the adoption of a national language by minority groups (indigenes or immigrants), many researchers have found a positive relationship between biological attributes and language, which together, offer many clues to the course of human population history. This is not surprising to those workers who study linguistic and biological differentiation. The processes underlying changes in either biology or language share many elements in common (Ruhlen 1994). When a group fissions into two or more subgroups, followed by the geographic isolation of descendant populations, differences in language and biology invariably develop. Of course, the underlying bases of linguistic and biological change are different – biological change is dictated by changes in a population's genetic make-up while linguistic change is dictated by extra-biological factors (i.e., our DNA does not condition what language we learn during our formative years, although genes are ultimately responsible for the pan-human neural template that allows language acquisition). However, both genes and words involve the transfer of information from one generation to the next and either form of information is subject to forces which bring about change. The evolutionary mechanisms underlying biological differentiation are genetic drift and natural selection, with mutations adding new but rare genetic variation. Genetic differentiation, in turn, can be obscured by various forms of migration, as well as by gene flow between populations, a process that leads to biological convergence. Sound shifts, changes in grammatical structure, the adoption of loan-words, and other linguistic modifications are brought about by processes broadly analogous to those responsible for biological differentiation. We know that some linguists concur with this view while others do not.

Given the processual parallels in the two fields, linguists borrow many terms from biology. They talk about genetic relationships between languages although the term 'genetic' should not be taken in the literal biological sense. If languages are said to be genetically related, this indicates they shared a common ancestral language at some time in the past. Just how long ago they shared this common ancestor, as measured by overall linguistic differences, is reflected in the use of terms that sometimes parallel those from the Linnaean hierarchy, such as language phylum or language family. Groups remotely related linguistically are sometimes placed in higher order Macro-Phyla. Although not derived from biology, the term dialect is used to distinguish the minor linguistic differences that develop in a few hundred years. Languages do evolve at a faster pace than biological characteristics. Linguists, for example, cannot find any 'genetic' relationship between the Indo-European and Caucasian language families (Ruhlen, 1987), although the common ancestry of groups within these two families is evident in many shared biological characteristics.

Temporally, the five-fold population classification we employ is bracketed by the early Neolithic and the so-called Age of Discovery which began c. AD 1500. Our five subdivisions do not presume to reflect human variation during the later Paleolithic (although it might in some cases). The emergence of food production ten to twelve thousand years ago in the Old World resulted in a massive geographic reshuffling of populations. Only the most remote areas were untouched by this process until recent times (e.g., Australia, Arctic). The Holocene has seen the expansion of Afro-Asiatics into North Africa, Indo-Europeans into India, Bantu-speaking populations into South Africa, Austronesians into the Pacific, Eskimos into Canada and Greenland, and so forth. These examples represent no more than the tip of the iceberg for major population movements and shifts over the past 10,000 years. During the last 500 years, the pace of movement accelerated to the point where many indigenous populations were replaced, peripheralized, and/or assimilated by biologically disparate populations in the Americas, Australia, parts of Africa, and northern Eurasia. Thus, while our groupings do attempt to reflect earlier Holocene movements (e.g., Europeans and Asiatic Indians are both considered Western Eurasian), recent colonization events are ignored. To illustrate, American, Canadian, South African, and Australian whites are linguistically and biologically tied to Western Eurasia rather than the geographic areas where they now reside, so they are treated accordingly. The five subdivisions we define for humankind and the major language families noted in the following areal descriptions are shown in Fig. 5.1.

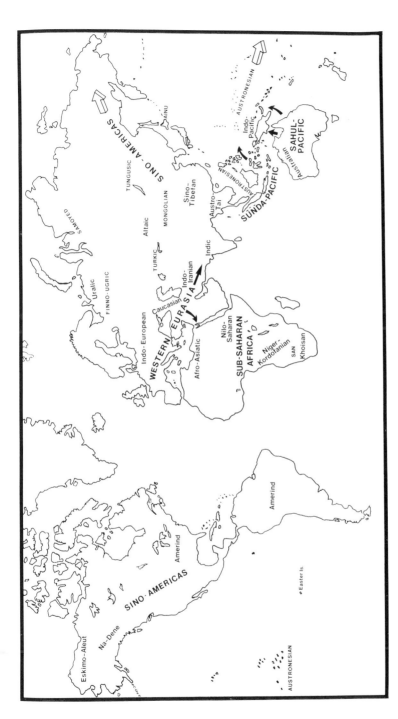

Fig. 5.1 Outline map of the world showing the five major subdivisions of humankind used for organizing data on world dental morphological variation. Also shown are major language phyla and families discussed in text and used in Tables 5.1 to 5.9.

The major subdivisions of humankind

Western Eurasia

It is difficult to coin a geographic term for this widely distributed set of populations traditionally called Caucasoids or Caucasians by natural historians and anthropologists. Excluding post-AD 1500 colonies, the geographic range of this grouping covers all of Europe and parts of Africa and Asia. More specifically, populations in this subdivision inhabit peninsular Europe, the British Isles, Scandinavia, European Russia, the Near and Middle East, North Africa, and much of the Indian subcontinent. The major language families represented in the subdivision are Indo-European, Afro-Asiatic, Caucasian, and Uralic. Populations within the first three families, despite the diversity of language, show many biological and historic ties. Uralic is exceptional as populations within one major branch, Finno-Ugrian, show biological ties to Europe while those in the Samoyedic branch exhibit closer ties to Asian populations.

Our primary characterization of this area is based on groups from: (1) Western Europe; (2) Northern Europe; and (3) North Africa. When comparative tables are constructed, it is possible to make finer subdivisions which include: (1) Europe (early: Neolithic to Iron Age); (2) Western Europe; (3) Eastern Europe (Slavic, Baltic, Armenian); (4) Indo-Iranian (Tadjik, Ossetian); (5) India (early: Neolithic to Iron Age); (6) Indic; (7) Caucasian; (8) Finnic-Permian (Estonians, Finns, Komi); (9) Ugrian (primarily Ob Ugrian); (10) Samoyedic (Nentsy); and (11) Afro-Asiatic. Forming groups 3, 4, 7, 8, 9, and 10 would not have been possible without the extensive data set compiled and described by A.A. Zubov and his colleagues (Zubov and Khaldeeva, 1979).

Sub-Saharan Africa

Earlier anthropologists referred to populations in this area as Negroid or Ethiopian. Coon (1965) divided Sub-Saharan Africans into two groups: Congoid and Capoid (Bushmen). Despite the great diversity of languages spoken in Africa, American linguist Joseph Greenberg (1963), the quintessential lumper, devised a classification that placed all Sub-Saharan African languages within one of three major phyla: (1) Niger-Kordofanian; (2) Nilo-Saharan; and (3) Khoisan. Most anthropologists feel that Khoisan-speaking groups were once widely distributed throughout the southern third of Africa but are now represented only by numerically small and peripheralized populations such as the Bushmen (San), Hadza, and Sandawe. The most widely distributed populations in Africa fall within the

Niger-Kordofanian phylum, whose speakers, including Bantu, extend over most of west and south Africa. Populations of the Nilo-Saharan phylum are found principally in East Africa, extending from Kenya and the reaches of the upper Nile northwest to the deserts of Sudan (Ruhlen 1987).

Our characterization of African dentitions focuses on: (1) West Africa; (2) South Africa; and (3) the San. The first and second groupings are comprised almost exclusively of samples from the Niger-Kordofanian language phylum. The San are considered separately given their unique history and many distinctive biological differences from other African populations (Coon, 1965; Garn, 1971).

Beyond the study of Irish (1993), whose data are used in our basic characterization, information on crown and root morphology is limited for African populations. For that reason, we employ only broad geographic headings in the comparative tables, e.g., East Africa, South Africa. When only South Africa is noted, the data are derived primarily from Bantu groups. Data on the San from sources other than Irish are listed in tables under the heading of their language family, Khoisan (see Tables 5.1–5.8).

Sino-Americas

In classical parlance, this geographic subdivision includes populations commonly known as Mongoloids or, in the case of American Indians, proto-Mongoloids. The geographic range of this subdivision today covers almost half of the earth's land surface, from the Ural Mountains in the west to the mouth of the Amazon River in the east. In the Old World, the region's north–south boundaries would extend from the Timor Peninsula in Siberia to the Yangtze River in south China. In the New World, these boundaries are set by northern Greenland and Tierra del Fuego at the southern terminus of South America.

Despite the biological and linguistic diversity shown by groups within this subdivision, they share many biological features in common. The primary language families in the Old World branch are Sino-Tibetan, Altaic, and Chukchi-Kamchatkan. Despite an exceptionally large number of New World languages, Joseph Greenberg (1987) reduced these to three primary families: Amerind, Na-Dene, and Eskimo-Aleut. Despite the sharing of many biological features by all populations of the New World and North and East Asia, working out linguistic relationships across the Bering Strait remains a formidable challenge (Ruhlen, 1987).

Because much time and effort has focused on the ties within and among Asian and Native American populations, it is possible to characterize eight major groupings under the Sino-Americas subdivision: (1) China-Mongolia; (2) Japan (Jomon); (3) Japan (recent); (4) Northeast Siberia; (5) South Siberia; (6) American Arctic (Eskimo-Aleuts); (7) Northwest North America (Indians); and (8) North and South America (Indians).

When adequate comparative data are available from other researchers, they are summarized under the headings of: (1) Sino-Tibetan; (2) Japanese; (3) Ainu; (4) Taiwan aborigine; (5) Altaic (Turkic speakers); (6) Altaic (Mongolian speakers); (7) Altaic (Tungusic speakers); (8) Eskimo-Aleut; (9) North American Indians; and (10) South American Indians. The work of Russian dental anthropologists allows the breakdown of Altaic groups into three major branches (Zubov and Khaldeeva, 1979; Ismagulov and Sikhimbaeva, 1989). As a point of geographical clarification, when we refer to East Asian groups in our trait by trait summaries, we are referring principally to Chinese and recent Japanese populations. North Asia is used in reference to the populations of northeast Siberia (but not South Siberia and Central Asia).

Sunda-Pacific

When sea levels were lowered during major Pleistocene glaciations, the relatively shallow Sunda continental shelf was exposed, linking mainland Southeast Asia to islands of the East Indian Archipelago (e.g., Sumatra, Java, Borneo, etc.). This geographic region is called 'Sundaland.' With rising sea levels during the Holocene, Sundaland was flooded and again divided into mainland and insular components. As peoples on or around the Sunda shelf are linguistically related to the far-flung populations of the Pacific, especially those of Polynesia, we combine terms to form our designation Sunda-Pacific.

In most classifications, the populations of mainland and insular Southeast Asia are lumped with East and North Asians into the Mongoloid category, although they are sometimes qualified as 'southern Mongoloids.' Workers have also referred to Polynesians as Oceanic Mongoloids. Working with only limited data, Blumenbach felt the distinction between groups in North and Southeast Asia was sufficient to warrant different labels – Mongolian and Malayan, respectively. We concur with Blumenbach that the biological differences between populations in the two regions are great enough to warrant separate geographic subdivisions. Additionally, the creation of two 'Mongoloid' subdivisions helps distinguish two major colonization events. That is, Native Americans (Eskimos, Aleuts, Indians) were derived from late Pleistocene stem populations in East and North Asia (Haeussler, 1996), while Polynesian ancestry can be traced, by archaeology, linguistics, and physical anthropology, to Southeast Asia (Bellwood, 1991; Green, 1994).

Populations of the Sunda-Pacific subdivision have been classified into four major linguistic families in a proposed Austric phylum: Miao-Yao, Austroasiatic, Daic, and Austronesian. The first three language families are spoken primarily in mainland Southeast Asia, including south China, with scattered pockets of Munda speakers (Austroasiatic family) in India. The far-flung Austronesian language family, with Malayo-Polynesian a principal division, is spoken by many peoples of insular Southeast Asia, including Taiwan aborigines. Polynesian and Micronesian languages fall within the Austronesian family as do many of the Melanesian languages (Ruhlen, 1987). We recognize that another over-arching phylum, Austro-Tai, embraces the same groupings (Benedict, 1975).

For our primary characterization, Sunda-Pacific is divided into four units: (1) Southeast Asia (early); (2) Southeast Asia (recent); (3) Polynesia; and (4) Micronesia. Samples (1) and (2) are composites with both mainland and insular components – they are divided along temporal rather than geographic lines. The inclusion of Micronesia in this subdivision should be qualified as the inhabitants of these coral atolls are of mixed ancestry. While the major colonization events were initiated from Southeast Asia, there is a discernible Melanesian contribution to some of the modern populations in Micronesia (Howells, 1973a).

To organize data from other workers, comparative samples are listed under the headings: (1) Southeast Asia; (2) Polynesia; and (3) Micronesia. While there is an adequate sampling of Polynesian groups, due largely to the efforts of Japanese dental anthropologists, relatively few studies are available for other Southeast Asian or Micronesian samples.

Sahul-Pacific

Australia, New Guinea, and Tasmania were linked during glacial periods by the exposed Sahul continental shelf to form the larger island continent of Sahulland. At no time during the Pleistocene was there a land bridge that connected Sundaland to Sahulland because of deep inter-island oceanic trenches. Anthropologists are confident that the peopling of Australia and New Guinea involved the use of watercraft some 40,000 to 60,000 years ago. During the late Pleistocene, rising sea levels separated Australia and New Guinea. The islands of Melanesia (e.g., New Britain, New Hebrides, Solomons), while never connected to Sahulland, were peopled by seafaring populations from this area at a very early date, some 30,000+ years ago (Green, 1994). The population history of Melanesia is, however, quite complicated as these islands have, in more recent times, been infused with a discernible Southeast Asian component. This is indicated in part by the spread of Austronesian languages throughout much of Melanesia. Many authors avoid this conundrum by classifying Melanesians, Polynesians, and Micronesians under separate headings (e.g., Garn, 1971) or include them all in a large Oceanic grouping. Melanesian heterogeneity belies any simple solution so we group them here with Australia and New Guinea more out of convenience than conviction.

The peoples of Australia have been isolated from other human populations for such a long period of time that they show no linguistic ties whatsoever to any population outside the continent. Within the Australian phylum, linguists divide 29 first-order groups under two broad headings: (1) Pama-Nyungan, with languages spoken throughout the lower two-thirds of the continent; and (2) non-Pama-Nyungan, which includes the remaining 28 first-order groups distributed in north-central and northwest Australia (Ruhlen, 1987). New Guinea groups are also characterized by a great diversity of languages. Many are placed in a broad Indo-Pacific phylum, although along the northern coasts there are many Austronesian speakers. Melanesia, likewise, has both Indo-Pacific and Austronesian components, sometimes on the same island. Relative to time depth, Indo-Pacific is far more ancient than Austronesian in this region. Greenberg's (1971) inclusion of Andaman Islanders in the Indo-Pacific phylum supports Howells' (1976) concept of an 'Old Melanesia' that would have dark-skinned, short-statured populations scattered across the islands of the East Indian Archipelago prior to the Holocene incursion of populations from the Southeast Asian mainland. It is the antecedent population of Old Melanesia, along with Australia, that we try to capture under the heading Sahul-Pacific (granting much movement of Austronesian peoples through New Guinea and Melanesia over the past 5000 years).

In our characterization, we list samples under the headings of: (1) Australia; (2) New Guinea; and (3) Melanesia. Many, if not most, authors include New Guinea as part of Melanesia but we prefer to retain this distinction given the unique position of both New Guinea and Melanesian populations. Comparative data provided by other workers are listed under the same three headings. Despite the seminal work on the Australian dentition by T.D. Campbell some 70 years ago, Sahul-Pacific groups have received only scant attention from dental morphologists so comparative data are limited, particularly for New Guinea. For this region, we know more about tooth size than tooth morphology (Brace, 1980; Brace and Hinton, 1981).

Characterization of dental variation

Do any dental traits show patterned geographic variation that can be used to distinguish one major subdivision from other subdivisions? Previous authors have used the relative frequencies of tooth crown traits to define broad complexes, notably the Mongoloid dental complex (Hanihara, 1968a), the Caucasoid dental complex (Mayhall *et al.*, 1982), and the Australian dental complex (Townsend *et al.*, 1990). Each of these complexes is defined by five to seven crown traits. Are these complexes valid and useful? If so, can they be expanded through the addition of more crown traits and, importantly, root traits? Can other dental complexes, beyond the three noted, be defined? These questions are addressed at the end of the chapter. We must initially lay the foundation by describing the geographic variation of individual traits.

Two different strategies are employed to portray morphological variation on a world scale. First, frequency data are presented for groups representing most of the major geographic subdivisions of humankind based largely on the observations of one of the authors (CGT). The total number of individuals he has scored for a wide array of crown and root traits now exceeds 25,000. This data set has the following advantages: (1) observations are all based on one set of standards and trait definitions; (2) inter-observer error is minimized; and (3) many of the traits surveyed have not been extensively studied by other workers.

Our characterization of crown and root trait variation includes data on 21 geographic groupings, studied primarily in skeletal series from a few hundred to a few thousand years old. Samples from 17 regions were studied by C.G. Turner II (cf. Turner, 1983a, 1986a, 1989), augmented by the work of Irish (1993) on North Africa, South Africa, and the San, and by Scott (Scott and Alexandersen, 1992; Scott *et al.*, 1992) on northern Europe. All three workers used the ASU dental anthropology scoring system (Turner *et al.*, 1991). The collections that make up the regional samples are listed and mapped in Appendix A1.

The second method we use to characterize worldwide dental variation is to extract comparative data from the literature, standardize the observations of different workers on a common scale, and summarize this variation through descriptive statistics. It will be evident that efforts to characterize dental variation are highly variable relative to our five subdivisions. Thanks to the intensive long-term projects of Russian dental anthropologists, led by A.A. Zubov, Western Eurasian summaries can be broken down on a very fine level. Due largely to Japanese dental anthropologists, including K. Hanihara, T. Sakai, M. Suzuki, T. Hanihara, and Y.

Mizoguchi, among others, extensive comparative data are also available for East Asian and Pacific populations. Their work, along with that of Zubov and his colleagues on Altaic populations, allows for detailed subdivisions in our summaries of Sino-American dental variation. Unfortunately, comparative data for Sub-Saharan Africans and both Sunda-Pacific and Sahul-Pacific groups are limited to relatively small numbers of samples. As Russian and Japanese dental anthropologists have studied primarily crown traits, comparative data on root traits are woefully lacking. Despite the limitations set by the scope of regional studies and differential emphasis on particular morphological traits, we still feel the observations of other workers should help the reader gauge the generality of our initial geographic characterizations.

For each morphological variant, the format used includes: trait name, tooth observed, classificatory scale, breakpoint for deriving trait frequency, worldwide range in frequencies, and figure number. The geographic variation for 16 crown traits and six root traits is graphically portrayed in the form of hi–lo charts where trait frequency is marked by a vertical line with a horizontal bar extending ± 2 standard errors (frequencies and samples sizes are provided in Appendix A2). In a few instances, Northern Europe and the San are not included in the graphs. For Northern Europe, Scott did not observe the full set of traits (frequencies in Appendix A2 are from comparable Northern European samples). For the San, Irish's sample sizes for root traits were often quite small, resulting in large standard errors.

Each graph is followed by a brief description of that trait's pattern of worldwide variation. After this characterization, we compare our findings to those of other researchers. For intensively studied traits, data from the literature are summarized for subgroups in each region by mean trait frequencies and standard deviations. As each summary table is based on information from 150–300 samples, the sources of data are provided as a separate bibliography in appendix B.

Crown traits

Winging

Tooth: upper central incisor; classification: Enoki and Dahlberg 1958; breakpoint: grade 1 (bilateral winging only); world range: 4.2–50.0%; Fig. 5.2.

Bilateral winging is rare (i.e., frequencies less than 10%) in Western Eurasia, Australia, and New Guinea and is only slightly more common in

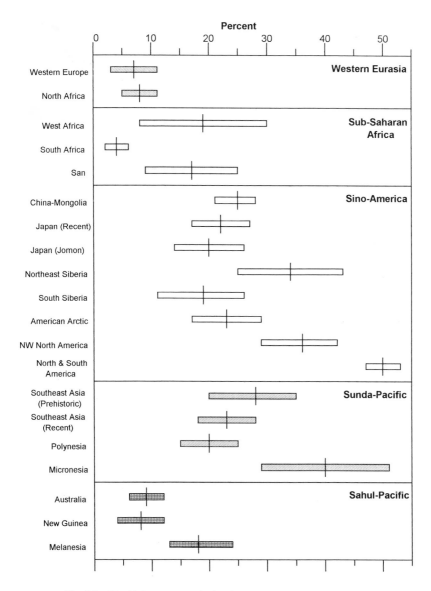

Fig. 5.2 World frequency variation in upper central incisor winging among the five major subdivisions of humankind (trait frequency represented by vertical line with horizontal bars denoting ± 2 standard errors).

Africa and Melanesia where frequencies fall between 10 and 20%. North, East, and Southeast Asians have winging frequencies in the 20 to 35% range. For New World groups, there are distinct differences between Eskimo-Aleuts (20%), Northwest North American Indians (35%) and North and South American Indians, who have the world's highest winging frequency (50%). Although Micronesians appear to be an outlier in a Sunda-Pacific context, their winging frequency is based on a small sample so this characterization may be deceiving.

Only limited comparative data are available for upper central incisor winging. Given its rarity in Western Eurasia, it is rarely mentioned in European and American dental anatomy texts and only occasionally scored by dental morphologists who study these populations. Scott (1973) and Mayhall *et al.* (1982) calculated frequencies of 5.1% and 4.3% for American and Canadian whites, respectively. For Asiatic Indians, Kaul and Prakash (1981) reported an incidence of 2.1% in the Jat while Sharma and Kaul (1977) found a somewhat higher frequency of 13.8% in Panjabi Hindus. Asiatic Indians residing in South Africa had a winging frequency of 1.0% (Scott, unpublished data).

Few African groups have been scored for winging. In modern populations, Haeussler *et al.* (1989) found this trait was more common in the San (19.6%) than in the Bantu (2.2%), a finding in accord with Irish's (1993) observations. It appears that Africa differs only slightly from Western Eurasia for the presence of this trait, although the San may be exceptional.

Workers who study populations in the Sino-American subdivision are in the best position to observe patterned variation in incisor rotation. Given the high frequency of winging in American Indians, it is not surprising that the earliest reference to this trait is associated with observations on Indian groups in Mexico – 'The two middle teeth of the Huichols are placed obliquely against each other, turning inward, making a symmetrical and not unpleasant break in the row of teeth' (Lumholtz, 1902:84–5). Enoki and Dahlberg (1958), who set up the classification system for this variant, worked with Japanese and American Indian populations, respectively, where the trait is expressed in about one of three individuals. Interestingly, a modern Ainu sample had a winging frequency of 34.5%, similar to that of other Japanese populations (Turner and Hanihara, 1977). In his classic study of the Indians of Pecos Pueblo, Nelson (1938:290) reported a winging frequency of 29.5%, adding that 'It seems scarcely possible that such a feature should have a significant racial distribution – yet one cannot fail to be impressed with the relatively large number of otherwise 'normal' (i.e., perfect) dentitions which exhibit mesio-palatal torsion of the upper centrals.' Scott (1973) arrived at an overall winging frequency of 27.3% for ten American Southwest Indian samples while another Southwest group, the Pima, had an incidence of 38.2% (Scott *et al.*, 1983). Wright (1941) observed winging in a South American Indian group, the Jivaro, who had a trait incidence of about 50%. Other South American Indians, including the Pewenche (Rothhammer *et al.*, 1968), Yanomama, and Makiritare (Brewer-Carias *et al.*, 1976) also have frequencies close to 50%. Although comparative data are limited for American Arctic groups, winging frequencies for modern Koniag Eskimos (Scott, 1994), prehistoric Kachemak and Koniag Eskimos (Scott, 1991b) and prehistoric St. Lawrence Island Eskimos (Scott and Gillispie, 1997) fell between 4.8 and 18.9%, supporting the notion that this trait is less common in Eskimo-Aleuts than American Indians.

For Sunda-Pacific groups, workers have reported winging frequencies of 11.0% for Easter Island (Turner and Scott, 1977), 18.1% for Ontong Java (Harris, 1977), and 24.6% for the Cook Islands (Yamada and Kawamoto, 1988). Winging was absent in small samples from Bali (Jacob, 1967) and Yap (Harris *et al.*, 1975).

Sahul-Pacific groups apparently have lower winging frequencies than Sunda-Pacific groups. Birdsell (1993) found a frequency of only 5.5% in his continent-wide survey of Australian aborigines. Among living Melanesians of the Solomon Islands, Harris (1977) reported mean winging incidences of 6.8% and 11.6% for Bougainville and Malaita Islanders, respectively. While these values are lower than the one tabled for Melanesian groups (i.e., 18.7%), they support the observation that this trait is relatively rare in Sahul-Pacific populations.

Our characterization and comparative data from the literature on bilateral winging variation support the following subdivisions:

(1) Low frequency groups (0–15%): Western Eurasia, Sub-Saharan Africa, Sahul-Pacific.
(2) Intermediate (15–30%): East and Central Asia, American Arctic, Sunda-Pacific.
(3) High frequency groups (30–50%): Northeast Siberia, Northwest North America, North and South America.

Shoveling

Tooth: upper central incisor; classification: Scott 1973; breakpoint: grades 3–6 (= Hrdlička's semi- and full-shovel classes; Zubov's 2+3); world range: 0–91.9%; Fig. 5.3.

Given the great attention shovel-shaped incisors have received, it is not

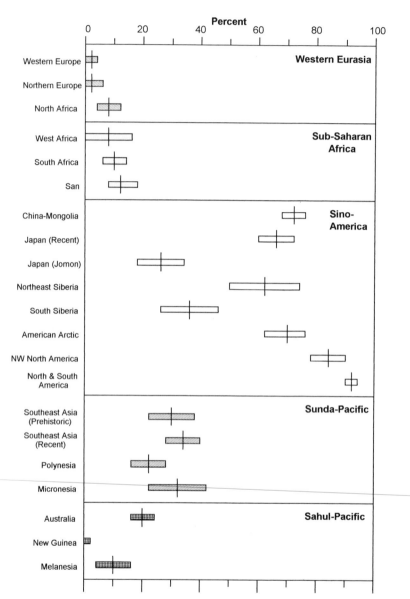

Fig. 5.3   World frequency variation in upper central incisor shoveling among the five major subdivisions of humankind (trait frequency represented by vertical line with horizontal bars denoting ± 2 standard errors).

Table 5.1. *Mean frequencies of shoveling (UI1) by geographic region and/or language family based on the observations of other workers (see Appendix B for references).*

Major region	Language/group	k	Mean %	SD
	Europe (early)	6	29.4	7.60
	Western Europe	13	7.6	4.67
	Eastern Europe (Slavs)	38	3.0	3.55
	Indo-Iranian	14	12.8	10.14
	India (early)	3	18.2	10.90
WESTERN EURASIA	Indic	9	12.6	9.55
	Caucasian	45	6.2	5.43
	Finnic-Permian	13	8.4	5.58
	Ugrian	2	53.7	1.63
	Samoyedic	9	47.4	11.89
	Afro-Asiatic	6	7.1	2.19
SUB-SAHARAN	East Africa	3	9.6	4.33
AFRICA	South Africa	3	11.0	4.81
	Khoisan	1	14.9	
	Sino-Tibetan	7	82.3	6.50
	Japanese	7	78.9	9.79
	Ainu	3	22.6	11.72
	Taiwan aborigine	3	79.0	7.21
SINO-AMERICAS	Altaic (Turkic)	37	36.0	19.84
	Altaic (Mongolian)	4	85.8	6.15
	Altaic (Tungusic)	2	60.0	35.78
	Eskimo-Aleut	10	72.7	16.23
	North American Indian	16	83.7	15.87
	South American Indian	4	78.7	20.32
	Southeast Asia	4	37.2	3.70
SUNDA-PACIFIC	Polynesia	6	29.7	8.20
	Micronesia	2	27.9	10.04
	Australia	2	6.5	9.19
SAHUL-PACIFIC	New Guinea	4	6.8	3.40
	Melanesia	11	5.4	5.02

k: number of samples; SD: standard deviation; UI1: upper central incisor.

surprising to find this trait clearly discriminates major regional groupings. With some minor exceptions, groups fall within one of three clusters for shoveling variation. At the low end, we find Western Eurasia, Africa, and Sahul-Pacific groups. An intermediate position is assumed by Sunda-Pacific populations. As Hrdlička (1920) and later workers have shown, for both frequency and degree of expression, shoveling sets East and North Asian and Native American populations far apart from all other human groups. In the Sino-Americas subdivision, Jomon and South Siberian groups stand out by having frequencies closer to those of Southeast Asia than North and East Asia.

Most comparative data available for shoveling are based on Hrdlička's (1920) four grade scheme. We have found that adding the semi- and full-shovel grades approximates the sum of grades 3–6 on the ASU scoring standard (Turner *et al.*, 1991) used to construct Fig. 5.3. Zubov's grades 2 and 3 (combined), used by most Russian and eastern European dental morphologists, also approximates our 3–6 breakpoint. Given these accommodations, other researchers have provided enormous amounts of data to test the generalizations stated above. These data are summarized in Table 5.1 for more than 280 samples.

Table 5.1 shows that almost half of the Western Eurasian groups have an incidence between 3 and 10% at the grade 3/semi-shovel breakpoint. This is true for western and eastern Europeans, Afro-Asiatics, Caucasians, and Finnic-Permians. Indo-Iranian and Indic groups show slightly more shoveling than Indo-European populations to the west. Two Uralic groups, Ugrians and Samoyeds, stand out clearly in Table 5.1, indicating their closer links to Asian peoples. Considering samples with more time depth (without language labels), it appears that shoveling was more common in western Europeans and Asiatic Indians dating from the Neolithic to the Bronze and Iron Ages than in modern samples from those areas.

Africans exhibit a mean frequency of 11% for distinct shoveling, an incidence close to or slightly higher than that of Western Eurasians. Hrdlička (1920) reports frequencies of 8.4% for American whites and 12.5% for American blacks. In assessing lingual fossa depth, Hanihara (1968a) found American blacks show slightly greater mean values (0.53 mm) than American whites (0.41 mm). Still, the difference is so subtle that shoveling would not be a useful trait to distinguish West Eurasians from Africans.

Sino-Americans show uniform frequencies of 70 to 85% for distinct shoveling. These summary values do not support the distinction shown in Fig. 5.3 between East and North Asians and American Indians. Regarding lingual fossa depth, Hanihara (1977) reported the following mean measurements: Pima Indians (1.2 mm) > Eskimos (1.13 mm) > Japanese (.99 mm). Although more data are required, we feel that shoveling expression is slightly more pronounced in American Indians than in East and North Asian populations. The notable exception to the general Sino-American pattern for shoveling is seen in Altaic groups and the Ainu. Turkic speakers from Central Asia have reduced frequencies of shoveling, more in line with Sunda-Pacific groups, while Mongolian and Tungusic populations show the elevated frequencies that typify their East Asian neighbors. The reduced shoveling frequency for the Ainu also makes them stand out in the context of recent Japanese and other East Asian and Native American populations.

Sunda-Pacific populations are distinct from Sino-American groups in having lower shoveling frequencies, although members of this cluster consistently exhibit more shoveling than groups from other major subdivisions. The mean values of 30 to 40% correspond closely to our initial characterization.

Sahul-Pacific groups show less shoveling than Sino-American and Sunda-Pacific populations, with frequencies approximating those of African and Western Eurasian populations. The low values for New Guinea and Melanesia shown in Fig. 5.3 are mirrored closely in Table 5.1 although the value of 6.5% for Australia contrasts our estimate of 20%. Richards and Telfer (1979) reported high shoveling frequencies in Australians (some samples over 90%), but they did not break their data down by grade of expression. Hanihara (1977) gives a shoveling frequency of 89.8% (breakpoint: lingual fossa depth > 0.50 mm) and a mean lingual fossa depth of 0.82 mm. This metric value is twice that given for American whites so it is unlikely that the grade 3/semi-shovel breakpoint frequencies would be equal for these two groups. It is probable that the Australian shoveling frequency falls in the 20–30% range, approximating that of Sunda-Pacific groups.

Our characterization and the summary data of other researchers show the following pattern for shoveling variation:

(1) Low frequency groups (0–15%): Western Eurasia (recent), Sub-Saharan Africa, Sahul-Pacific.
(2) Intermediate groups (20–50%): Sunda-Pacific, Western Europe and India (prehistoric), Samoyeds, South Siberia, Central Asia, Jomon, Ainu.
(3) High frequency groups (60–90%): East and North Asia, Americas.

Double-shoveling

Tooth: upper central incisor; classification: Turner *et al.*, 1991; breakpoint: grades 2–6; world range: 0–70.5%; Fig. 5.4.

While labial marginal ridging, or double-shoveling, is not entirely independent of the lingual marginal ridging characteristic of shoveling, the pattern of geographic variation for these two traits is not the same. Double-shoveling is by far the most common in Sino-Americans, but the

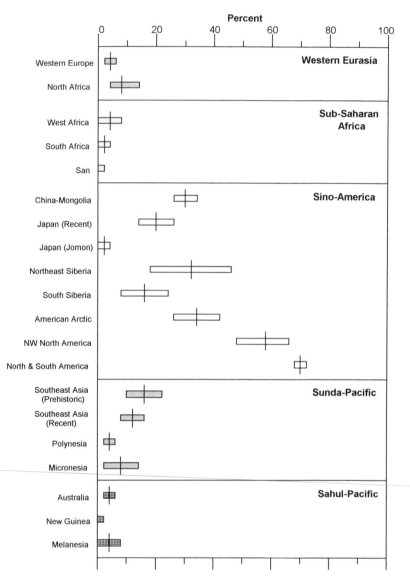

Fig. 5.4 World frequency variation in upper central incisor double-shoveling among the five major subdivisions of humankind (trait frequency represented by vertical line with horizontal bars denoting ± 2 standard errors).

two American Indian groups stand out with frequencies much higher than those of North and East Asians. For the rest of the world, double-shoveling is relatively rare. The frequencies for all other groups, excluding two Southeast Asian samples, fall below 10%.

Because double-shoveling is rare in Western Eurasia and Africa, workers have seldom reported trait frequencies for populations from these regions. While no European data are available, the Pushtun of India have a double-shoveling incidence of 9.7% (Sakai *et al.*, 1969). For Africans, Haeussler *et al.* (1989) reported very low frequencies for both San (2.0%) and South African Bantu (0.0%) samples.

The few workers who have described double-shoveling in East Asian and Native American populations support our characterization of these groups. The trait attains a frequency of 42.2% in the Bunun, a Taiwan aboriginal group (Manabe *et al.*, 1991), and 50.5% in modern Japanese (Suzuki and Sakai, 1965). For American Indians, the Lengua of Paraguay (Kieser and Preston, 1981) and Peruvian Indians (Goaz and Miller, 1966) have frequencies of 61.3% and 60.5%, respectively.

For Sunda-Pacific populations, Sakai (1975) reported a double-shoveling frequency of 15.9% for Hawaiians while Yamada and Kawamoto (1988) found an incidence of 9.7% in Cook Islanders. Although more data are needed, these values conform to our characterization of Southeast Asian and Pacific populations. The absence of data on Sahul-Pacific groups may reflect the rarity of this trait in Australian, New Guinea, and Melanesian populations.

Our figure and limited observations in the literature support the following characterization of upper central incisor double-shoveling:

(1) Low frequency groups (0–15%): Western Eurasia, Sub-Saharan Africa, Sahul-Pacific, Sunda-Pacific.
(2) Intermediate (20–40%): East and North Asia, American Arctic.
(3) High frequency groups (55–70%): American Indians.

Interruption grooves

Tooth: upper lateral incisor; classification: Turner *et al.*, 1991; breakpoint: grade 1 (total frequency); world range: 10.4–65.0%; Fig. 5.5.

While not rare in any major regional grouping, lateral incisor interruption grooves are least common in Sub-Saharan African and Sahul-Pacific groups where frequencies fall between 10 and 20%. The trait is most common in Sino-Americans where frequencies vary from 45 to 65%, with Native Americans and the Jomon at the high end of the range. The

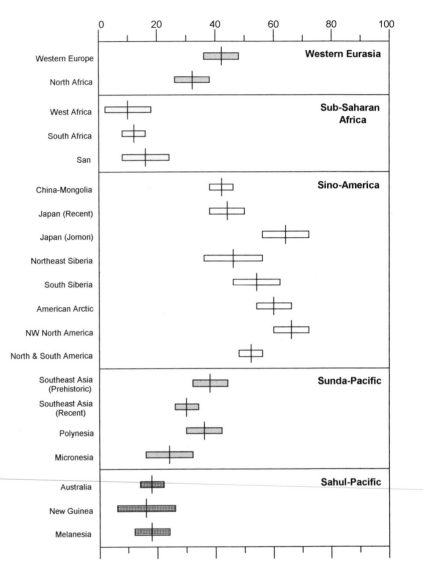

Fig. 5.5   World frequency variation in upper second incisor interruption grooves among the five major subdivisions of humankind (trait frequency represented by vertical line with horizontal bars denoting ± 2 standard errors).

frequencies for Western Eurasians and Sunda-Pacific groups are slightly, but consistently, lower than those of East and North Asian populations and their derivatives.

Few comparative data are available on upper lateral incisor interruption grooves. Workers who study casts and wax-bite impressions seldom deal with this trait. Even those who study skeletons and loose teeth, where the trait is easier to score, rarely include these grooves in their trait inventories. For Western Eurasia-India, all comparative data come from Brabant (1971) and Lukacs (1987, 1988). Brabant (1971), who referred to this trait as 'coronal-radicular grooves,' reported frequencies ranging from 6.3–10.7% for French, Swiss, and Belgian Neolithic samples, and a somewhat higher incidence of 12 to 20% for French Megalithic samples. For prehistoric India and Pakistan, Lukacs (1987, 1988) found interruption groove frequencies of 35.0% and 51.6% for Chalcolithic and Neolithic samples, respectively. The figures provided by Lukacs correspond to our characterization of Western Eurasia while Brabant's European frequencies are much lower.

More data are available on interruption grooves for Sino-American groups. One primary comparative check is Matsumura (1990) who reported frequencies for the protohistoric Kofun period in Japan (39.9%), recent Japanese (22.4%), the Jomon (60.0%), and recent Ainu (48.5%). His frequency for the Ainu is similar to the 42.6% reported by Turner and Hanihara (1977) for the same group. The high incidence for the Jomon corresponds well with the frequency we illustrate for Japan (Jomon) in Fig. 5.5. For Native Americans, frequencies close to 50% were found by Scott and Gillispie (1997) for St. Lawrence Island Eskimos (48.6%) and by Sciulli *et al.* (1984) for prehistoric Ohio Valley Indians (46%). Although subtle subregional distinctions are not discernible, data provided by other researchers support our view that this trait attains its highest frequencies in Sino-Americans.

Comparative data on interruption grooves are nil for Sub-Saharan African, Sahul-Pacific, and Sunda-Pacific populations. It is thus not possible to independently corroborate our characterization of these groups for lateral incisor interruption grooves.

Based primarily on Fig. 5.5, a characterization of UI2 interruption grooves would place the subdivisions in the following order:

(1) Low frequency groups (10–20%): Sub-Saharan Africa, Sahul-Pacific.
(2) Intermediate groups (20–40%): Western Eurasia, Sunda-Pacific.
(3) High frequency groups (45–65%): Sino-Americas.

Mesial canine ridge (Bushmen canine)

Tooth: upper canine; classification: Turner *et al.*, 1991; breakpoint: grades 1–3 (total frequency); world range: 0–35.1%; Fig. 5.6.

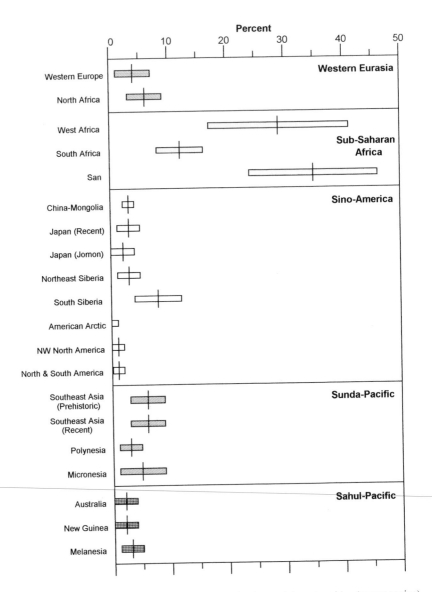

Fig. 5.6   World frequency variation in the mesial canine ridge (upper canine) among the five major subdivisions of humankind (trait frequency represented by vertical line with horizontal bars denoting ± 2 standard errors).

This trait, called the 'Bushmen canine' by Morris (1975), shows a distinctive pattern of geographic variation. It is moderately common in Sub-Saharan Africans, ranging from 12 to 35%, with the San showing the highest frequency. In groups throughout the remainder of the world, the trait does not exceed 10%. Although rare in general, it is least common in the Americas.

Almost all comparative data for the mesial canine ridge comes from Morris (1975) who reported complete trait absence (0.0%) for the following groups: Western Eurasia (American whites; Asiatic Indians in South Africa); Sino-Americas (San Francisco Chinese, Papago Indians); Sahul-Pacific (Solomon Islanders). For African groups, he found frequencies of 7.6% for Meroitic Nubians, 9.4% for Central Sotho Bantu, and 43.1% for the San, all of which correspond closely to the portrayal shown in Fig. 5.6. Sakuma *et al.* (1991) found an incidence of 6.3% in another Bantu group, the Chewa. Manabe *et al.* (1991) report a 'Bushmen canine' frequency of 7.0% for the Bunun of Taiwan. This value is more in line with our figure for Southeast Asians than North and East Asians, although the distinction is subtle.

The pattern of variation of the mesial canine ridge can be characterized in the following manner:

(1) Very rare (0–3%): Sino-Americas, Sahul-Pacific, Polynesia.
(2) Rare (4–7%): Western Eurasia, Sunda-Pacific.
(3) Moderate (12–35%): Sub-Saharan Africa.

Odontomes (premolar occlusal tubercles)

Tooth: upper and lower first and second premolars; classification: Turner *et al.*, 1991; breakpoint: 1 (total frequency); world range: 0.0–6.5%; Fig. 5.7.

Odontomes, while rare in all populations, show a distinctive pattern of geographic variation with three discernible clusters. This trait is very rare if not absent in most Western Eurasian, African, and New Guinea populations. It is also uncommon in South Siberia and the Jomon. Odontomes occur most frequently in other Sino-American groups with natives of northern North America (both Indians and Eskimo-Aleuts) having the highest frequencies. Australian, Melanesian, and Sunda-Pacific populations fall between these extremes.

It is hard to observe that which is not there. This truism may account for the dearth of observations on odontomes in Western Eurasian and African populations. Given their experience in examining Eskimo dentitions, Mayhall *et al.* (1982) were at least stimulated to look for this trait on the

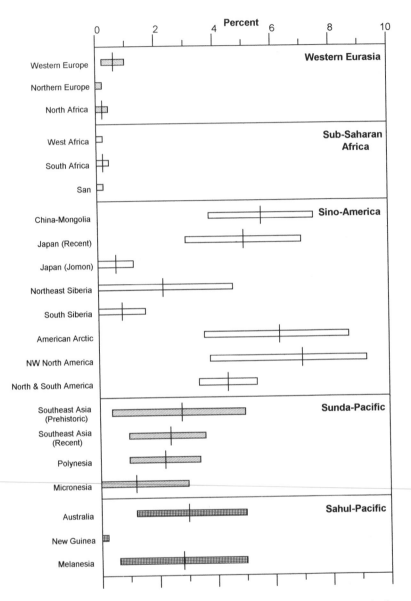

Fig. 5.7 World frequency variation in premolar odontomes among the five major subdivisions of humankind (trait frequency represented by vertical line with horizontal bars denoting ± 2 standard errors).

premolars of Canadian whites and reported, not surprisingly, a frequency of 0.0%.

Most comparative observations on odontomes come from those groups where the trait is most common: North and East Asians and Native Americans. Other authors support our observation that odontomes occur most frequently in the northern reaches of North America. Merrill (1964) reported that 4.6% of 650 Alaskan Indians, Eskimos, and Aleuts exhibited one or more odontomes. For central Canadian Eskimos, Curzon *et al.* (1970) and Mayhall (1979) found odontome frequencies of 3.0% and 1.4%, respectively. Scott (1991b, 1994) gave frequencies of 2.4% for prehistoric Kodiak Islanders and 4.4% for living Koniag Eskimos. An exceptionally high odontome frequency of 17.6% was observed in prehistoric St. Lawrence Island Eskimos (Scott and Gillispie, 1997). Merrill (1964) noted that a dentist in Wainwright, Alaska found a frequency of 15% in sixty Eskimo children. These high values probably approach the upper limit for this trait. For American Indians, Escobar *et al.* (1977) arrived at an odontome frequency of 0.74% for the Queckchi of Guatemala but added that frequencies of 0.0, 1.0, 1.1, 3.2, 3.5, and 6.3% were reported for Guatemalan Indians by other researchers.

Odontome frequencies reported for East Asian groups tend not to be as high as those shown in Fig. 5.7. Kato (1937) and Sumiya (1959) gave frequencies of 1.09% and 1.88% for two Japanese series while Lau (1955) found a frequency of 1.55% in a large Chinese sample. Manabe *et al.* (1991) reported a comparable frequency of 1.1% in a Taiwan aboriginal tribe. As for Sunda-Pacific groups, Reichart and Tantiniram arrived at an incidence of 1.01% in more than 5000 Thai premolars (in Escobar *et al.*, 1977).

In his paper on occlusal anomalous tubercles on premolars, Merrill (1964:491) concluded 'it seems safe to assume that this anomaly affects only members of the Mongolian race.' This statement now requires some qualification. Odontomes are certainly most common in East Asian and Native American populations, but the frequencies in Australians and Melanesians (non-'Mongolians') match or exceed those of Southeast Asians. The trait does occur, albeit rarely, in most human groups.

It is granted that odontomes, or tuberculated premolars, are uncommon in all human groups. We therefore characterize world variation under different terms than those used for traits found in higher frequencies:

(1) Near absence groups (0–1%): Western Eurasia, Sub-Saharan Africa, New Guinea, Jomon, South Siberia.
(2) Very rare (1–3%): Australia, Melanesia, Sunda-Pacific, Northeast Siberia.
(3) Rare (4–7%): East Asia, Americas.

Hypocone (absence)

Tooth: upper second molar; classification: Turner *et al.*, 1991; breakpoint: grades 0–1 (frequency of 3-cusped UM2; equivalent to 3 on Dahlberg scale); world range: 3.3–30.6%; Fig 5.8.

For the past 40 years, most workers have followed the Dahlberg scale (4, 4–, 3+, and 3) to classify upper molar cusp number. Prior to the adoption of this standard, cusp number was commonly reported as: (1) 4 or 3, or (2) 4, 3 or 4, and 3. The six grade scale of Turner *et al.* (1991) has two grades, 0 and 1, which workers using other scales would record as 3. Our goal is to reduce all comparative data to a common ground of hypocone absence so our graphs and summaries are for 3-cusped upper second molars. While most traits are presented in terms of presence rather than absence frequencies, the hypocone and hypoconulid are major cusps of the molar crown that show a trend to evolutionary reduction in the hominid lineage. Our focus is on loss rather than retention of these major cusps. For the upper molars, this is represented by the 3-cusped form (hypocone loss) while in the lower molars, we focus on the 4-cusped form (hypoconulid loss). These data can be readily converted to hypocone and hypoconulid presence frequencies by subtracting from 1.

Upper molar cusp number, determined by the presence or absence of the hypocone, shows a relatively limited frequency range on a world scale. Three-cusped upper second molars are least common in Sub-Saharan African and Sahul-Pacific groups who have frequencies of less than 10%. The frequencies for Southeast Asians, East Asians, and American Indians are only slightly higher. Europeans show elevated frequencies of hypocone loss (20–25%), although this is less evident in North Africans. The American Arctic group shows the greatest amount of hypocone reduction with 3-cusped upper second molars in excess of 30%. The only other Sino-American group to approximate this condition is Northeast Siberia.

In contrast to the mesial ridge of the upper canine or premolar odontomes that have been observed by very few researchers, there are extensive comparative data for hypocone variation, especially from Europe, Asia, and the Americas. Upper molar cusp number has been a member of the 'standard trait battery' for at least 100 years. Hence, it is more efficient to provide summaries for each major geographic subdivision rather than individual sample frequencies. Unfortunately, not all areas are equally represented – a long-standing problem for much of human biology, not just surveys of dental morphology.

The summary of 3-cusped upper second molars presented in Table 5.2 supports our main views for hypocone variation around the world.

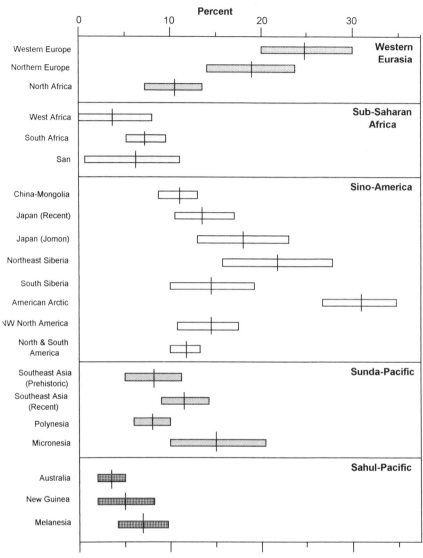

Fig. 5.8   World frequency variation in 3-cusped upper second molars among the five major subdivisions of humankind (trait frequency represented by vertical line with horizontal bars denoting ± 2 standard errors).

Western Eurasians, arranged under 12 geographic, linguistic, and/or temporal categories, show 3-cusped UM2 frequencies ranging, for the most part, between 22 and 34%. The only exceptions are the two early samples from Europe and India with frequencies of less than 20%.

Table 5.2. *Mean frequencies of 3-cusped upper second molar (hypocone absence) by geographic region and/or language family based on the observations of other workers (see Appendix B for references).*

Major region	Language/group	k	Mean %	SD
	Europe (early)	11	19.4	14.96
	Western Europe	10	32.1	11.73
	Eastern Europe (Slavs)	35	26.6	9.32
	Indo-Iranian	11	27.4	12.70
	India (early)	3	13.6	15.71
WESTERN EURASIAN	Indic	7	22.7	12.25
	Caucasian	45	33.2	13.21
	Finnic-Permian	13	27.1	14.66
	Ugrian	2	33.7	11.10
	Samoyedic	8	22.6	12.15
	Afro-Asiatic (early)	3	24.6	9.52
	Afro-Asiatic	5	27.5	16.00
SUB-SAHARAN	South Africa	1	0.0	
AFRICA	Khoisan	1	0.0	
	Sino-Tibetan	2	19.4	16.55
	Japanese	5	15.9	3.08
	Ainu	1	13.3	
SINO-AMERICAS	Taiwan aborigine	3	8.6	2.73
	Altaic (Turkic)	41	26.5	8.51
	Eskimo-Aleut	9	31.6	7.43
	North American Indian	12	14.9	8.83
	South American Indian	3	24.1	10.79
	Southeast Asia	2	8.0	7.99
SUNDA-PACIFIC	Polynesia	8	11.8	8.05
	Micronesia	1	0.0	
	Australia	2	1.8	2.48
SAHUL-PACIFIC	New Guinea	4	4.9	3.80
	Melanesia	11	25.5	9.53

k: number of samples; SD: standard deviation.

Comparative data for Sub-Saharan African populations are scanty for hypocone variability. The two samples in Table 5.2 are from the classic works of Drennan (1929) and Shaw (1931) on the Bushmen and Bantu, respectively. They do support our position that 3-cusped UM2 are rare in these populations, but more samples need to be examined.

For Sino-American groups, the summarized frequencies correspond in most details to those presented in Fig. 5.8. That is, East Asian and North American Indian populations show moderate levels of hypocone loss (*c.* 15%) while American Arctic groups have frequencies twice as high (*c.* 30%). The elevated frequency of 3-cusped UM2 in Turkic-speakers is also noteworthy, although we have no direct comparison to this group in our characterization.

Sunda-Pacific groups, excepting Polynesians, are not well represented in Table 5.2. However, the frequencies derived for early Southeast Asians and Polynesians show good correspondence to our characterization. That is, at a frequency of 10%, these groups fall between East Asians, with slightly higher frequencies, and Sahul-Pacific groups, with slightly lower frequencies.

Sahul-Pacific populations, while not well represented, show an interesting pattern of variation in hypocone loss. Australian and New Guinea groups show little hypocone reduction on UM2 ($<5\%$), consistent with our findings. Melanesians, in contrast, appear to have significantly higher frequencies of 3-cusped UM2. This discrepancy could be due to any number of factors, but it should be noted that Bailit *et al.* (1968) and Harris (1977), whose observations form the basis of the Melanesian frequency of 25.5%, examined living populations. Our frequency of 7.5% was based on skeletal series. As a final note, there is no way we can reconcile any of our observations, or those of other workers, with the data presented by Birdsell (1993) on living Australian aborigines. In his table on upper second molar cusp number, Birdsell (1993:107) gives a frequency of 86.3% for the category of 'Few [*sic*] than four cusps (percent).' This figure should not be taken literally. Although Birdsell does not describe his methods of observation, he mentions upper molars with $3\frac{1}{2}$ cusps, which we presume he included in the category of 'fewer than four cusps.' In fact, it is possible that any cusp reduction on UM2 (4−, 3+ on Dahlberg scale; 1, 2, 3, 4? on ASU scale) led him to conclude the tooth was not 4-cusped.

Based on our findings and the summary table (Table 5.2), the pattern of variation in hypocone loss on the upper second molars is not particularly revealing. The trait shows the following pattern of geographic variation:

(1) Low frequency groups (0–10%): Sub-Saharan Africa, Australia, New Guinea.
(2) Intermediate (10–20%): Sunda-Pacific, East Asia, Jomon, American Indian, North Africa, Melanesia?.
(3) High frequency groups (20–35%): Europe, India, Northeast Siberia, American Arctic.

Carabelli's trait

Tooth: upper first molar; classification: Dahlberg 1956; breakpoint: grades 5–7 (tubercle and cusp forms only); world range: 1.9–36.0%; Fig. 5.9.

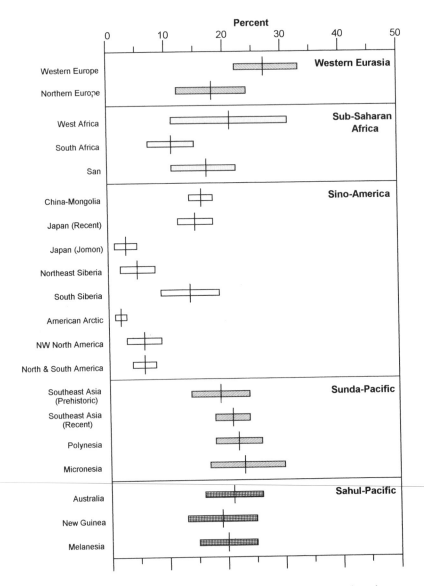

Fig. 5.9   World frequency variation in Carabelli's trait (tubercle and cusp forms only; upper first molar) among the five major subdivisions of humankind (trait frequency represented by vertical line with horizontal bars denoting ± 2 standard errors).

No dental variable has stimulated more attempts at classification than Carabelli's trait (Dietz, 1944; Shapiro, 1949; Kraus, 1951; Dahlberg, 1956; Zubov, 1968; Scott, 1973; Alvesalo *et al.*, 1975; Mizoguchi, 1977; Townsend and Brown, 1981b). Unfortunately, this makes it unduly complicated and error prone to standardize data from different sources when authors use a variety of different standards. All standards have to accommodate a wide range of trait expression, from slight to distinct single furrows, double furrows, pits, partial cusp outlines, complete cusp outlines lacking free apexes, small to moderate tubercles with free apexes, and the classic free-standing cusps (with independent centers of calcification) that rival the hypocone in size.

In the early literature, manifestations below the level of distinct tubercle or cusp were not often counted as Carabelli's cusps. From the 1940s on, however, workers started taking the full Carabelli continuum into account and established new standards, or used those developed by other workers, which included four to ten grades of trait expression. Reducing data based on variable standards to a common comparative ground is not an easy task (cf. Mizoguchi, 1993). While total trait frequency would seem to be the solution to this conundrum, the highly variable treatment of small grooves, furrows, pits, and partial or complete cusp outlines indicates this value, considered across observers, is of limited value. Summarizing data based on a grade 2 breakpoint, a method followed in the publications of C.G. Turner II, A.A. Zubov, and their co-workers, is not feasible for a world summary because many workers combine all grades below the level of tubercle/cusp. For these reasons, our summary of Carabelli's trait variation focuses on the frequencies of distinct tubercles and cusps, or grades 5–7 on the Dahlberg scale and small plus pronounced tubercles on the Kraus (1951) scale.

When just tubercle and cusp forms are considered, the pattern of geographic variation in Carabelli's cusp is not particularly striking (Fig. 5.9), an observation that might surprise the general student of physical anthropology. At this level, Western Eurasian and all Sunda-Pacific and Sahul-Pacific groups have frequencies close to 20%, while Africans fall at or below this value. The only groups that are clearly outliers are Northeast Siberians, the Jomon, and Native Americans who have tubercle-cusp forms in frequencies of 5% or less. East Asians and South Siberians fall between the two extremes at about 15%.

The mean frequencies for distinct tubercles and cusps on the Carabelli continuum in worldwide groups are presented in Table 5.3. As most authors consider Carabelli's cusp a Caucasoid trait, it is perhaps surprising to find that most of the Western Eurasian subgroups have frequencies in

Table 5.3. *Mean frequencies of Carabelli's trait (UM1; tubercle and cusp forms only) by geographic region and/or language family based on the observations of other workers (see Appendix B for references).*

Major region	Language/group	k	Mean %	SD
	Europe (early)	6	4.4	2.82
	Western Europe	15	22.1	8.22
	Eastern Europe (Slavs)	35	20.1	6.89
	Indo-Iranian	11	17.6	7.36
	India (early)	3	8.6	9.31
WESTERN EURASIA	Indic	7	22.8	11.33
	Caucasian	45	16.8	6.62
	Finnic-Permian	13	16.4	5.06
	Ugrian	2	19.7	3.25
	Samoyedic	8	9.2	5.18
	Afro-Asiatic	5	30.1	10.37
SUB-SAHARAN AFRICA	East Africa	7	12.3	5.44
	South Africa	4	13.9	5.41
	Khoisan	2	20.0	4.45
	Sino-Tibetan	2	15.6	7.28
	Japanese	10	10.6	4.12
	Ainu	5	8.2	2.73
SINO-AMERICAS	Taiwan aborigine	4	9.6	8.56
	Altaic (Turkic)	40	8.0	3.87
	Eskimo-Aleut	8	1.9	3.54
	North American Indian	20	6.6	4.89
	South American Indian	4	18.0	8.51
SUNDA-PACIFIC	Southeast Asia	3	17.2	8.64
	Polynesia	9	15.9	9.7
SAHUL-PACIFIC	Australia (living)	4	18.4	1.86
	Australia (skeletal)	4	3.2	2.65
	New Guinea	4	18.0	7.78
	Melanesia	11	10.3	3.93

k: number of samples; SD: standard deviation; UM1: upper first molar.

the 15 to 20% range. The exceptions are the prehistoric samples from Western Europe and India and the Samoyeds with frequencies of less than 10% and Afro-Asiatics with an exceptionally high frequency of 30%.

The mean frequency of Carabelli's cusp forms in Sub-Saharan Africa of 12.5% approximate the figures we report for South Africans and the San. Cusp forms appear to be somewhat less common in Africa than Western Eurasia, although Turner and Hawkey (1995) have proposed the opposite.

For Sino-Americans, the East Asian groups show mean cusp frequencies of 10 to 15%. The reduction of cusp forms in New World groups parallels our results to the point of showing Carabelli's cusp is less common in

Eskimo-Aleuts than American Indians. The Ainu, probable mixed descendants of the Jomon, have a cusp frequency of 8.2%. The elevated cusp frequencies in the four South American Indian samples are likely an artifact of observational differences.

For Sunda-Pacific and Sahul-Pacific groups, mean Carabelli's cusp frequencies fall slightly below our values, but the differences are minor. We place these groups at 20% for cusp forms while summarized data from other workers fall mostly in the 15 to 20% range. The most enigmatic finding for either of these two groupings was the exceptionally low frequency of Carabelli's cusp in Australian skeletal samples. As a rule, recent skeletal samples and modern populations from the same area (assuming genetic propinquity) show similar morphologic trait frequencies. In collating the Australian data, the variance in frequencies among eight samples was surprisingly high. It became evident that all the low frequencies were based on skeletal observations, while the high and uniform frequencies of cusp forms were made on dental casts. Our Australian frequency of 21.4% cusp forms (n = 332) was based on skeletal material, but is much closer to the values derived from modern Australian samples. It is not clear why other workers dealing with Australian aboriginal skeletons found so few cusp forms. We would have to know their tolerance limits for scoring the trait in the presence of wear on the protocone, a process that can easily obscure the presence of tubercles and cusps.

Although Carabelli's cusp does not discriminate as clearly among human groups as previous workers surmised (cf. Garn, 1971; Brues, 1977), the pattern of variation, in terms of cusp forms, is as follows:

(1) Low frequency groups (0–10%): North Asia, Eskimo-Aleuts, American Indians, Jomon, Ainu, (prehistoric Europe and India?).
(2) Low intermediate (10–15%): East Asia.
(3) High intermediate (15–20%): Sub-Saharan Africa, Sunda-Pacific, Sahul-Pacific.
(4) High frequency groups (20–30%): Western Eurasia.

Cusp 5

Tooth: upper first molar; classification: Harris 1977; breakpoint: 1–5 (total frequency); world range: 10.4–62.5%; Fig. 5.10.

With few exceptions, cusp 5 variation divides the world into three major clusters. Western Eurasian and Sino-American groups have frequencies ranging from 10 to 25%. Sunda-Pacific groups have intermediate frequen-

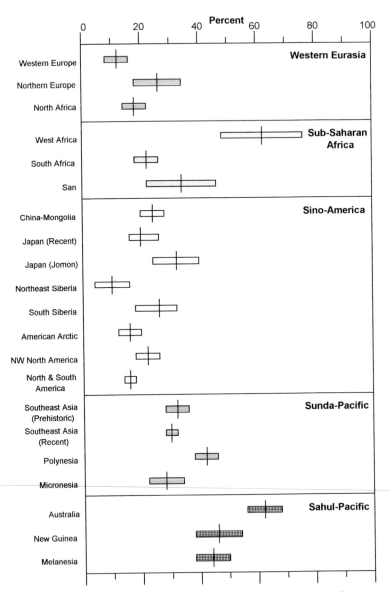

Fig. 5.10   World frequency variation in cusp 5 (upper first molar) among the five major subdivisions of humankind (trait frequency represented by vertical line with horizontal bars denoting ± 2 standard errors).

cies (30–40%) while Sahul groups, led by Australia, have the world's highest frequencies (45–60%). African groups are not easily characterized; South Africans fall in the low frequency cluster, the San fall with the Sunda-Pacific groups, and West Africans share with Australia the highest frequency of cusp 5 in the world. In all likelihood, Africans belong to the intermediate or high frequency cluster as the South African and San frequencies are based on breakpoints of grade 2 rather than grade 1 (Irish, 1993). The higher West African sample frequency is based on a grade 1 breakpoint.

Few workers provide comparative data for cusp 5 of the upper first molar. A useful survey of this trait's variation is, however, provided by Kanazawa *et al.* (1992) who observed the distal accessory tubercle (= cusp 5) in several groups using Moiré contourograms. These authors provide frequency data for nine samples representing all five of our subdivisions. For Western Eurasia, they found low frequencies in the Dutch (11.5%) and Asiatic Indians (6.3%). Their single African sample, the Bantu, had a frequency three times as high at 30.0%. However, this frequency was matched or exceeded by two North Asian groups (Japanese 30.3%; Yami 32.5%), one Sunda-Pacific group (Cook Islanders 35.0%), and one Sahul-Pacific group (Australians 36.4%). Falling between the low and high frequency groups were the Ainu (21.7%) and Eskimos (19.8%). These findings correspond to aspects of our characterization, but there are some exceptions. For example, we find a consistent, if not major, distinction between Sino-Americans and both Sunda-Pacific and Sahul-Pacific populations.

Other workers who have scored cusp 5 have found low frequencies in American whites (Scott, 1973: 10.4%) and the Ainu (Turner and Hanihara, 1977: 10.8%) and high frequencies in Polynesians (Harris, 1977: Ontong Java – 34.0%; Ulawa – 32.1%), Melanesians (Harris, 1977: Bougainville Island – 35.9%; Malaita Island – 43.1%), and Taiwan aborigines (Manabe *et al.*, 1991: Bunun – 35.6%). Low to intermediate frequencies have been reported for American Indians (Scott, 1973: American Southwest Indians – 18.5%; Scott *et al.*, 1983: Pima Indians – 19.9%) and Eskimos (Scott, 1991b: prehistoric Kachemak – 12.2%, prehistoric Koniag – 13.3%; Scott, 1994: modern Koniag – 15.7%). Most intriguing are the observations of Birdsell (1993) who found this trait, which he calls the 'Musgrave cusplet' (after the Musgrave Ranges in South Australia), to be extremely common in living Australian aborigines. In these groups, it is frequently pronounced in size and even bifurcated to form a 'double Musgrave cusplet.' Unfortunately, Birdsell does not present frequencies for each upper molar but combines all three molars to arrive at a cusp 5 frequency of

87.9%– an impressive figure by any standard. While Kanazawa *et al.* (1992) found Australians at the high end of the world range for this trait, they were not shown to be distinctly different from other Asian and Pacific groups. Birdsell's observations provides some weight to our view that Australians show an unusually high frequency of this trait. Townsend *et al.* (1990) would concur as they include a high frequency of cusp 5 as part of their Australian dental complex.

The geographic variation in cusp 5 of the upper first molar can be summarized in the following manner:

(1) Low frequency groups (10–25%): Western Eurasia, Sino-Americas.
(2) Intermediate (30–40%): Sunda-Pacific, Sub-Saharan Africa?.
(3) High frequency groups (45–60%): Sahul-Pacific, Sub-Saharan Africa?.

Enamel extensions

Tooth: upper first molar; classification: Turner *et al.*, 1991; breakpoint: 2–3; world range: 0–54.6%; Fig 5.11.

For enamel extension frequencies, all Western Eurasian, Sub-Saharan African, and Sahul-Pacific groups, along with Micronesians and the Jomon, exhibit frequencies of less than 10%. At the other extreme, East and North Asian and Native American populations have trait frequencies in the 45 to 55% range. Southeast Asian and Polynesian groups, along with South Siberians, fall between the low and high clusters with frequencies of 20 to 35%.

Enamel extensions are most readily scored in skeletons or extracted teeth. As many workers deal with casts or wax-bite impressions, comparative data on enamel extensions are limited. In their summary of European groups, Brabant and Ketelbant (1975) listed consistently low enamel extension frequencies for Denmark (6.5%), Norway (1.8%), Greece (3.1%), Belgium (1.6%), and North American whites (1.8%). These frequencies correspond closely to our characterization of Western Eurasians. At the other extreme, Nelson (1938) found an enamel extension frequency of 34% (all molars) for the Indians of Pecos Pueblo. For the upper first molar, Pedersen (1949) gave a frequency of 36.7% for 490 Greenlandic Eskimo skulls. Leigh, who had earlier described enamel extensions in California Indians (Leigh, 1928), said of the Chamorros of Guam, 'On the teeth of this Mongoloid tribe I found the enamel streaming into the bifurcation, particularly of the inferior second and third molars, but to a shorter

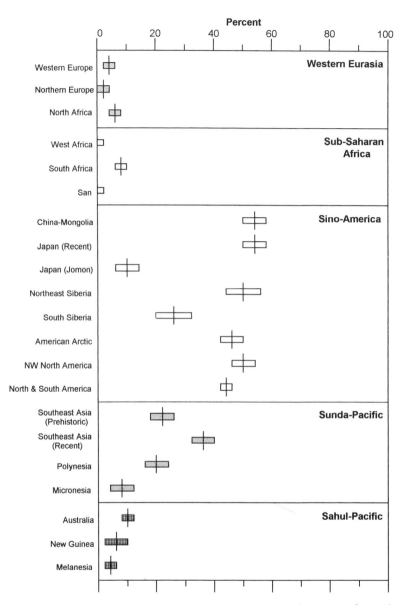

Fig. 5.11 World frequency variation in enamel extensions (upper first molar) among the five major subdivisions of humankind (trait frequency represented by vertical line with horizontal bars denoting ± 2 standard errors).

distance than with the Americans' (Leigh, 1929:455). In terms of percentage occurrence, his Chamorro sample shows 4% marked, 16% medium, 62% slight, and 18% absent. By our scale, his Guam frequency would be 20%. In a study of Hawaiian skulls, Chappel (1927) observed that 94% exhibited some form of enamel extensions. It is likely that only 20 to 30% of these expressions would be of the medium and marked variety. According to Leigh, who had experience observing both Pacific and American Indian populations, the latter groups exhibited higher frequencies and more pronounced expressions of enamel extensions.

Data from the literature supports our initial characterization of enamel extension variation:

(1) Low frequency groups (0–10%): Western Eurasia, Sub-Saharan Africa, Sahul-Pacific, Jomon.
(2) Intermediate (20–30%): Sunda-Pacific, South Siberia.
(3) High frequency groups (40–60%): East and North Asia, Americas.

Hypoconulid (absence)

Tooth: lower first and second molars; classification: Gregory 1916; breakpoint: 0 (4-cusped LM1 and LM2); world range: 0.0–10.0% and 4.4–84.4%; Figs. 5.12 and 5.13.

In most human populations, 4-cusped lower first molars are rare. They are most common in Western Eurasian populations. Groups from other geographic areas have an incidence of less than 1% although South Siberia (3.1%), New Guinea (4.5%) and Melanesia (1.9%) are interesting exceptions.

In contrast to the limited variability of 4-cusped lower first molars, loss of the hypoconulid on the lower second molars is common in many populations. Four-cusped lower second molars are most common in Western Eurasians (65–85%), distinguishing groups in this area from all other populations. The San, along with Northeast Siberians, the three Native American groups, and Australians, by contrast, seldom exhibit 4-cusped lower second molars (< 10%). China-Mongolia, recent Japan, and the Jomon show somewhat higher frequencies of hypoconulid loss (10–25%) than Northeast Siberians and New World groups. South Siberia is the most obvious outlier in a Sino-American context with an incidence greater than 50%. Sunda-Pacific groups, along with South Africans, have frequencies that cluster around 30%. Interestingly, New Guinea and Melanesia have high frequencies of 4-cusped lower second molars (50–60%), a marked contrast to the low frequency in Australians.

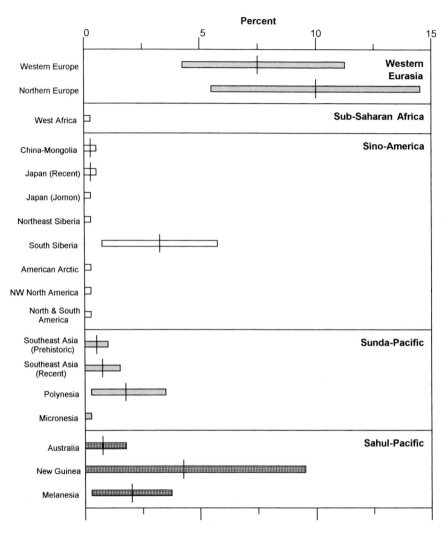

Fig. 5.12   World frequency variation in 4-cusped lower first molars among the five major subdivisions of humankind (trait frequency represented by vertical line with horizontal bars denoting ± 2 standard errors).

A significant fraction of the voluminous literature on lower molar cusp number is summarized in Table 5.4 for both lower first and second molars. There are interesting contrasts between our characterizations of trait variation in Figs. 5.12 and 5.13, so these are reviewed for each regional subdivision.

Our characterization of Western Eurasia appears conservative relative

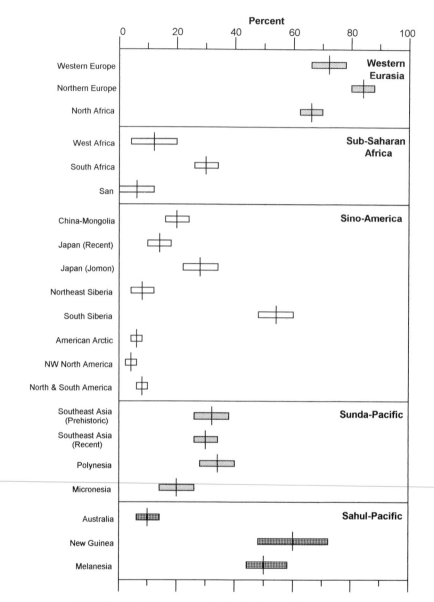

Fig. 5.13   World frequency variation in 4-cusped lower second molars among the five major subdivisions of humankind (trait frequency represented by vertical line with horizontal bars denoting ± 2 standard errors).

Table 5.4. *Mean frequencies of 4-cusped LM1 and LM2 (hypoconulid absence) by geographic region and/or language family based on the observations of other workers (see Appendix B for references).*

Major region	Language/group	LM1			LM2		
		k	Mean %	SD	k	Mean %	SD
WESTERN EURASIA	Europe (early)	6	6.8	4.20	6	94.6	4.56
	Western Europe	23	15.6	10.90	24	92.6	6.42
	Eastern Europe (Slavs)	41	9.3	4.70	41	87.1	6.03
	Indo-Iranian	15	13.4	6.97	14	82.0	9.53
	India (early)	3	19.1	10.40	3	88.2	5.34
	Indic	5	15.1	5.42	6	84.4	13.23
	Caucasian	45	17.1	7.02	45	91.9	4.08
	Finnic-Permian	15	12.7	7.20	15	81.6	9.09
	Ugrian	2	3.7	5.23	2	73.8	18.31
	Samoyedic	7	3.4	2.99	9	56.2	17.31
	Afro-Asiatic (early)	3	13.6	7.43	3	82.9	4.77
	Afro-Asiatic	10	14.2	7.12	5	94.6	5.12
SUB-SAHARAN AFRICA	East Africa	7	4.0	4.65	7	75.1	5.43
	South Africa	2	0.6	0.85	3	42.7	11.93
	Khoisan	2	0.4	0.57	3	16.6	7.92
SINO-AMERICAS	Sino-Tibetan	3	5.0	3.65	5	62.1	21.68
	Japanese	1	1.3		10	40.5	9.02
	Ainu	1	3.2		3	54.0	11.95
	Taiwan aborigine	3	3.7	2.90	4	55.1	31.47
	Altaic (Turkic)	50	4.3	4.37	50	64.9	16.63
	Altaic (Mongolian)	3	3.1	5.37	3	44.1	21.80
	Altaic (Tungusic)	2	9.4	3.82	2	32.9	6.86
	Eskimo-Aleut	5	3.3	4.29	8	26.2	13.04
	North American Indian	12	0.3	0.49	17	36.8	12.37
	South American Indian	4	0.6	1.20	4	39.8	30.19
SUNDA-PACIFIC	Southeast Asia	1	0.0		3	50.4	13.05
	Polynesia	5	1.9	1.35	10	44.4	11.61
	Micronesia	2	0.0		2	47.3	48.43
SAHUL-PACIFIC	Australia	5	2.4	2.22	7	44.2	21.77
	New Guinea	4	14.9	11.83	4	84.2	5.73
	Melanesia	11	8.6	5.22	11	61.4	14.71

k: number of samples; SD: standard deviation; LM1: first lower molar; LM2: second lower molar.

to the higher frequencies reported by other authors. The frequencies of 4-cusped LM1 in this region are largely in the 10 to 20% range. Early Europe shows a lower incidence as do Samoyeds and Ugrians. Four-cusped LM2 frequencies also tend to be higher, with most groups in the 80 to 95% range. Neolithic-Megalithic samples, with a mean of 94.7%, had already attained a high frequency of quadrate lower second molars. Within this Western Eurasian context, Samoyedic and Ugrian speakers show fewer 4-cusped LM2 paralleling the pattern indicated by their LM1 cusp number.

Africans show much less hypoconulid loss on the LM1 than Western Eurasians. The three African groups have frequencies of 5% or less. However, for LM2, our characterization appears conservative. East Africans have quadrate LM2 in frequencies of 75%, only slightly lower than those of Western Eurasians. Moreover, there is distinctive regional variation within Africa as South Africans have 4-cusped LM2 in a frequency only half that of East Africans. The San provide yet another contrast with one of the lowest 4-cusped LM2 frequencies in the world. The samples used by Irish (1993) to characterize South Africans came almost exclusively from Niger-Kordofanian-speaking populations. The 30% incidence we use in Fig. 5.13, taken from Irish, approximates the means for the South African grouping that was made up of three Bantu samples. The Nilo-Saharan populations of East Africa may have higher frequencies of quadrate LM2 than South African Niger-Kordofanians, such as the Bantu.

Our characterization indicates that groups in the Sino-American subdivision have almost no 4-cusped lower first molars. By contrast, other authors have found a low but consistent frequency of about 3% for North and East Asian and Eskimo-Aleut populations. The frequencies of less than 1% for North and South American Indians approximate our values. For LM2, we characterized East Asians as having more 4-cusped forms than Native Americans, and this pattern holds in Table 5.4. However, our total frequencies are lower than the summary figures which show East and North Asians in the range of 35 to 65% while Native Americans fall between 25 and 40%.

We characterized Sunda-Pacific populations as having very low frequencies of 4-cusped LM1, and this is supported by other workers. For LM2, our frequencies for Southeast Asians and Polynesians are relatively high (30–35%) but are somewhat lower than those shown in the summary table (45–50%).

The large-toothed populations in the Sahul-Pacific subdivision show exceptionally high frequencies of hypoconulid reduction for both LM1 and LM2. Even our relatively low values hinted that New Guinea and

Melanesian populations were regionally exceptional in terms of 4-cusped LM1. The summarized frequencies of other workers at 9 to 15% support this distinction, and even fall within the Western Eurasian range. Australians do not show as much hypoconulid loss on LM1 as their Pacific neighbors, but they do have 4-cusped LM1 in frequencies higher than Sunda-Pacific groups. For LM2, New Guinea and Melanesia again go against regional trends, showing exceptionally high frequencies of 4-cusped forms, almost rivaling those of Western Eurasians. The Australian frequency of about 40% for 4-cusped LM2 is also much higher than our characterization indicates.

Although the frequencies we provide in Figs. 5.12 and 5.13 differ from the summary values in Table 5.4, the differences are principally in magnitude, not order of arrangement. Keeping this in mind, geographic variation in lower molar cusp number shows the following pattern:

A. Lower first molar (4-cusped forms)
(1) Low frequency groups (0–3%): Sub-Saharan Africa, Sino-Americas, Sunda-Pacific, Australia.
(2) Intermediate (5–10%): New Guinea, Melanesia, prehistoric Europe.
(3) High frequency groups (10–20%): Western Eurasia.
B. Lower second molar (4-cusped forms)
(1) Low frequency groups (10–30%): San, Americas.
(2) Low intermediate (30–60%): South Africa, East Asia, North Asia, Altaic (Mongolian/Tungusic), Sunda-Pacific, Australia?.
(3) High intermediate (60–80%): New Guinea, Melanesia; East Africa, Altaic (Turkic).
(4) High frequency groups (> 80%): Western Eurasia.

Groove pattern

Tooth: lower second molar; classification: Gregory 1916; breakpoint: Y pattern; world range: 7.6–71.9%; Fig. 5.14.

Despite the attention this trait has received, the retention of a Y groove pattern on LM2 does not exhibit a pattern of geographic variation that clearly differentiates human populations. North and East Asians, Native Americans, Sunda-Pacific groups, and Australians show the lowest frequencies of the Y pattern on LM2 where it ranges from 10 to 20%. Western Eurasians, Melanesians, and the Jomon show only slightly higher frequencies (20–30%). Sub-Saharan African and New Guinea populations show the highest frequencies of Y pattern retention with the San exhibiting the remarkably high frequency of 70%.

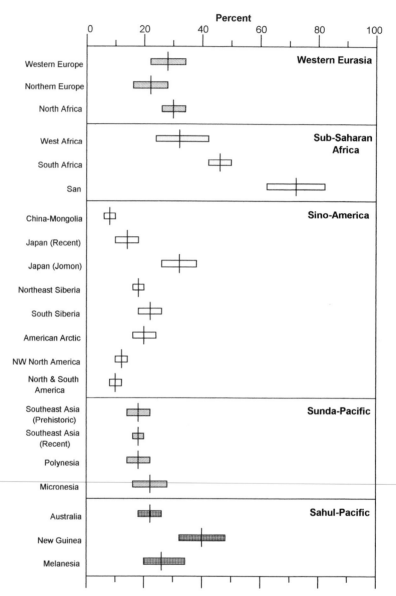

Fig. 5.14   World frequency variation in Y groove pattern (lower second molar) among the five major subdivisions of humankind (trait frequency represented by vertical line with horizontal bars denoting ± 2 standard errors).

Table 5.5. *Mean frequencies of Y groove pattern (LM2) by geographic region and/or language family based on the observations of other workers (see Appendix B for references).*

Major region	Language/group	k	Mean %	SD
WESTERN EURASIA	Europe (early)	5	19.0	8.77
	Western Europe	19	12.4	9.58
	Eastern Europe (Slavs)	35	6.5	3.49
	Indo-Iranian	12	10.0	7.25
	India (early)	3	27.8	3.99
	Indic	5	7.0	9.00
	Caucasian	45	12.8	6.09
	Finnic-Permian	14	9.6	7.58
	Ugrian	2	1.5	2.12
	Samoyedic	8	3.5	1.91
	Afro-Asiatic (early)	3	6.8	2.39
	Afro-Asiatic	6	8.6	3.24
SUB-SAHARAN AFRICA	East Africa	5	26.7	4.59
	South Africa	2	38.4	20.58
	Khoisan	2	68.7	24.96
SINO-AMERICAS	Sino-Tibetan	4	10.3	8.25
	Japanese	5	3.5	1.15
	Taiwan aborigine	5	8.7	9.17
	Altaic (Turkic)	45	6.0	4.30
	Eskimo-Aleut	11	20.7	9.80
	North American Indian	13	11.9	7.09
	South American Indian	4	7.6	2.62
SUNDA-PACIFIC	Southeast Asia	1	10.4	
	Polynesia	6	18.3	15.66
	Micronesia	1	16.7	
SAHUL-PACIFIC	Australia	4	9.1	9.09
	New Guinea	4	3.6	3.50
	Melanesia	11	39.4	15.12

k: number of samples; SD: standard deviation; LM2: lower second molar.

The summary data from other workers, shown in Table 5.5, are at some odds with our characterization. With but few exceptions, Western Eurasians show Y frequencies in the range of 6 to 12%. The early samples from Europe and India of 19.0 and 27.8% more closely approximate our values. Exceptionally low frequencies of the Y pattern are evident in Samoyedic and Ugrian samples.

Sub-Saharan African groups stand out for Y pattern retention on LM2 compared to other world populations. The summarized frequencies for East and South Africa and the San parallel those of Fig. 5.14. The San, for reasons that are not clear, are a world outlier for this variable.

Y-pattern variation among Sino-Americans does not clearly distinguish them from Western Eurasians. Excluding Eskimo-Aleuts, who have elevated frequencies of *c.* 20%, other samples in this grouping fall between 5 and 10%. Comparative Sunda-Pacific samples are limited, but Polynesians appear to have higher Y frequencies than East Asian and Native American populations.

For Sahul-Pacific groups, Australians do seem to have a relatively low incidence of the Y pattern compared to Melanesians, but our characterization is at odds with the values for New Guinea. Table 5.4 would indicate that Australians and New Guinea groups fall at the low end of the Western Eurasian and Sino-American ranges while Melanesia would fall in the middle of the African range.

Although it is not possible to reconcile completely the two sets of independent data on LM2 Y pattern variation, a tentative characterization would be:

(1) Low frequency groups (5–20%): Western Eurasia, Sino-Americas, Sunda-Pacific, Australia.
(2) Intermediate (25–40%): East and South Africa, Melanesia, New Guinea.
(3) High frequency group (60–70%): the San.

Cusp 6

Tooth: lower first molar; classification: Turner, 1970; breakpoint: 1–5 (total frequency); world range: 4.7–61.7%; Fig. 5.15.

On a worldwide scale, cusp 6 (*tuberculum sextum*) is a very common trait. Only in Western Eurasian and New Guinea populations, along with the San, is the trait relatively infrequent (5–15%). South Africans and South Siberians have slightly higher frequencies at *c.* 20%. In all other North and East Asian, Native American, Southeast Asian, Australian, and Pacific groups, cusp 6 frequencies fall between 32 and 62% with the majority of groups within the bounds of 40 to 55%. Australians occupy the high end of this frequency range, which is interesting given the incidence of only 15% in neighboring New Guinea.

Our characterization matches, in broad outline and many details, the summarized observations on more than 300 world samples (Table 5.6). Western Eurasians are distinctive in having consistently low frequencies of cusp 6, mostly in the 5 to 10% range. Sub-Saharans Africans have only slightly higher frequencies (10–20%). Excluding Altaic-speakers, who show moderate cusp 6 frequencies, other North and East Asian and Native

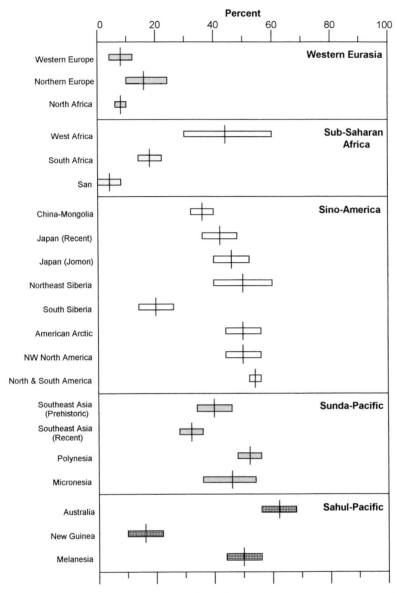

Fig. 5.15 World frequency variation in cusp 6 (lower first molar) among the five major subdivisions of humankind (trait frequency represented by vertical line with horizontal bars denoting ± 2 standard errors).

Table 5.6. *Mean frequencies of cusp 6 (LM1) by geographic region and/or language family based on the observations of other workers (see Appendix B for references).*

Major region	Language/group	k	Mean %	SD
	Europe (early)	2	1.9	0.14
	Western Europe	20	3.7	5.80
	Eastern Europe (Slavs)	35	3.1	2.95
	Indo-Iranian	14	4.0	2.58
	India (early)	3	9.5	2.90
WESTERN EURASIA	Indic	4	6.0	4.68
	Caucasian	45	2.2	2.44
	Finnic-Permian	15	3.9	3.10
	Ugrian	2	5.5	1.84
	Samoyedic	7	7.9	5.28
	Afro-Asiatic	15	11.2	12.10
SUB-SAHARAN AFRICA	East Africa	4	8.6	10.42
	South Africa	2	16.1	0.85
	Khoisan	2	18.8	4.10
	Japanese	13	27.0	6.72
	Ainu	5	22.9	3.56
	Taiwan aborigine	2	39.6	11.03
	Altaic (Turkic)	50	11.2	5.67
SINO-AMERICAS	Altaic (Mongolian)	2	17.9	1.34
	Altaic (Tungusic)	2	15.9	1.06
	Eskimo-Aleut	11	28.6	27.10
	North American Indian	15	36.8	17.73
	South American Indian	5	22.2	18.72
SUNDA-PACIFIC	Southeast Asia	3	17.1	5.72
	Polynesia	9	52.0	15.99
SAHUL-PACIFIC	Australia	7	52.3	22.19
	New Guinea	4	5.4	4.90
	Melanesia	10	38.9	11.43

k: number of samples; SD: standard deviation; LM1: lower first molar.

American groups have frequencies in the 25 to 40% range. Polynesians and Australians have exceptionally high frequencies of cusp 6 ($>50\%$), followed not too distantly by Melanesians. Although the summarized incidence for New Guinea is lower than the value we report, the much reduced frequency for this group compared to Australians and Melanesians parallels our tabulation.

Cusp 6 LM1 variation on a world scale could be summarized as follows:

(1) Low frequency groups (0–10%): Western Eurasia.
(2) Low intermediate (10–20%): Sub-Saharan Africa, South Siberia, Altaic-speakers, New Guinea.

(3) High intermediate (30–50%): North and East Asia, Americas, Melanesia.
(4) High frequency groups (>50%): Polynesia, Australia.

Cusp 7

Tooth: lower first molar; classification: Turner 1970; breakpoint: 1–4 (excludes grade 1A); world range: 3.1–43.7%; Fig. 5.16.

Cusp 7 (*tuberculum intermedium*) is found in low and uniform frequencies throughout the world (5–10%), although there is one important exception. Contrasting all other human groups, Sub-Saharan African populations exhibit cusp 7 in relatively high frequencies (25–45%). Melanesians, with a frequency of 12%, are the only group outside of Africa with an incidence greater than 10%.

Summarized world values for more than 270 samples, shown in Table 5.7, confirm this succinct characterization of cusp 7 variation. This is a trait common in Africa and rare, or at least uncommon, in all other human groups. In this instance, it is difficult to even trichotomize variation as there is no distinct pattern evident among the low frequency groups:

(1) Low frequency groups (0–10%): Western Eurasia, Sino-Americas, Sunda-Pacific, Sahul-Pacific.
(2) High frequency groups (25–40%): Sub-Saharan Africa.

Deflecting wrinkle

Tooth: lower first molar; classification: Turner *et al.*, 1991; breakpoint: grade 3; world range: 4.9–39.5%; Fig. 5.17.

The geographic pattern of deflecting wrinkle variation is not particularly distinctive. This trait is relatively uncommon in Western Eurasians and the Jomon (< 10%), but is not radically higher in East Asians or Sunda-Pacific and Sahul-Pacific groups (10–25%). Native Americans, along with Northeast Siberians, are the only groups that show frequencies exceeding 30%.

Compared to the summarized observations of other workers (Table 5.8), our characterization is conservative. The basic dichotomy of Western Eurasian groups versus all other world populations is maintained, but for all geographic areas, mean trait frequencies exceed those of Fig. 5.17. That is, almost all Western Eurasian groups fall in the 10 to 20% range, while all other populations have trait frequencies varying between 25 and 50%. Making some adjustments for contrasting results, the deflecting wrinkle shows the following pattern of variation on a world scale:

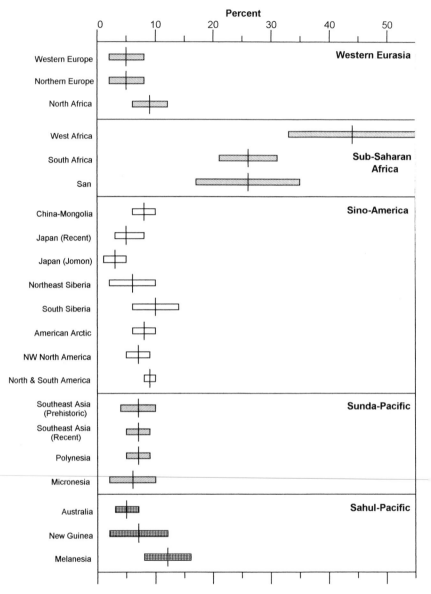

Fig. 5.16   World frequency variation in cusp 7 (lower first molar) among the five major subdivisions of humankind (trait frequency represented by vertical line with horizontal bars denoting ± 2 standard errors).

Table 5.7. *Mean frequencies of cusp 7 (LM1) by geographic region and/or language family based on the observations of other workers (see Appendix B for references).*

Major region	Language/group	k	Mean %	SD
	Western Europe	7	4.2	1.91
	Eastern Europe (Slavs)	37	3.5	1.70
	Indo-Iranian	13	1.6	2.05
	India (early)	3	6.9	2.70
WESTERN EURASIA	Indic	3	5.8	5.10
	Caucasian	45	3.1	2.76
	Finnic-Permian	14	3.0	2.68
	Ugrian	2	1.7	0.78
	Samoyedic	8	5.5	6.07
	Afro-Asiatic	11	13.6	7.74
SUB-SAHARAN AFRICA	East Africa	5	12.3	6.84
	South Africa	2	42.9	40.16
	Khoisan	2	27.8	10.54
	Japanese	12	4.5	2.06
	Ainu	5	4.7	3.40
	Taiwan aborigine	2	5.4	5.02
	Altaic (Turkic)	49	5.2	3.08
SINO-AMERICAS	Altaic (Mongolian)	2	9.2	1.63
	Altaic (Tungusic)	1	10.0	
	Eskimo-Aleut	9	5.9	7.05
	North American Indian	15	4.7	2.61
	South American Indian	1	5.5	
	Southeast Asia	3	5.8	5.90
SUNDA-PACIFIC	Polynesia	9	6.0	3.33
	Micronesia	1	19.2	
SAHUL-PACIFIC	Australia	4	5.6	3.80
	Melanesia	9	8.4	4.29

k: number of samples; SD: standard deviation, LM1: lower first molar.

(1) Low frequency groups (5–15%): Western Eurasia.
(2) Intermediate (20–35%): Sub-Saharan Africa, East Asia, Altaic, Sunda-Pacific.
(3) High frequency groups (35–55%): Sahul-Pacific?, North Asia, and Americas.

Distal trigonid crest

Tooth: lower first molar; classification: Turner *et al.*, 1991; breakpoint: 1 (presence); world range: 0.0–18.7%; Fig. 5.18.

This trait is relatively rare throughout the world. Western Eurasian,

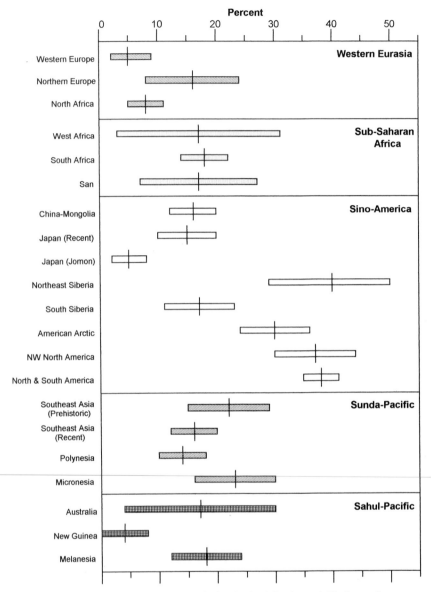

Fig. 5.17   World frequency variation in the deflecting wrinkle (lower first molar) among the five major subdivisions of humankind (trait frequency represented by vertical line with horizontal bars denoting ± 2 standard errors).

Table 5.8. *Mean frequencies of the deflecting wrinkle (LM1) by geographic region and/or language family based on the observations of other workers (see Appendix B for references).*

Major region	Language/group	k	Mean %	SD
	Western Europe	5	9.7	13.91
	Eastern Europe (Slavs)	37	7.2	5.09
	Indo-Iranian	14	14.9	7.24
	India (early)	2	13.2	2.56
WESTERN EURASIA	Indic	4	16.9	7.04
	Caucasian	45	9.6	5.03
	Finnic-Permian	14	12.7	5.90
	Ugrian	2	15.9	1.77
	Samoyedic	9	24.8	8.78
	Afro-Asiatic	2	5.6	2.26
SUB-SAHARAN AFRICA	East Africa	1	28.6	
	South Africa	2	32.7	19.30
	Khoisan	2	50.0	40.31
	Japanese	12	27.6	9.36
	Ainu	5	27.5	14.43
	Taiwan aborigine	2	40.2	12.45
SINO-AMERICAS	Altaic (Turkic)	50	23.6	6.75
	Altaic (Mongolian)	2	26.2	12.59
	Altaic (Tungusic)	2	31.4	2.76
	Eskimo-Aleut	5	55.5	12.33
	North American Indian	12	49.2	15.33
SUNDA-PACIFIC	Southeast Asia	2	30.3	17.82
	Polynesia	7	24.7	11.06
SAHUL-PACIFIC	Australia	1	41.1	
	Melanesia	9	38.5	8.23

k: number of samples; SD: standard deviation; LM1: first lower molar.

Sub-Saharan African, Sunda-Pacific, and Sahul-Pacific populations have frequencies of less than 10%. In most cases, the frequencies are closer to 5%. Only within the Sino-American subdivision do we find groups with trait incidences greater than 10%, notably recent Japan, Northeast Siberia, and the American Arctic. Other Sino-American groups parallel the low frequencies shown by groups in other parts of the world.

Comparative data on the distal trigonid crest are extremely limited except for those provided by Russian dental anthropologists for European and Asian populations. The summary frequencies available are in general accord with our findings (Table 5.9). This trait is rare in Europe and somewhat more common in North Asians, but apparently not East Asians. Only Mongolians and Eskimo-Aleuts have trait occurrences greater than 20%.

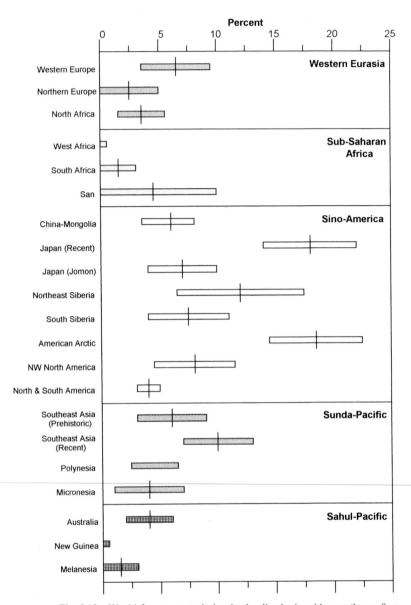

Fig. 5.18 World frequency variation in the distal trigonid crest (lower first molar) among the five major subdivisions of humankind (trait frequency represented by vertical line with horizontal bars denoting ± 2 standard errors).

Table 5.9. *Mean frequencies of the distal trignoid crest (LM1) by geographic region and/or language family based on the observations of other workers (see Appendix B for references).*

Major region	Language/group	k	Mean %	SD
WESTERN EURASIA	Eastern Europe (Slavs)	39	2.1	1.98
	Indo-Iranian	14	4.1	3.06
	Caucasian	45	5.6	3.69
	Finnic-Permian	13	2.0	2.95
	Ugrian	2	5.0	2.76
	Samoyedic	9	9.0	7.41
	Afro-Asiatic	2	10.8	1.77
SINO-AMERICAS	Japanese	4	6.7	10.01
	Taiwan aborigine	1	6.8	
	Altaic (Turkic)	50	16.1	8.03
	Altaic (Mongolian)	3	23.5	7.56
	Altaic (Tungusic)	2	10.7	3.75
	Eskimo-Aleut	2	30.9	3.46
SUNDA-PACIFIC	Southeast Asia	1	6.9	
	Polynesia	3	10.6	4.78

k: number of samples; SD: standard deviation; LM1: first lower molar.

The distal trigonid crest shows the following pattern of geographic variation:

(1) Low frequency groups (0–10%): Western Eurasia, Sub-Saharan Africa, Sunda-Pacific, Sahul Pacific.
(2) Intermediate (10–20%): Altaic (Turkic), Northeast Siberia; East Asia.
(3) High frequency groups (20–30%): Altaic (Mongolia), American Arctic.

Root traits

As most dental morphologists have studied plaster casts or wax-bite impressions, it was not possible to summarize the frequency of root traits in tables due to a paucity of observations. As the following characterizations show, root variants show as much patterned variation as crown traits. We strongly urge workers to observe these variables whenever possible.

Upper premolar root number

Tooth: upper first premolar; classification: Turner, 1981; breakpoint: 2-rooted UP1 (also includes uncommon 3-rooted variants); world range: 4.9–66.7%; Fig. 5.19.

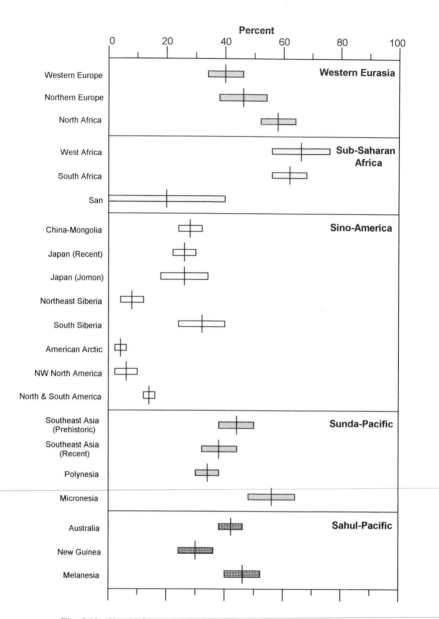

Fig. 5.19 World frequency variation in 2-rooted upper first premolars among the five major subdivisions of humankind (trait frequency represented by vertical line with horizontal bars denoting ± 2 standard errors).

With frequencies close to 65%, Sub-Saharan Africans show more 2-rooted upper first premolars than any other world population. North Africans show a frequency slightly lower than South Africans and about 10% higher than Europeans. Sahul-Pacific and Sunda-Pacific groups approximate Western Eurasia for this trait with frequencies varying around 40%. Sino-Americans, as a group, have lower frequencies than all other world populations, but there is a dichotomy in this subdivision. China-Mongolia, recent Japan, and the Jomon have frequencies in the 20 to 30% range while the lowest frequencies in the world (5–15%) are found among Northeast Siberians and all Native Americans. As with many variables, South Siberia is exceptional in the context of Sino-American groups, falling between East Asia on the one hand and Western Eurasia on the other.

For Sub-Saharan Africans, both Shaw (1931) and Barnes (1969) found high frequencies of 2-rooted UP1. For the Teso, an East African group, Barnes gave an incidence of 73.9%. For 2- and 3-rooted UP1, Shaw reported a frequency of 66.0% for the Bantu; if his category of '2 partly fused roots' is added, the rate of occurrence goes up to 82.6%. Greene (1967) also found 2-rooted UP1 to be common in North Africa; he gave frequencies of 62.2%, 68.0%, and 81.3% for Meroitic, X-group, and Christian samples, respectively. Frequencies of *c.* 70%, almost as high as those of Africans, were reported for European samples by Bennejeant and Visser (in Brabant, 1964). Campbell's (1925) observations on 2-rooted UP1 in Australian Aboriginals (56% or 67%, depending on how '2 partly fused roots' are counted) would place this group in the Western Eurasian range, somewhat below that of Africans.

In ancient and recent East Greenland Eskimos, Pedersen (1949) was impressed by the high number of 1-rooted UP1 which provided a clear contrast to European dentitions. His frequencies of 2-rooted UP1 of 5.1% and 9.0% closely approximate those shown in Fig. 5.19. Scott (1991b) found frequencies almost as low in prehistoric Kachemak (14.5%) and Koniag (11.4%) samples from Kodiak Island, Alaska. Nelson (1938), one of the few to observe this trait in American Indians, reported an incidence of 13.6% for the Indians of Pecos Pueblo. This would be elevated to 33.4% if the cases of '2 partly fused roots' were added to his '2 or 3 roots' category.

While comparative data on upper premolar root number are limited, the observations of other workers follow the same basic pattern as our characterization:

(1) Low frequency groups (5–15%): North Asia, Americas.
(2) Low intermediate (20–30%): East Asia, Jomon.

(3) High intermediate (30–60%): Western Eurasia, Sunda-Pacific, Sahul-Pacific.
(4) High frequency groups (>60%): Sub-Saharan Africa.

Upper molar root number

Tooth: upper second molar; classification: Turner *et al.*, 1991; breakpoint: 3-rooted UM2; world range: 37.4–84.5%; Fig. 5.20.

Sub-Saharan Africans and Australians show the highest frequencies of 3-rooted upper second molars in the world with rates of occurrence between 80 and 85%. At the other extreme, New World groups, notably those from the American Arctic and Northwest North America, have frequencies half as great at 35 to 40%. Western Eurasians and East Asians show no tight clusters but fall in the intermediate range of 50 to 70%. Southeast Asians, Melanesians, Micronesians, and North Africans have slightly higher frequencies (70–80%), but New Guinea and Polynesia are outliers by Pacific standards with frequencies of 50 to 55%.

Although comparative data are few, the observations of other workers agree with our basic characterization. Campbell (1925) found the exceptionally high incidence of 95.8% for 3-rooted UM2 in Australian aborigines. The figure for the Bantu given by Shaw (1931) of 76.8%, although lower than Australians, is close to the values shown in Fig. 5.20 for West and South Africans. Three European samples studied by Visser, Hjelmman, and Fabian average 56.6% for 3-rooted UM2 (in Brabant, 1964), also in accordance with our figures for Western Eurasia. At the other extreme, American Arctic samples have very low frequencies of 3-rooted UM2. Pedersen (1949) reported a frequency of 23.7% for East Greenland Eskimos, while Scott (1991b) found frequencies of 30.7% and 31.3% in two prehistoric Alaskan Eskimo samples.

The overall pattern of variation shown by 3-rooted UM2 can be summarized as follows:

(1) Low frequency groups (35–45%): American Arctic, Northwest North America.
(2) Low intermediate (50–70%): Western Eurasia, East Asia, North and South American Indian, Polynesia, New Guinea.
(3) High intermediate (70–80%): North Africa, Southeast Asia, Micronesia, Melanesia.
(4) High frequency groups (>80%): Sub-Saharan Africa, Australia.

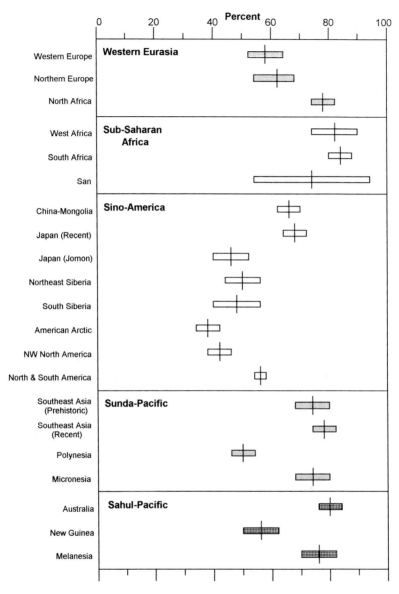

Fig. 5.20 World frequency variation in 3-rooted upper second molars among the five major subdivisions of humankind (trait frequency represented by vertical line with horizontal bars denoting ± 2 standard errors).

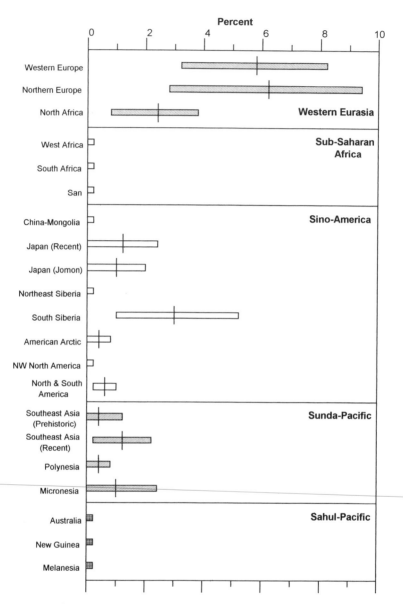

Fig. 5.21   World frequency variation in 2-rooted lower canines among the five major subdivisions of humankind (trait frequency represented by vertical line with horizontal bars denoting ± 2 standard errors).

Lower canine root number

Tooth: lower canine; classification: Alexandersen 1962, 1963; breakpoint: 2-rooted LC; world range: 0–6.1%; Fig. 5.21.

Double-rooted lower canines are rarely found outside of Europe. This root variant is effectively absent in Sub-Saharan African and Sahul-Pacific groups and occurs in a frequency of 1% or less in North and East Asian, Native American, and Sunda-Pacific groups. This is basically a Western Eurasian trait; in Europe, it attains a frequency of 5–6% while in North Africa the incidence is slightly above 2%. South Siberia is the only other regional grouping where the population frequency exceeds 2%.

At this point, there are few data to summarize beyond those found in Alexandersen's (1963) review of 2-rooted lower canines. He made observations on one Danish Neolithic sample and two medieval samples and listed the frequencies for six other European countries. For these nine samples, the mean incidence of 2-rooted LC is 6.3% (range: 4.9–10.0%) – almost identical to the frequencies illustrated in Fig. 5.21. Pal (1972) observed this trait in one Asiatic Indian sample and reported a frequency of only 0.5%. Observations on additional samples from the Indian subcontinent are needed to determine if this Western Eurasian trait, like all others, extends to populations of the eastern division of Indo-European languages. No 2-rooted lower canines were found in Nelson's (1938) Pecos Pueblo sample nor did Scott (1991b) find any in two Kodiak Island samples. Pedersen (1949), however, did report an incidence of 1.3% for East Greenland Eskimos while Shaw (1931) found a frequency of 1.6% in the Bantu.

Although rare in general, 2-rooted LC is considered a European marker. It can be characterized in terms of:

(1) Low frequency groups (0–1%): Sub-Saharan Africa, Sino-America, Sunda-Pacific, Sahul-Pacific.
(2) Intermediate (2–4%): North Africa, South Siberia.
(3) High frequency groups (>5%): Europe.

Tomes' root

Tooth: lower first premolar; classification: Turner *et al.*, 1991; breakpoint: 4–7; world range: 0–38.7%; Fig. 5.22.

Multiple rooted lower premolars are much less common than two-rooted upper premolars. Tomes' accessory root on the lower first premolar exceeds 20% only in Africa, Australia, and Southeast Asia. This trait is relatively rare in Western Eurasia (<10%). Although the difference is subtle, Tomes' root is generally more common in Sunda-Pacific groups

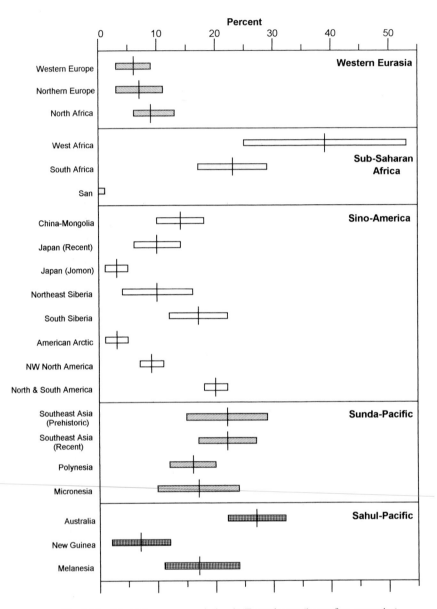

Fig. 5.22 World frequency variation in Tomes' root (lower first premolar) among the five major subdivisions of humankind (trait frequency represented by vertical line with horizontal bars denoting ± 2 standard errors).

than in North and East Asians and Native Americans. Melanesia parallels Sunda-Pacific groups, but the trait is rare in New Guinea.

Shaw (1931) reported a frequency of 37.0% for Tomes' root in the Bantu that supports our view the trait is most common in African populations. Unfortunately, Campbell (1925) did not provide a frequency for Australian aborigines, although he does mention Tomes' root in passing. Two disparate frequencies on Tomes' root are reported for Europeans: 4% by Bennejeant and 24.4% by Visser (in Brabant, 1964). The second frequency is clearly high by our standards although the first corresponds to our European characterization. For East Greenland Eskimos, Pedersen (1949) gave a frequency of 1.4% while Nelson (1938) found an incidence of 10.1% for the Indians of Pecos Pueblo. These findings are in accord with our position that the trait is more common in American Indians than Eskimos.

While more data are needed, the tentative pattern of Tomes' root variation is:

(1) Low frequency groups (0–10%): Western Eurasia, Jomon, American Arctic, New Guinea.
(2) Low intermediate (10–15%): North and East Asia, Northwest North America.
(3) High intermediate (15–25%): Sunda-Pacific, Melanesia, South Siberia, North and South American Indian.
(4) High frequency groups (> 25%): Sub-Saharan Africa, Australia.

Lower first molar root number

Tooth: lower first molar; classification: Turner, 1971; breakpoint: 3-rooted LM1 (3RM1); world range: 0–31.1%; Fig. 5.23.

A supernumerary distolingual root on the lower first molar (3RM1) reaches a frequency of 20 to 30% in some Sino-American groups (North and East Asians and Eskimo-Aleuts). This trait distinguishes them from all other human populations, including others in the Sino-American subdivision. South Siberians, the Jomon, and North and South American Indians all have 3RM1 frequencies of less than 10%. Northwest North American Indians, with a frequency of about 15%, fall between the two extremes. This trait occurs only rarely in Western Eurasian, Sub-Saharan African, and Sahul-Pacific populations (0–5%). While not common in Sunda-Pacific groups (8–15%), it is more common in Southeast Asia and Polynesia than in any other area outside of North and East Asia and northern North America.

In contrast to other root traits, 3RM1 has captured the imagination of

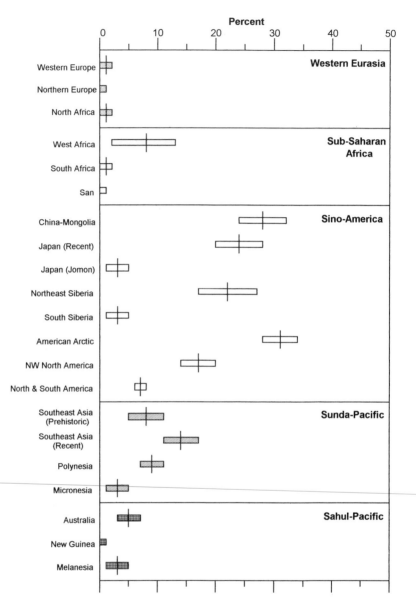

Fig. 5.23 World frequency variation in 3-rooted lower first molars among the five major subdivisions of humankind (trait frequency represented by vertical line with horizontal bars denoting ± 2 standard errors).

dental researchers who have provided valuable comparative data on this trait's distribution, including some based on X-ray studies. In a classic study, Tratman (1938) was among the first to show a distinct dichotomy between Asians and Europeans for 3RM1 frequencies. His mean frequency for several southern Chinese samples is 9.8% with a rate of occurrence somewhat higher for Malay (11.2%) and Javanese (14.8%) samples. Eurasian hybrids in Malaya have a 3RM1 frequency of 6.2%, intermediate to European and Southeast Asian populations. Hochstetter (1975) found Micronesians had this extra root in a frequency of 14.3%.

For Sino-American groups, de Souza-Freitas *et al.* (1971) found a 3RM1 frequency of 17.8% for individuals of Japanese descent living in Brazil. Pedersen (1949) considered this trait a striking feature of the East Greenland Eskimo dentition, given its rarity in comparative European material, but he reported an incidence of only 13.2% for prehistoric and 10.0% for modern Greenlanders. St. Lawrence Island Eskimos, by contrast, have a 3RM1 incidence of 32.2% (Scott and Gillispie 1997) while Kachemak and Koniag samples from Alaska have frequencies of 18.8 and 21.5% (Scott 1991b). A good comparison to our Northwest North American Indian sample is provided by Somogyi-Csizmazia and Simon (1971) who found a frequency of 15.6% in a mixed sample of Athabaskans and Algonkins.

For Western Eurasians, workers have reported 3RM1 frequencies ranging from 1.0 to 4.3% in European and European-derived samples (Brabant and Ketelbant, 1975; Curzon, 1973; de Souza-Freitas *et al.*, 1971). Younes *et al.* (1990) observed this supernumerary root in Saudi Arabians and Egyptians and, depending on method of observation, gave frequencies of 2.1, 2.3, or 3.6% for Saudis and 0.7, 1.4, or 1.6% for Egyptians. In his Southeast Asian survey, Tratman (1938) found an incidence of only 0.3% in Asiatic Indians. Lukacs (1983, 1988) found higher frequencies in Neolithic (5.6%) and Iron Age (5.0%) samples from Pakistan.

A basic geographic characterization of 3RM1 frequencies includes the following divisions:

(1) Low frequency groups (0–5%): Western Eurasia, Sub-Saharan Africa, Jomon, South Siberia, Sahul-Pacific.
(2) Intermediate (5–15%): Sunda-Pacific, American Indian.
(3) High frequency groups (>20%): North and East Asia, American Arctic.

Lower second molar root number

Tooth: lower second molar; classification: Turner *et al.* 1991; breakpoint: 1-rooted LM2; world range: 3.6–39.8%; Fig. 5.24.

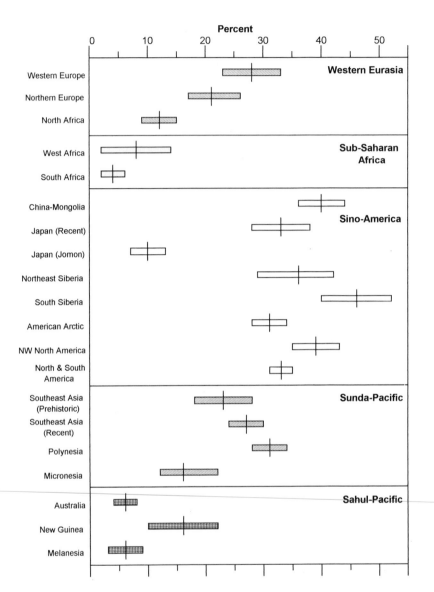

Fig. 5.24   World frequency variation in 1-rooted lower second molars among the five major subdivisions of humankind (trait frequency represented by vertical line with horizontal bars denoting ± 2 standard errors).

Single-rooted lower second molars are most commonly found in Sino-American groups where trait frequencies fall in the 30 to 40% range. The only outlier is the Jomon sample with an incidence of less than 10%. Unseparated LM2 roots are least common in Sub-Saharan Africans, Australians, and Melanesians (0–10%). European and Sunda-Pacific groups show frequencies scattered mostly between 20 and 30%. North Africans show a reduced frequency of 1-rooted LM2 compared to Europeans, tending more in the Sub-Saharan African direction.

Campbell (1925) and Shaw (1931) found very few 1-rooted LM2s in their studies of Australian aboriginal (1.7%) and Bantu (4.4%) dentitions. Much higher frequencies have been reported for American Arctic groups. Pedersen (1949) found an incidence of 22.2% in East Greenland Eskimos, while Scott (1991b) reported frequencies of 28.2% and 32.7% for Kachemak and Koniag samples. Nelson (1938) reported a frequency of 30.4% for the Indians of Pecos Pueblo. For Europeans, Taviani and Visser gave a 1-rooted LM2 frequency of 25.7% (in Brabant, 1964).

Single-rooted lower second molars show the following pattern of geographic variation:

(1) Low frequency groups (0–10%): Sub-Saharan Africa, Jomon, Australia, Melanesia.
(2) Low intermediate (10–20%): North Africa, Micronesia, New Guinea.
(3) High intermediate (20–30%): Europe, Southeast Asia, Polynesia.
(4) High frequency groups (>30%): North and East Asia, South Siberia, Americas.

Regional characterizations

On a world scale, crown and root traits show distinctive patterns of geographic variation. This is evident from the graphs, tables, and summary characterizations for each trait. Now, we reverse emphasis to characterize trait variability within our five subdivisions. For each major grouping, we are most interested in the traits that make them stand out on the world stage. We also note which variables do not distinguish a subdivision from other subdivisions. This characterization is based on relative, not absolute, trait frequencies so all crown and root polymorphisms are viewed in terms of low, intermediate, and high frequencies.

Regional divisions

Western Eurasia

Groups in this region are distinguished from other world populations primarily by two traits: 4-cusped lower first and second molars and 2-rooted lower canines. Carabelli's cusp and 3-cusped upper second molars are also at the high end of world variation, but are nearly equaled by groups in other regions. Although five traits show intermediate frequencies compared to other groups (UI2 interruption grooves, LM2 Y pattern, 2-rooted UP1, 3-rooted UM2, and 1-rooted LM2), Western Eurasians fall at the low end of the world scale for more than half the traits described (winging, shoveling, double-shoveling, mesial canine ridge, odontomes, cusp 5, enamel extensions, cusp 6, cusp 7, deflecting wrinkle, Tomes' root, and 3-rooted LM1). In this respect, we concur with Mayhall *et al.* (1982) and Zubov and his associates who feel Western Eurasians, or Caucasoids, are characterized more by trait absence or rarity than trait elaboration.

Sub-Saharan Africa

Africans occupy an extreme position on a world scale for three crown traits and four root traits. The crown traits, in fact, are rather distinctive indicators of the African dentition: cusp 7, the mesial canine ridge, and LM2 Y pattern. For root traits, African populations show the highest occurrences of differentiated roots (i.e., high frequencies of 2-rooted UP1, 3-rooted UM2, and Tomes' roots and a low frequency of 1-rooted LM2), coupled with the minimal addition of extra roots (i.e., 2-rooted LC, 3-rooted LM1). In addition to retaining normative root numbers, Africans also show low frequencies of 3-cusped UM2 and 4-cusped LM1 and LM2, indicating minimal reduction or simplification of the major cusps of the upper and lower molars. For most other crown traits, Sub-Saharan Africans, like Western Eurasians, fall at the low end of the world range (e.g., winging, shoveling, double shoveling, interruption grooves, odontomes, enamel extensions), although some assume intermediate positions (Carabelli's cusp, cusp 5, cusp 6, deflecting wrinkle).

Sino-Americas

East and North Asians and Native Americans exhibit greater dental morphological elaboration than populations from any other geographic area. In addition to their well documented distinction for shovel-shaped incisors, Sino-Americans occupy the high end of the world range for many other crown traits, including winging, double shoveling, interruption grooves, odontomes, enamel extensions, cusp 6, and the deflecting wrinkle. For root traits, Sino-Americans, in contrast to Sub-Saharan Africans, show the highest levels of undivided roots, resulting in a high frequency of 1-rooted LM2 and low frequencies of 2-rooted UP1 (especially Native Americans) and 3-rooted UM2. Coupled with the relatively common occurrence of unseparated roots, another distinguishing variable of this region is the addition of a supernumerary root – 3-rooted LM1. For groups within this subdivision, the relatively few traits found at the low end of the world scale are: mesial canine ridge, cusp 5, cusp 7, 4-cusped lower molars, and LM2 Y pattern. An intermediate position is assumed by Carabelli's cusp (although extremely low in Native Americans), 3-cusped UM2 (although very high in the American Arctic), 4-cusped LM2, 2-rooted LC, and Tomes' root.

Sunda-Pacific

For most crown and root traits, groups from Southeast Asia, Polynesia, and Micronesia fall in the middle range of world variation (e.g., winging, shoveling, double shoveling, interruption grooves, odontomes, cusp 5, enamel extensions, 3-cusped UM2, 4-cusped LM2, LM2 Y pattern, deflecting wrinkle, 2-rooted UP1, 3-rooted UM2, 2-rooted LC, Tomes' root, 3-rooted LM1, and 1-rooted LM2). Sunda-Pacific groups have no trait in an exceptionally high frequency that sets them apart from other world groups. Although they occupy the high end of the range for Carabelli's cusp and cusp 6, they share this position with Western Eurasians or Sahul-Pacific groups, respectively. The few traits that occupy the low end of the frequency range for these groups (i.e., mesial canine ridge, cusp 7, and 4-cusped LM1) are also shared by other regional populations. To succinctly summarize the Sunda-Pacific pattern of dental variation, one might say they are core rather than peripheral.

Sahul-Pacific

Any characterization of this area has to be qualified by noting that some distinctive differences exist between Australian, New Guinea, and other island Melanesian populations. For a number of traits, the three groups cluster closely together but, for others, they show pronounced differences. As with Sunda-Pacific groups, the populations from this region show few high frequency traits which distinguish them on a global scale. They occupy the high end of the world range for cusp 5, Carabelli's cusp, and cusp 6, but only cusp 5 might set them apart, albeit slightly, from other groups. In contrast to the Sunda-Pacific cluster, Sahul-Pacific groups have many more traits at the low end of the world scale (i.e., winging, shoveling, double shoveling, interruption grooves, mesial canine ridge, 3-cusped UM2, enamel extensions, cusp 7, 2-rooted LC, 1-rooted LM2, and 3-rooted LM1). However, they do occupy an intermediate position for several traits (i.e., odontomes, 4-cusped LM1 and LM2, but highly variable, LM2 Y pattern, deflecting wrinkle, 2-rooted UP1, Tomes' root, and 3-rooted UM2).

Dental complexes

Our characterization of crown and root trait variation for the five principal subdivisions of humankind parallels, in some respects, the efforts of earlier workers who defined dental complexes for Mongoloids (Hanihara, 1968a), Caucasoids (Mayhall *et al.*, 1982), and Australians (Townsend *et al.*, 1990). Their conclusions as to which traits characterize each of these major groups are supported by our data in most details. One difference between these complexes and our characterizations is that we include more traits, not only additional crown traits but also root traits for the first time. In this sense, we augment those complexes defined previously. A more noteworthy difference is in geographic perspective. The three complexes noted above focused on single regions. We have tried to view the variation within each major region relative to all other regions.

What does it all mean? an example from the real world

In this chapter, we have summarized dental data on thousands of individuals from hundreds of samples. This numerical scale is far removed from the realities of the laboratory and field where workers often have to deal with small samples of 20 or so individuals, especially when these are archaeologically-derived skeletal remains. To place small samples into a world context, we have found an interesting case study, published in a regional journal, that helps illustrate how dental characterizations can be used to assess the biological relationships of earlier human populations. The study we draw from, written more than 40 years ago, includes observations on 12 crown and root traits that can be calibrated on the scales used to construct Figs. 5.2 to 5.24. The author who made the original observations (to be named later) had no preconceptions about what dental traits he might find in his sample because only a few had been hitherto documented in a wide range of human populations. He simply recorded the

Table 5.10. *Crown and root trait frequencies in small archaeologically-derived 'mystery sample'; examine frequency profile and determine which geographic region this sample most likely came from (see text for explanation).*

Trait phenotype (tooth)	Affected/n	Percent
Semi- and full shoveling (UI1)	0/7	0.0
3-cusped UM2	5/10	50.0
Carabelli's cusp (UM1)	3/9	33.3
4-cusped LM1	1/11	9.1
4-cusped LM2	7/9	77.8
Y-groove pattern (LM2)	2/7	28.6
Cusp 7 (LM1)	1/11	9.1
2-rooted UP1	4/6	66.7
3-rooted UM1	6/7	85.7
Tomes' root (LP1)	1/8	12.5
3-rooted LM1	0/12	0.0
1-rooted LM2	0/9	0.0

n: number in sample; UI1: upper central incisor; UM1/UM2: upper first/second molars; LM1/LM2: lower first/second molars; UP1: upper first premolar; LP1: lower first premolar.

presence or absence of features to the best of his ability and, from what we can discern, he was quite able. Most of the data we extracted from this article were not in tables but were embedded in the text.

Before going on, we ask the reader to examine the trait frequencies shown in Table 5.10. Although the author observed the dentitions of 17 skeletons, no trait was scored in more than 12 individuals nor fewer than six. The percentage of scorable traits/individual ranged from 35 to 70%, not an unusual circumstance when working with skeletal remains. In this sample, three traits have no affected phenotypes so their frequencies are 0.0%. The remaining nine traits vary between 9.1 and 85.7%.

Given the characterizations in this chapter, which major subdivision of humankind was this sample most likely drawn from? Taking one trait at a time, and using the process of elimination, you can arrive at a reasonable answer without mathematical machinations and biological distance estimates. Our assessment of this sample, hereafter referred to as sample X, is presented below for each of the 12 traits.

(1) Shoveling: the total absence of semi- and full shoveling, albeit in seven individuals, effectively rules out East Asian, North Asian, and all Native American populations as a source for sample X.

Sunda-Pacific affiliation, while unlikely, cannot be ruled out. All other subdivisions remain viable relatives on the basis of shoveling.

(2) Three-cusped upper second molars: a 50% frequency of 3-cusped UM2 makes it unlikely that sample X hails from Sub-Saharan African or Sahul-Pacific groups. Sunda-Pacific groups, while unlikely, are not so far removed as to be eliminated from consideration. The group that has an incidence closest to 50% is American Arctic, but the absence of shoveling in sample X has already ruled out most Sino-American groups.

(3) Carabelli's cusp: as one-third of sample X exhibits this trait, we can eliminate once again North Asians and Native Americans as potential relatives. This trait frequency would not eliminate any other specific group, but some appear to be more viable candidates than others.

(4) Four-cusped lower first molars: this trait is rare on a world scale and it is exhibited by one of nine individuals in sample X. For this trait alone, we would not eliminate any single subdivision, although Sub-Saharan African, Sunda-Pacific, and Sino-American (except South Siberian) groups are certainly the least likely to show any 4-cusped LM1.

(5) Four-cusped lower second molars: over three-quarters of the individuals in sample X have 4-cusped LM2. The groups least likely to have a frequency of this magnitude include Sub-Saharan Africans, Sino-Americans (excepting South Siberians), Southeast Asians, Polynesians, and Australians.

(6) Y groove pattern of lower second molar: a frequency of 28.6% for this trait in sample X would not eliminate any world subdivision as a potential relative, although one Sub-Saharan African group, the San, could be ruled out.

(7) Cusp 7: one of 11 individuals (9.1%) expresses this trait. A frequency of less than 10% characterizes almost all human populations. The only subdivision that could possibly be eliminated on the basis of this low trait frequency is Sub-Saharan Africa.

(8) Two- and three-rooted upper first premolars: four of six individuals exhibit this trait, an unlikely finding for any Sino-American group, especially those of North Asia and the Americas. No other group could be eliminated from consideration based on this trait's world distribution.

(9) Three-rooted upper second molars: a frequency of 85.7% for this trait only allows the conclusion that the sample probably does not

hail from North Asia or the Americas. Beyond that, no group can be eliminated.

(10) Tomes' root: one of eight individuals exhibited a Tomes' root, an occurrence rate that falls in the middle of an already limited frequency range. No group could be eliminated on this basis, although Sub-Saharan African affiliation is least likely.

(11) Three-rooted lower first molars: the absence of this trait in 12 observable individuals tells us only that it is unlikely that the sample originates from East Asia, North Asia, or the American Arctic, but this has already been made clear by other variables.

(12) One-rooted lower second molars: out of nine observable individuals in the sample, no 1-rooted LM2 were found. Trait absence, once again, makes Sino-Americans (except the Jomon) an unlikely relative. Sunda-Pacific and European groups have moderate frequencies of 1-rooted LM2, but the difference is not sufficient to remove these groups from the category of prospective relatives.

From the foregoing, we can conclude with a high degree of certainty that this sample was not derived from an East Asian, North Asian, or Native American population. Sub-Saharan African affinity also appears unlikely based on several traits. Sunda-Pacific populations, with their characteristic intermediacy, cannot be eliminated on the basis of most traits, but their low frequencies of 4-cusped LM1 and LM2 are at some remove from those of sample X. Australians, but not New Guinea and Melanesian populations, can also be eliminated because of their low frequencies of 4-cusped lower molars. New Guinea and Melanesia can be eliminated on the basis of their very low 3-cusped UM2 frequencies which stand in sharp contrast to the 50% reported for sample X. This leaves one major subdivision that shows crown and root trait frequencies paralleling, in the main, those of the 'mystery sample' – Western Eurasians.

Western Eurasians are characterized by: (1) high frequencies of 3-cusped UM2, Carabelli's cusp, and 4-cusped LM1 and LM2; (2) low frequencies of shoveling, cusp 7, Tomes' root, and 3-rooted lower first molars; and (3) intermediate frequencies of LM2 Y groove pattern, 2- and 3-rooted UP1, 3-rooted UP2, and 1-rooted lower second molars. Sample X shows this Western Eurasian dental pattern almost perfectly, despite its small size.

Mystery sample unmasked: the skeletal remains represented in Table 5.10 come from the site of Alaca Hoyuk in Turkey. Eleven of 17 skeletons date to the Copper Age while six come from the subsequent Chalcolithic and Bronze Ages. A site in Turkey dating to several thousands of years ago would likely contain a population of Western Eurasian affiliation. M.S.

Senyurek (1952) provided the observations that were published in the journal *Turk Tarih Kuruma Belleten*. To specify which division of Western Eurasian (e.g., Indo-European, Afro-Asiatic, or Caucasian) is represented is more difficult, given the small sample size. With larger samples and more traits, such distinctions might be made.

There are many archaeological circumstances where recognizing major biological subdivisions based on small samples could be quite useful, especially at or near the boundaries of these subdivisions. Archaeological, skeletal, and linguistic data indicate that North Africa and India were invaded by Western Eurasians. Austronesians moved into the already populated islands of New Guinea and Melanesia and the uninhabited islands of Polynesia and Micronesia. North Asians colonized (i.e., no resistance from indigenous groups) the Americas, and so forth. The timing of these events could, under the right circumstances, be resolved by a handful of teeth. Regarding more controversial theories, including 'American Indians in the Pacific' (Heyerdahl, 1952) or the colonization of South America from Oceania, teeth could also serve as an important line of evidence to reject or support particular hypotheses. On a more whimsical note, some of the fanciful scenarios for early Western Eurasian exploration or migration to the New World (e.g., the lost tribes of Israel, Phoenicians, Welshman, etc.) could be settled by teeth, assuming (1) they were ever in the New World and (2) their remains are ever found. The Vikings pose no such problem; they clearly brought and left many Western Eurasian teeth in Greenland and a few in North America (although not yet recovered). The problems surrounding the movements of people through space and time abound in the anthropological and historical literature – in many instances, teeth would have a tale to tell about whether or not migrations actually occurred and the timing and impacts of such migrations. The next two chapters document these possibilities in more detail.

6 Establishing method and theory for using tooth morphology in reconstructions of late Pleistocene and Holocene human population history

Introduction

Today, biological anthropology encompasses subfields that focus on various aspects of primate and human variation, adaptability, and evolution. Even with numerous methodological and theoretical developments over the last 150 years, one of the major objectives of biological anthropology predates Charles Darwin, evolutionary theory, and the discovery and/or recognition of primate and hominid fossils. During the first half of the nineteenth century, the natural historians who laid the foundations for biological anthropology were concerned with human variation. Their avowed purpose was to systematize the 'varieties of man' (i.e., racial classifications) and provide insights into the course of human history. These early classifications were seldom free of assumptions about racial superiority, and most confused the causal relationships between, and independence of, biological, linguistic, and cultural variables. Despite the growth and diversity of modern biological anthropology, the field has not abandoned one of its initial goals – reconstructing human population history through the analysis of biological variation.

From the early 1800s to 1950, research in human variation dealt primarily with visible traits (e.g., skin color) and body dimensions (anthropometry for the living, osteometry for skeletons). Less emphasis was placed on other traits and systems such as dermatoglyphics, serology, and the dentition. In 1950, the immunologist W.C. Boyd opined that traditional studies of human variation emphasizing measurements and qualitative traits of unknown inheritance were passé. He advocated the use of serological characters (e.g., ABO, Rh, and MN blood group antigens) to resolve anthropological, historical, and evolutionary problems because they: (1) could be objectively observed with little or no interobserver error;

(2) were not alterable by environmental factors during development; (3) had simple and known modes of inheritance; (4) were not subject to the action of natural selection (i.e., nonadaptive); and (5) did not mutate at a high rate.

Although Boyd explicitly discussed blood group genes in the context of human racial classifications, his primary interest was nonetheless in historical problems: 'If the anthropologist is not content with the current popular classification of men into 'races' and nations, it is partly because he hopes to find something more fundamental, such as a classification which will tell him something of the history of the human race' (Boyd, 1950:15), in other words, an evolutionary classification. In essence, Boyd was providing a rationale to substitute genetic markers for the quantitative and qualitative traits with complex and unknown modes of inheritance used traditionally by anthropologists to address classificatory and historical problems. Boyd's urging, in conjunction with the discovery of new genetic polymorphisms, stimulated a great increase in the number of articles devoted to serological profiles of human populations. The compendia of gene frequencies made possible by these surveys attest to the scope and high level of activity in this research area from 1950 to the present day (Mourant, 1954; Mourant *et al.*, 1976; Steinberg and Cook, 1981; Roychoudhury and Nei, 1988).

Today, geneticists do not set 'racial classification' as a goal of their studies, but they use the same rationale as Boyd to justify the use of genetic markers for phylogenetic and historic reconstructions. While it is natural for workers to concentrate on the strengths of their particular approach, it should be evident that one limitation of classical genetic markers is the synchronic perspective they provide. They are limited to one 'slice of time' – namely, the moment when blood samples were collected in the field. Efforts to blood-type human bone in the 1950s and 1960s, aimed to meet this limitation, met with little success despite an initial flush of optimism. Boyd (1950:17) even commented to the effect that 'If we are interested merely in human taxonomy, we shall want characters varying from one geographical locality to another, but expressed, as nearly as may be, in the same way in each individual possessing them, and which are controlled by known genes. On the other hand, if we are primarily interested in human prehistory, we shall want characteristics which can be identified in ancient human remains.' While recognizing the potential significance of skeletal studies in prehistory, Boyd's statement was a prelude to a critique of osteometry rather than a promise for diachronic studies on human skeletal material. Assessing issues in prehistory using tooth morphology, observable in both the living and the dead, had been little explored by 1950.

At the time Boyd's book was published, craniometry remained the primary anthropological tool for assessing population relationships. In his chapter on incompletely analyzed genetic characteristics, the only dental trait Boyd listed was 'absent teeth.' That same year, Lasker (1950) addressed the potential significance of tooth morphology in racial and historical studies, but he had few good examples to draw on to illustrate this promise. That task was, however, taken up by later researchers, and we are now in a position to consider how dental morphology can shed light on anthropological and historical problems.

The literature in biological anthropology shows that researchers have one of two primary orientations: historical or processual. These are not mutually exclusive orientations as history is a product of process, and differing processes operate as histories unfold. Still, those in the processual camp often chide historical reconstructions as old-fashioned throwbacks to '19th century anthropology' (cf. Marks, 1995). Interestingly, criticism is less vocal toward those who use 'high tech' methods to reconstruct history, with the recent emphasis on mitochondrial DNA (mtDNA) the outstanding example. The analysis of mtDNA variation does promise to shed light on anthropological problems, but these problems are historical, not processual. Process is acknowledged by mtDNA geneticists (e.g., D loop sequence is selectively neutral, nucleotides mutate at a faster rate than those of nuclear DNA, genetic bottlenecking can have significant consequences on variation, etc.), but the aim of this research remains primarily historical. To illustrate, many of the early publications on mtDNA variation focused on the temporal and spatial origins of anatomically modern humans and the timing and number of migrations into the New World. These are the same historical problems shared with workers who dig up bones and stones, measure and observe skulls and teeth, and make proto-language reconstructions.

What sets historically-oriented research in human variation apart from processually-oriented research is the former's focus on population origins and relationships. Human history (including all of prehistory) is a massively complex and interwoven story, replete with examples of population movements, amalgamations, splittings, and extinctions. Throughout the Pleistocene and Holocene epochs, human groups have migrated to previously uninhabited regions (i.e., primary colonizations) or moved into settled lands, with some combination of population replacement and amalgamation. Historical questions abound in every geographic area of the world on the when, where, why, how, and who of these movements. Classic anthropological issues include the colonization of previously empty lands such as Arctic Eurasia, the New World and all the hundreds of islands of the Pacific basin. In Africa, Asia, and Europe, continents with temporally deep fossil records, scores of questions remain on the origins and relationships of small populations that appear to be remnants of once larger population systems. A few examples include the Paleoasiatics of Northeast Siberia (e.g., the Yukaghir), Oceanic Negritos of mainland and insular Southeast Asia, Tasmanians, the Basques of the French and Spanish Pyrenees (a linguistic island in a sea of Indo-European speaking populations), and the Bushmen (San) of South Africa. Reconstructing population expansions are as challenging as they are interesting. Consider, for example, the origins and dispersal of Indo-European speakers into Europe and India; the Bantu expansion into South Africa; and the dispersal of Austronesian speakers to hundreds of settled and uninhabited islands in the Pacific basin. There are many other such historical problems that generate great interest (and controversy), not only within anthropology but throughout the scientific and lay community.

The primary anthropological utility of tooth morphology is in historical rather than processual studies. Some researchers have addressed crown and root traits in the context of selection (see next section), gene flow, and genetic drift, but these efforts contribute to rather than supplant the major aim of most studies – elucidating the origins, relationships, and microevolution of populations at the regional, continental, and global levels. If this is the primary aim of dental morphologists, how do dental traits meet Boyd's criteria for choosing variables for historical reconstructions? First, regarding objectivity and replicability in scoring dental phenotypes, some workers have expressed reservations (cf. Sofaer *et al.*, 1972a) about interobserver concordance, but the use of well-defined standards, coupled with experience and caution, greatly diminishes observational problems, as noted in chapter 2. We note also that Boyd made no mention that serologists do get false positives and false negatives in blood typing. Second, crown and root traits are not entirely immune from environmental influences during development (chapter 3), but the changes wrought appear as minor and random deviations that do not significantly impact population frequencies. The influence of most exogenous factors can be clearly recognized in the dentition, as in hypoplasias and twinned teeth. Subtle alterations of individual phenotypes aside (e.g., between monozygotic twins), no worker has demonstrated the high level of plasticity in tooth morphology that Boas (1912) found for anthropometric traits. Third, dental traits do not appear to have simple autosomal dominant or recessive modes of inheritance, like Boyd's blood groups, but their genetic basis, albeit polygenic, has been well established through twin and family studies. Moreover, the long-standing debate over the susceptibility of monogenic traits to drift and founder effect suggests that polygenic traits should be favored for historical reconstruction. Dental morphologists cannot convert trait frequencies into gene frequencies, but the incidence and expression of phenotypes in populations does reflect underlying genetic variation as hundreds of studies have demonstrated. Fourth, there is some debate over whether or not crown and root traits are subject to the effects of selection, but patterns of variation within recent historical periods appear to be primarily a product of random processes (i.e., genetic drift and founder effect) rather than genetic adaptation. Fifth, there is no evidence to suggest that a high mutation rate affects dental phenotypes. In fact, the evolutionary conservatism of the dentition is well known – many traits expressed on the teeth of modern humans have been observed in hominid fossils more than two million years old and some traits are homologous to those of Oligocene and Miocene fossil hominoids.

Before we discuss studies that have used dental morphology for historical reconstructions, three additional considerations should be reviewed. First, we address the issue of adaptation. Do crown and root traits have any measurable effect on differential mortality or fertility; that is, are they strongly adaptive or only weakly so? Second, most biohistorical problems are solved using multiple variables and multiple samples. How do workers measure relative relationships among a set of populations? In this regard, the methods of estimating biological distance to reach conclusions about affinity are briefly introduced. Third, at what level of biological organization or within-species differentiation can dental variation best be used to assess affinity? Some workers claim they are useful only at distinguishing the major subdivisions of humankind, i.e., geographic races (Palomino *et al.*, 1977). Are there studies at the sub-continental, regional, or local levels that suggest otherwise? Secondary to this question is the degree to which relationships indicated by dental morphology correspond with affinity assessments based on other biological systems. According to the 'hypothesis of nonspecificity', similarity/dissimilarity matrixes derived from different types of data sets should show congruence for a common array of samples (Sokal and Sneath, 1963). That is, do teeth, genes, measurements, fingerprints, and the like, give concordant results? Provided with this background, the reader has some context for the discussions and findings in chapter 7 on tooth morphology and population history.

Adaptation and dental morphology

For some crown and root traits, between-group differences are modest, but, for others, they are striking. Why and how did these differences develop? Why do East and North Asians and derivative populations in the

Americas exhibit such high frequencies and pronounced expressions of shovel-shaped incisors? Why do Sub-Saharan Africans hold a near world monopoly on cusp 7? Why do Australians have inordinately high frequencies of cusp 5 and cusp 6? The 'why' questions on dental variation could be extended to most traits and geographic subdivisions.

Dental traits, like other variables of interest to biological anthropologists, vary not only between but also within populations. In other words, they are polytypic (between) and polymorphic (within). In evolutionary biology, there is a long-standing debate over the factors that maintain polymorphisms within populations. 'Selectionists' believe polymorphisms occur for a reason. It may not be clear how a particular polymorphic trait is related to differential fertility and/or mortality within a population, but its existence is taken as an indicator that it contributes in some way, either directly or indirectly, to genetic fitness. At the other end of the spectrum are the 'neutralists' who feel the majority of polymorphisms have no reason for being that can be translated into fitness components. Neutralists view many polymorphisms as a secondary consequence of minor changes in the genetic code that confer neither advantages nor disadvantages to an organism. When new genes are introduced into populations by mutations, chance rather than selection determines whether or not they increase in frequency or are simply lost. From the neutralist's vantage, frequency variation through time and space for most polymorphic traits reflects chance events and accidents of history, not the actions of natural selection.

Earlier, we remarked that the dentition is a crucial element in survival for those animals who use teeth to procure and/or process food. Animals who lose their teeth prematurely through trauma or infection are severely handicapped from nutritional, energetic and even defensive-predation standpoints. While the dentition as a whole, and teeth as individual units, are no doubt quite important, how is overall importance translated down to the level of the constituent components of crowns and roots? Can we reckon differential fitness within human populations to the level of extra ridges, cusps, and supernumerary roots? In other words, are polymorphic crown and root traits adaptive or nonadaptive?

When workers speculate on the potential adaptive significance of dental traits, it is usually in the context of how specific features enhance a tooth's strength and durability. Shovel-shaped incisors have frequently been perceived from such a structural standpoint. For an incisor with shoveling, Dahlberg (1963b:244) noted that 'Such a tooth generally has added mass as well as two supporting struts on the margins similar to the design of the builder's 'I' beam.' Although he gave no exact figures, he added that incisor loss through trauma was less common in Japanese populations, with their moderate to pronounced shoveling, than in Western Eurasians, who ordinarily show minimal expressions or complete absence of shoveling. In a like manner, Cadien (1972:208) commented 'Undoubtedly, the shovel-shape form results in greater structural strength for the incisor.'

In his monograph on shovel-shaped incisors, Mizoguchi (1985) concluded this variable has some adaptive significance. Analyzing 148 samples from around the world, he found that shoveling frequencies showed a significant positive association with latitude and significant negative associations with average annual temperature and average temperature of the coldest month. In other words, shoveling is a character that is associated with peoples of the far North. Rather than viewing shoveling as an isolated variable, Mizoguchi considered it in the context of other craniofacial characteristics such as facial flatness, pinched nasal bones, sagittal keeling, thickened tympanic plates, etc., which are common in North Asians and Eskimos. Hylander (1977) has developed a strong case to argue that these features, at least in Eskimos, act to generate or dissipate pronounced vertical occlusal forces. Mizoguchi (1985:110) concluded that 'shovelling should be considered basically as part of the facial structure associated with powerful biting forces.' This sentiment is echoed by Guthrie (1996) who feels shoveling and other morphological features that characterize the northern Mongoloid dentition (e.g., deflecting wrinkle, protostylid) parallel the increased dental morphological complexity that evolved rapidly in fauna of the Mammoth Steppe (e.g., in microtines, horses, and bison) during the late Pleistocene. Given the rigors of the North, especially during the Pleistocene, shoveling was a potential contributor to overall fitness. However, proto-Europeans also inhabited northern latitudes during this period, and they lacked shoveling for the most part. An alternative explanation is that shoveling drifted to high frequencies in groups on the northern population edge of East Asia – a much more likely possibility given the isolated mountainous environments of Asia compared to the more steppe-like genetically-webbed European habitat.

Although shovel-shaped incisors have received the most attention, other crown traits have also been viewed in structural-functional terms. In discussing the common occurrence of Carabelli's trait in Europeans, Dahlberg (1963b:244) commented that 'More than 40% of them have the advantage of the Carabelli's cusp in the upper molars, giving additional substance and size to the pestle-like protocone in its crushing contacts with the mortar-like central fossae of the lower molars.' Also regarding this trait, Cadien (1972:211) noted 'Cusps that do not reach the occlusal surface of the tooth cannot play a very direct role in mastication, which makes the function of the cusps uncertain. However, any structure that adds to the tooth material (dentin included) may increase the strength of the tooth, which may confer a selective advantage.' In his world analysis of Carabelli's trait variation, Mizoguchi (1993) also found this polymorphism may be influenced by local selective factors. Specifically, he correlated trait frequencies with anthropometric, climatic, and economic variables, and found significant positive associations with average annual temperature, average temperature in the hottest month, milking, and agriculture. He concluded that the long history of food production in the Near East and Europe led to decreased buccolingual diameters of the molars but 'the still remaining intensive biomechanical demand for the reduced first molar caused the strengthening of the first molar in the buccolingual direction through the enlargement of Carabelli's trait' (Mizoguchi, 1993:50). However, Carabelli's trait is rather common in pre-agricultural groups in Europe and the Middle East (i.e., Neanderthals and anatomically-modern Cro-Magnons).

For cusp 5 (UM) and cusp 6 (LM), traits noted to be in high frequency among aboriginal Australians, Townsend *et al.* (1986) feel that 'It is possible that the metaconule, and perhaps C6 on mandibular molars, provide additional enamel bulk in a region normally subjected to early wear. In a population such as the Yuendumu Aboriginals where the masticatory system was used vigorously, this may have been of selective value.' These comments may apply to other supernumerary cusps and ridges that are often the first crown components to exhibit enamel wear facets. In earlier populations, attrition would always win the battle in the end, wearing off cusps to produce a featureless planed surface, but accessory cusps and ridges may have served as the first line of defense to slow down this inexorable process. While these traits may indeed serve a function, it nonetheless remains difficult to move from inferred function to a demonstration of fitness.

When the subject of adaptability is addressed in the context of tooth morphology, it is usually to the effect that surficial features provide some advantage in terms of enhancing function, inhibiting wear, preventing tooth loss, etc. In a few instances, some trait expressions may be deleterious to their possessors. Enamel extensions have no obvious masticatory function – they do not participate in occlusion, inhibit wear, or add mass to a tooth. Since the periodontum attaches to the cementum of the root and not to enamel, an enamel extension toward the bifurcation of the roots also involves an apical displacement of the periodontal attachment (Masters and Hoskins, 1964). Tsatsas *et al.* (1973) feel enamel extensions are implicated in the formation of periodontal pockets, a precursor to periodontal disease. This association has not, however, been conclusively established. Premolar odontomes are another trait for which disadvantages seemingly outweigh advantages. Odontomes have both enamel and dentine components, but the pulp also projects into the tubercle about half the time. The location of odontomes in the center of the premolar crown, in conjunction with their conical shape, invites rapid wear and even breakage. In either case, pulpal necrosis and premature tooth loss are possible consequences. Any modest addition an odontome makes to crown mass is more than offset by the increased risks of tooth loss. Alaskan dentists, well aware of this problem, often 'drill and fill' unworn and unbroken odontomes as a preventative measure.

The adaptive advantages that might be associated with variations in root number have invited less speculation than ridges and extra cusps of the crown. One exception is the possible advantage associated with 3RM1, an accessory disto-lingual root on the lower first molar (Hylander, 1977; Turner, 1987; Guthrie, 1996). While relatively rare on a world scale, 3RM1 is common in East and North Asians, and reaches its highest frequencies among Eskimos and Aleuts of the American Arctic. This extra root provides a secure anchor for the lower first molar, as Alaskan dentists who practice in Eskimo villages readily attest. They are often forced to break the accessory root before a lower first molar can be successfully extracted. Turner (1987:311) adds 'The occurrence of the supernumerary third root on lower first molars would surely help hold this important tooth in proper occlusion for more years that would be the case without it.' While the root appears to 'serve a purpose' in a high-attrition environment, the role of selection in its maintenance remains speculative.

While most discussions of dental traits in an adaptive context focus on strength and durability, another consideration is sexual selection. We might ask, for example, if any dental trait plays a role in attracting potential mates? For most crown traits, which are partially or completely hidden from view, any effects from sexual selection seem unlikely. Incisor winging, however, is directly visible to prospective mates. When groups artificially mutilate their teeth as a sign of status and group affiliation or to attract members of the opposite sex, the upper incisors are the most commonly altered teeth (Prologue). Whether or not winging serves as a biological analogue to intentional incisor modification is only conjectural. If this polymorphism had any impact on function, it would more likely be negative than positive because the incisors are rotated out of normal occlusal alignment with their opponents. Given that winging, like shoveling, is most common in Asia and the Americas, the functional explanations for the presence of shoveling appear to run counter to the concurrent

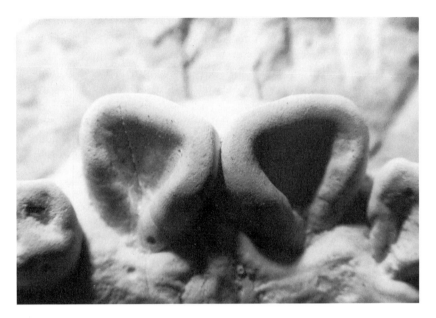

Fig. 6.1   The co-occurrence of pronounced upper central incisor winging and shoveling in an American Indian dentition. The well-developed marginal ridges of shovel-shaped incisors would strengthen the tooth and slow the rate of crown wear, but these advantages appear offset when this trait is found in association with winging. Winging has no obvious adaptive function and its role in sexual selection remains obscure.

expression of incisor winging (Fig. 6.1). However, one additional complicating factor is that incisor rotation permits relatively large teeth to be retained in small jaws. At this time, we do not know if incisor winging functions to enhance tooth longevity or attract mates, or, alternatively, is simply neutral in terms of biological fitness.

From a paleontological standpoint, minor variants of the crowns and roots are the primary subjects of dental evolutionary change (Butler, 1982). During the course of primate evolution, subtle changes in surficial morphology are reckoned in millions of years. The addition or subtraction of cusps and ridges may well be due to natural selective agencies operating over many hundreds of generations. In the short run, however, it appears that within and between group variation in crown and root traits is largely the result of chance processes. When subpopulations of a larger population exhibit inter-village differences, founder effect, genetic drift, and systems of mating are the primary mechanisms affecting this variability. At higher levels of differentiation, involving longer periods of isolation and significant contrasts in environmental conditions (e.g., South Africans vs. North

Asians), selective agencies may have played a role in generating patterns of dental variation. However, the role of selection has never been demonstrated for any single dental trait, including shoveling.

Geneticists face the same adaptive conundrum in determining if and/or how specific genes are acted upon by natural selection? The selective advantage conferred by hemoglobin S has been cited so many times in textbooks, it has become the 'melanic moth' of anthropological genetics. Heterozygotes with one normal (A) and one sickle-cell (S) gene have higher relative and absolute fitness values than either homozygote (AA or SS) in a malarial environment. This gene is maintained as a balanced polymorphism through the action of both differential mortality and differential fertility. Other polymorphisms, including G6PD-deficiency, the Duffy blood group system, thalassemia, and other hemoglobinopathies, also appear to be subject to balancing selection in response to malaria. After the 'malarial polymorphisms,' researchers have found it difficult to demonstrate the effects of selection on other genes, despite massive and valiant efforts toward this end, especially for the ABO and MN blood group systems.

In their recent analysis of genetic variation on a continental and global scale, Cavalli-Sforza *et al.* (1994) were primarily interested in reconstructing human population history. Given this interest in history and not the evolution of individual genes, they acknowledge, as Boyd (1950) did, that selectively neutral genes provide the most suitable data for historical reconstructions. As malarial polymorphisms are decidedly not neutral, they were omitted from their analysis. Based on the variance in gene frequencies among 490 world samples, it also appears that additional loci are subject to stabilizing selection, disruptive selection, or some combination of the two. Despite indications that selection operates on some genetic polymorphisms other than those associated with malaria, the authors conclude that 'Considering the total number of regions and alleles tested, the deviations from neutrality of classical markers detected by this test are not overwhelming' (Cavalli-Sforza *et al.*, 1994:120). In other words, these researchers do not feel the effects of selection on a small proportion of loci in their overall data set seriously bias historical reconstructions.

We take much the same position on dental polymorphisms that Cavalli-Sforza *et al.* (1994) adopt for genetic polymorphisms. We would indeed remove dental traits with selective effects from an analysis, if any could be identified. Clearly, there are no dental morphological parallels to hemoglobin S and other 'malarial polymorphisms.' For this reason, we find it unnecessary to omit any variable in a review of dental differentiation. With a focus on recent human history, this should present no significant problem even if some variables, such as shoveling, are influenced by low levels of selection. Our explicit assumption is that the divergence in dental trait frequencies among populations has resulted principally from the chance processes of founder effect and genetic drift and the effects of population structure. Given the phylogenetic history of most crown and root traits, mutations likely contribute little if anything to normal dental variation, excepting perhaps rare traits with no fossil history such as the 'Uto-Aztecan premolar' (Morris *et al.*, 1978). In analyzing relationships among recent populations, the only directional evolutionary process likely to alter trait frequencies is gene flow. As with serological genes or other genetically determined phenotypes, hybridization results in a convergence of trait frequencies. In general, however, the impact of gene flow on historical reconstruction is greater at the local and regional levels than at the continental and global levels. With the great increase in human movements across the globe in the past few centuries, recent populations (skeletal or living) are also more likely to exhibit the effects of gene flow than can be seen in skeletal samples dating to earlier times. In summary, we find very little evidence to suggest any significant adaptive changes in dental morphology for the time period of this book, namely the last 20,000 years.

Biological distance

Early nineteenth-century natural historians used biological data to infer relationships among human populations. While some were overly preoccupied with racial classifications, the aim of many was to understand human movements, mixtures, and modifications, or, in short, the biological course of human history. During this era, workers had access to information on a limited suite of biological variables that included skin color, hair color and form, eye color, and nose form. Populations in a particular area were characterized for each of these traits in normative or modal terms (e.g., brown skin, black hair, brown eyes, narrow noses). In the mid-nineteenth century, this short list was augmented by what became the 'all-important' cephalic or cranial index (i.e., head breadth/head length × 100) of A. Retzius. In some respects, the popularity of this index ushered in the 'age of measurements.'

Two important developments occurred in anthropology between 1870 and 1910. First, research in biometrics and statistics stimulated many biologists and early physical anthropologists to think in terms of population samples rather than individual types, even though typological thinking retained some popularity in anthropology until the mid-twentieth century. Workers who thought in terms of populations, characterized samples by measures of central tendency (e.g., mean) and dispersion (e.g., variance) rather than by normative types (i.e., mean = 165 cm vs. medium stature). Second, with the rise of anthropological societies in the 1870s, there was a rush to develop more 'measures of man.' By the turn of the twentieth century, several international congresses had been convened to define and standardize dozens of anatomical landmarks and hundreds of measurements for anthropometric and osteometric studies. In short order, the mass of measurement data on living and skeletal samples outstripped the abilities of researchers to derive insights into historical relationships, the original goal of the monumental efforts to quantify human variation.

With ever expanding volumes of measurement data on human subjects from around the world, disentangling relationships on the basis of similarities and differences between populations became a challenging mathematical problem. It was tedious and only minimally informative to compare two or more populations one variable at a time across a long list of measurements. Some workers circumvented the problem by focusing on a small handful of indices that were considered especially useful in discriminating between groups (cf. Dixon, 1923). Both approaches had limitations, leaving much data unanalyzed and problems unresolved. Workers clearly needed some statistical method to determine how geographically-removed groups were related. The issue that had to be addressed was how to assess multiple variables simultaneously to arrive at one summary value that encapsulated the overall difference between any two individuals or groups.

The quantitative methods that were eventually developed to summarize between-group differences are referred to as 'distance statistics.' Initially emphasizing the analysis of quantitative variables, pioneers who developed such statistics include F. Heinke, J. Czekanowski, K. Pearson, P.C. Mahalanobis, P.J. Clark, L.S. Penrose, and J. Hiernaux (Constandse-Westermann, 1972). The most elementary measure of distance is Czekanowski's DD, derived by finding the average of the absolute differences in multiple quantitative variables between two individuals or two groups. This measure does not, however, account for: (1) the differential contribution of 'small' and 'large' traits to a distance value, e.g., nasal breadth vs. cranial length; (2) sample size; (3) sample variances; (4) correlated variables; or (5) tests of statistical significance. To correct some of these deficiencies, Pearson (1926) developed the 'coefficient of racial likeness' (CRL) to gauge whether or not the overall difference (or distance) between any two groups was statistically significant (i.e., were two groups sampled from the same population?). While the CRL gained popularity for a time, it also had its deficiencies. Many of these problems, including adjustments for correlated variables, were rectified by the Generalized Distance (D^2) statistic of Mahalanobis (1936). However, to calculate Generalized Distance, a worker starts with individual measurements rather than sample means and standard deviations. Moreover, in pre-computer days, this statistic was computationally tedious. In response to these problems, Penrose (1954) reverted to the earlier ideas of Heinke, Zarapkin, and Pearson to develop 'size and shape' distance statistics. These distances could be derived using sample summaries readily available in the literature and were also easier to calculate. Today, between-group differences in measurement variables are assessed by some combination of these traditional distance statistics and multivariate techniques, especially principal coordinates, principal components, and factor analysis (cf. Howells, 1989).

The historical development of distance statistics for measurement variables mirrors the preoccupation of early twentieth century anthropologists with anthropometric and osteometric data. Comparable statistics for qualitative traits received little attention until the 1950s, coincident with the rapid growth and increased popularity of serological genetics and nonmetric traits in anthropological research. In contrast to earlier methods that focused on differences in means between groups, qualitative methods involve comparisons of frequencies or proportions. Paralleling Czekanowski's DD statistic, the most elementary method to derive distances for qualitative traits is to square the difference between two sample frequencies, sum the squared differences across all variables, and divide by the total number of variables. Spuhler (1954) called the square root of this derived number the 'coefficient of relationship.' Since that time, statisticians and geneticists, including A.H. Edwards, L.L. Cavalli-Sforza, C.A.B. Smith, L.D. Sanghvi, V. Balakrishnan, J.C. Gower, and M. Nei, have developed a wide array of techniques for estimating biological distance comparing populations for gene or phenotype frequencies. Many of these methods involve variants of the chi-square statistic, angular transformations of frequencies, or kinship coefficients (cf. Constandes-Westermann, 1972; Weiner and Huizinga, 1972).

Most distance values used by dental anthropologists are measures of dissimilarity. That is, two samples with identical trait frequencies would have a pairwise distance coefficient of 0.0. As deviations increase from zero, dissimilarity increases, resulting in ever larger distance coefficients. Distances are computed among three or more groups (a distance value between just two groups would have no context for comparison) and all pairwise distance values are arrayed in matrix form where values above and below the diagonal are reversed images (in many cases, only values below the diagonal are presented as in Appendix A3). The upper limits of a distance value vary relative to the particular statistic used, but, in all cases, when many pairwise distances are shown in a matrix, the smallest values are associated with groups showing the greatest overall similarity across all traits while the largest values indicate the most dissimilar or divergent groups.

Biological distance is used ordinarily in the context of affinity assessment. It has nothing to do with geographic distance except in the sense that neighboring groups are likely to be more closely related than groups separated by hundreds or thousands of miles. With the extensive movements of human populations over the past 500 years, there are, of course, many exceptions to this rule as 'biologically distant' groups are now often geographically proximate neighbors.

Any biological distance value is a relative measure of relationship. Two distance values are comparable only when based on a common set of variables and distance statistic. Five additional aspects of biological distance should be noted: (1) similarity, as indicated by a small inter-group distance value, is assumed to indicate a close biological relationship and recent common ancestry while larger values indicate a distant relationship and more remote common ancestor; (2) between group divergence is caused primarily by the stochastic processes of genetic drift and founder effect; (3) gene flow, which leads to convergence, can alter or even erase a pattern of phylogenetic relationships; (4) distance values based on many variables are more reliable than those based on only a few variables (Livingstone, 1991); and (5) in most analyses, the traits used to estimate distance are weighted equally.

Although the development and use of different distance statistics has proceeded with some controversy, and various authors are quick to point out the advantages of one particular method and the weaknesses of alternative approaches, we follow the prudent view of Constandse-Westermann (1972:137) who notes 'The high correlation generally found between the various coefficients creates the impression that the importance of finding *the* [author emphasis] statistically perfect way of calculating the biological distance should not be overstated.' More recently, Cavalli-Sforza *et al.* (1994:30) aver 'In general, the distances calculated by different formulas are always highly correlated.' We concur with both of these statements. When the same data sets are analyzed using different distance statistics, the corresponding distance matrixes tend to be highly correlated. Workers who specialize in nonmetric cranial and dental traits often employ C.A.B. Smith's 'Mean Measure of Divergence' (MMD), but this is just one of many useful distance measures appropriate for the analysis of phenotypic trait frequencies. When we review the analyses of other workers, a number of different distance techniques are alluded to but all perform basically the same function – to determine relative degrees of pairwise dissimilarities and/or similarities.

When only three groups (A,B,C) are compared in a distance analysis, the resultant matrix includes three values corresponding to all possible pairwise comparisons: AB, AC, and BC. Analyses of more groups, for example, 5, 10, and 15 samples, give matrixes with 10, 45, and 105 pairwise distance values, respectively. Analyses involving more than 15 groups yield very large matrixes, making it difficult to evaluate the overall pattern of individual pairwise distance values. Stimulated to a large extent by research in numerical taxonomy (cf. Sokal and Sneath, 1963), graphical methods have been developed to reduce the complexities of a large distance matrix to two or three dimensions. The most popular method of reducing distance values to two dimensions is cluster analysis that produces tree diagrams. In any tree diagram, or dendrogram, groups that cluster together on one branch typically have the smallest pairwise distance values. That is, groups with similar trait frequencies cluster together while groups with disparate frequencies fall on different branches.

When values on the X, Y, and Z axes are derived for each group in a sample array through principal components analysis, factor analysis, or multidimensional scaling, they can be plotted in either two (X,Y) or three (X,Y,Z) dimensions to provide a visual representation of 'biological distance.' Such ordination methods provide an alternative or complementary method to cluster analysis for graphically depicting biological distances among groups. Two groups with a small pairwise distance value will have similar coordinates and plot closely together. Conversely, groups with the largest pairwise distance values ordinarily fall at opposite ends of a coordinate plot, mirroring their greater relative distance.

Some workers distinguish genetic distance from morphological or phenetic distance, depending on whether gene frequencies or phenotype frequencies are used for analysis. Given this distinction, distance estimates based on crown and root trait frequencies are phenetic. Phenetic distances have been shown to be powerful and robust in assessing population affinity when the phenotypes used in their estimation have a strong heritable component. When this condition is met, it is still debatable as to whether genetic or phenetic distances are the better indicator of population relationships. Genetic and phenetic data often yield similar distance matrixes, but this is not always true.

Levels of differentiation

In chapter 5, we illustrated how sundry crown and root traits show patterned geographic variation in frequencies among high-order groups from five major subdivisions of humankind. Differences in trait frequencies are also evident at lower levels of differentiation. At the lowest levels are differences between individuals and between families within a single population (cf. Alt *et al.*, 1995). While interesting in some forensic and archaeological contexts (see Epilogue), research in dental morphology does not focus ordinarily on individual or family level distinctions, but rather on population or group differences.

The lowest level of differentiation at the group level is between sub-populations (e.g., villages) within a larger population that are tied together by bonds of language, kinship, mate exchange, geography, etc. Many authors refer to this as microdifferentiation although, to be consistent with other terms, we use the phrase local differentiation to describe subpopulation variation. In many instances, sub-populations studied in this context fall within groups at tribal levels of political organization. For example, the Yanomama of Venezuela number about 10,000 individuals residing in dozens of villages, any one of which rarely exceeds 200 inhabitants (Chagnon, 1972). Inter-village variation is thus synonymous with intratribal variation.

The next level of differentiation occurs between populations at the regional level. Individual populations within a region may exhibit different forms of population structure. Some small populations, for example, exist as geographically circumscribed isolates. Larger groups might include subpopulations, as the Yanamamo noted above, or show a continuous distribution through space, with genetic neighborhoods but no concentrated settlements. The focus at this level is in comparisons between populations rather than subpopulations. Regions are generally defined by geography, culture, or some combination thereof, and are sub-continental in scope. They may, in some instances correspond to what Wissler (1917) and Kroeber (1939) refer to as culture areas. For example, in the American Southwest culture area, population units are represented by self-defined tribes such as Zuni, Pima, Papago, Hopi, and Navajo.

Transcending the analysis of between-group variation within circumscribed geographic areas is differentiation among regional groups in broader continental contexts. This is effectively intra-continental, or, more simply, continental differentiation. At this level, contrasts are made between widely dispersed regional groups. In North America, comparisons might be made between Indian groups of the American Southwest, Great

Basin, Northwest Coast, and Sub-Arctic, and Eskimos and Aleuts of the Arctic. In the Pacific, comparisons could be made between different island populations in Polynesia, Melanesia, and Micronesia. We recognize there is heterogeneity within these larger regional groupings, but, in most instances, there is more biological cohesion and shared history within regions than between regions. Recent immigrants to a region can usually be discerned on the basis of biological and linguistic differences (e.g., Athapaskans in the American Southwest, Polynesian outliers in Melanesia).

Moving up the ladder of divergence to its highest infraspecific rung, the last level of differentiation is inter-continental or global in scope where comparisons are made between major human geogenetic subdivisions (e.g., Africans, Asians, Europeans, Australians, Native Americans). Global level analyses of this sort have a long history in anthropology (e.g., see papers in Count 1950), and are the focus of many human genetic and craniometric studies today (cf. Howells, 1973b, 1989; Nei and Roychoudhury, 1982, 1993; Cann *et al.*, 1987; Cavalli-Sforza *et al.*, 1988, 1994; Relethford, 1994; Relethford and Harpending, 1994; Chen *et al.*, 1995; Hanihara, 1996).

Palomino *et al.* (1977) considered dental morphological variation at four levels of differentiation, differing but slightly from the one outlined above. Their first two levels, village and village cluster, correspond to what we refer to singularly as local differentiation. Their next level, that of tribe, is roughly equivalent to our continental designation while the fourth level, race, refers to global differentiation. Using frequency data on six dental traits compiled from four separate sources, Palomino *et al.* (1977) calculated within and between group variance components and failed to find a significant relationship between coefficients of diversity and their four levels of differentiation. In other words, their analysis did not show the expected pattern of lowest diversity among villages, with increasingly larger diversity coefficients going from village cluster to tribe and, finally, to race. While discouraged by this finding, they had modest success in assessing biological distance at the tribal (= continental) level. They concluded 'there are some general features in the tooth morphology which make them useful indicators of biological distance when one is concerned only with differentiation at a broad level' (Palomino *et al.*, 1977:67).

How well does tooth morphology gauge population relationships at the four defined levels of differentiation? We concur with Palomino *et al.*, (1977) that dental traits are quite effective at discriminating between populations at high levels of differentiation. This was demonstrated, in part, in chapter 5 and is further elaborated in chapter 7. Another issue, however, is whether or not dental morphological variation is sufficiently sensitive to estimate affinity within (local differentiation) and between (regional differentiation) populations. One complication in making this determination is that there is no absolute standard for assessing biological distance among closely related groups. Some workers consider distances based on simple genetic markers as 'the standard' to be met by estimates based on other types of biological variables. Others, however, feel that gene frequency variation at the local and regional levels is too subject to the vagaries of founder effect and genetic drift to provide the definitive portrayal of relative population relationships. This problem is discussed by Howells (1977), and also Simmons (1976) who concluded, after 35 years of blood group genetic research, that gene frequency data provided no clues as to the origins of Australian aborigines. As this issue is not fully resolved, we review studies that determine to what extent patterns of relationships indicated by dental traits agree with patterns estimated by genetic markers and other biological variables. In a few cases, authors also correlate dental distances with non-biological variables such as geographic distance and language.

In only a few instances have dental morphologists been afforded the opportunity to study dental variation within populations, i.e., between sub-populations of a single population system. The largest studies of this type were conducted among the Yanomama tribe of South America (Brewer-Carias *et al.*, 1976) and Solomon Islanders (Harris, 1977). In a sample of more than 800 Yanomama, Brewer-Carias *et al.* (1976) estimated biological distance among seven villages using eight crown traits. The primary aim of the authors was to determine the level of congruence between distance matrixes based on crown traits and those based on simple genetic markers, noting that 'in the absence of any correspondence one would be forced to conclude the microdifferentiation was more noise than signal' (Brewer-Carias *et al.*, 1976:12). Using two different methods, these workers found significant, albeit moderate, positive correlations between dental and genetic distances. They concluded 'the pattern of microdifferentiation suggested by dental traits is roughly comparable to that projected by other types of biological traits whose genetic basis is more firmly established' (Brewer-Carias *et al.*, 1976:13).

Harris (1977) observed 44 dental variables in 1217 individuals distributed among 14 villages on Bougainville Island in the Solomon Islands chain. He correlated distance matrixes based on crown morphology with those derived from simple genetic markers, dermatoglyphics, anthropometrics, odontometrics, geography, and language. In contrast to the Yanomama study, he found no significant correlation between distance matrixes derived from dental morphological and genetic data. Moreover, neither male nor female distance matrixes showed any correlation with distances based on dermatoglyphic variables or geographic distance. The male dental distance matrix did, however, correlate positively with distances based on anthropometric and odontometric traits and language. Female dental distances, by contrast, showed a significant correlation with distances based on odontometric variation but with no other type of biological variable, language, or geographic distance.

Dental analyses of the Yanomama and Solomon Islanders suggest that crown traits exhibit patterns of intrapopulation variation that are only moderately congruent with other measures of relationship indicated by biology, language, and geography. One possible explanation for this finding is the conservative pattern of differentiation evident in crown trait frequencies among closely related sub-populations. For example, Harris (1977) used the MMD to calculate distances among his 14 village samples for 44 dental traits and 19 alleles. The mean inter-village MMD from the matrix of dental distances was 0.030 while the comparable mean distance for genetic markers was 0.110. In other words, the average pairwise genetic distance was three to four times larger than the mean dental distance. Brewer-Carias *et al.* (1976) used different methods to estimate dental and genetic distances so a direct comparison of their results is not possible. However, we computed the mean variance for eight dental traits and eight alleles across seven Yanomama villages and found that the dental traits (mean frequency = 0.482) had an average variance of 0.0164 while the genetic markers (mean frequency = 0.634) had a mean variance almost twice as great at 0.0301. Thus, in the two largest studies of intra-tribal/inter-village dental microdifferentiation, crown traits exhibit less frequency divergence than simple genetic markers. This supports the contention of Sofaer *et al.* (1986) who say 'the expression of each variable is probably influenced by several genes, presumably resulting in a slower rate of biological differentiation and therefore smaller trait frequency differences between groups than might be expected for single locus polymorphisms.' Although further studies of microdifferentiation should be encouraged, dental traits might prove to be only minimally useful for assessing relationships among subgroups of a population because of their slower rates of differentiation. As the analysis of micro-taxonomic variation in crown and root traits has never been a major goal of dental researchers, this observation does not pose a serious problem to those who use dental data to assess more distant relationships among populations separated for longer periods of time. The next question is whether or not crown traits can effectively discriminate between distinct yet closely related populations that reside within well defined geographic regions.

Sofaer *et al.* (1972a) were among the first to compare populations for both genetic markers and crown morphology to determine if these different types of variables revealed the same pattern of relationships. They made dental observations on three extant Indian tribes in the American Southwest, the Pima, Papago, and Zuni. Based on language, geography, and gene frequency variation, the Pima and Papago were expected to be more similar to one another than to the linguistically and geographically disparate Zuni. Using a battery of ten crown traits, the authors could not affirm this predicted triangular relationship. However, when they re-analyzed distances by removing six and then eight traits with the highest estimates of intraobserver error, the dental distances more closely conformed to the pattern of relationships expected on the basis of relative genetic differences. With some hesitation, the authors concluded 'This change, towards a relative closeness between the Pima and Papago, and their mutual distance from the Zuni, the fundamental relationship shown by geography and by the genetic estimates, suggests that tooth morphology has the potential of providing moderately good discrimination, even on the fine level on which the three Indian tribes are related' (Sofaer *et al.*, 1972a:364).

More extensive studies of American Southwest Indian dental variation were conducted by Scott, Dahlberg, and their collaborators (Scott, 1973; Scott and Dahlberg, 1982; Scott *et al.*, 1983, 1988) who observed more than 2500 individuals distributed among 11 living tribes from five language families: Uto-Aztecan (Pima, Papago, Hopi), Hokan (Quechan or Yuma, Mohave, Maricopa, Pai Pai), Athapaskan (Navajo, Apache), Tanoan (Tewa) and Zunian (Zuni). In this geographic area, extensive anthropological research among both prehistoric and living groups provides some basis for predicting the pattern of between-group relationships. For example, Uto-Aztecan, Hokan, Tanoan, and Zunian speakers likely have considerable time depth in the Southwest (>2000 years), but Athapaskans are recent migrants from Canada who arrived in northern New Mexico and northeastern Arizona only a few hundred years ago. Although both the Navajo and Apache have intermarried with long-standing inhabitants of the area, their northern roots are indicated clearly by language and simple genetic markers. The Navajo and Apache should thus be more similar to one another than they are to indigenous populations in the Southwest. Another expectation is a close similarity between two neighboring and closely allied Uto-Aztecan speaking tribes of the southern Arizona desert, the Pima and Papago. A third Uto-Aztecan group, the Hopi, reside on the Colorado Plateau in northeastern Arizona. Given their geographic separation and greater linguistic differences, the Hopi should not be as close to either Pima or Papago as these two groups are to one another. Hokan groups in the Southwest have not been thoroughly characterized for genetic markers although, on linguistic and geographic grounds, these populations, which reside mostly along the Colorado River in western Arizona, might also be expected to show greater similarities to one another based on their shared language and geographic proximity. It is difficult to make any specific predictions about the Zuni who are culturally allied with Western Pueblos but are linguistically isolated from other groups in the Southwest.

In a study focusing on the Pima Indian dentition, Scott *et al.* (1983) found many points of agreement between the pattern of relationships derived from crown morphology and relationships predicted by other evidence, especially language and anthropometry. That is, the Pima and Papago grouped closely together in a three-dimensional ordination and their next closest relationship was to another Uto-Aztecan group, the Hopi. Three Athapaskan samples (eastern Navajo, western Navajo, San Carlos Apache) also plotted together although the central Navajo were closer to the Hopi, perhaps indicating greater gene flow in the Navajo residing in closest proximity to the Hopi. The two Hokan groups, Yuma and Mohave, plotted closely together and showed only a distant relationship to other Southwest Indian groups. The two remaining Puebloan populations, Zuni and Tewa, were outliers in the context of this Southwest tribal array. With but few exceptions, these Southwest Indian populations show a consistent relationship between language affiliation and dental morphological variation.

In a subsequent study including additional Hokan-speaking samples, Scott *et al.* (1988) found further evidence for an association between language families and relationships indicated by dental morphology. Again, Pima and Papago as well as Navajo and Apache plotted closely together in a 3-dimensional ordination, with Hopi closest to their Uto-Aztecan affines (Pima, Papago) and Zuni as an outlier (Fig. 6.2). The four Hokan groups were found to be closer to one another than to other Southwest tribes, although Quechan and Mohave are slight outliers. Although serological profiles were not available for all the Southwest Indian samples observed for crown morphology, Scott *et al.* (1988) also found that for the Papago, Pima, Zuni, Maricopa, Navajo, and Apache, the ordinations based on genetic data were mirrored closely by those based on crown morphology. The first axis (i.e., eigenvector) distinguished the Pima, Papago, Maricopa, and Zuni from the two Athapaskan groups, Apache and Navajo. The second axis separated the Pima, Papago, and Maricopa, who remain tightly clustered, from the linguistically isolated

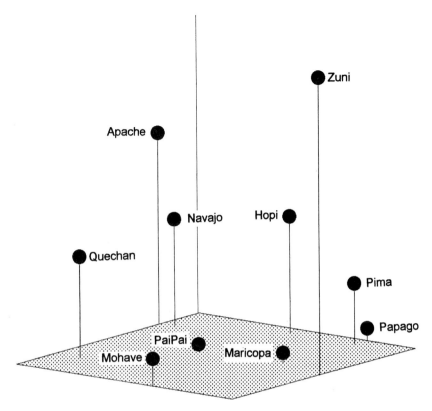

Fig. 6.2 Three-dimensional ordination of 10 American Southwest Indian tribes based on crown trait frequency variation. Language families represented: Uto-Aztecan (Pima, Papago, Hopi), Hokan (Maricopa, PaiPai, Mohave, Quechua), Athapaskan (Navajo, Apache), and Zunian (linguistic isolate).

Zuni. When additional genetic data become available for other Southwest Indian populations, it will be possible to examine in greater detail the correlation between dentally and genetically derived distance values.

As noted earlier, Harris (1977) found that dental morphological variation among Bougainville Islanders was not strongly correlated with other biological variables or linguistic differences at the inter-village level. However, when he performed an analysis based on language groups rather than villages, the results were more positive. To accomplish this, he grouped 14 Bougainville villages into four non-Austronesian (two Northern Papuan, two southern Papuan) samples and one Austronesian sample. He also expanded his sample array to include three Austronesian groups from Malaita Island in the southern Solomons and two Polynesian outliers from Ulawa and Ontong-Java.

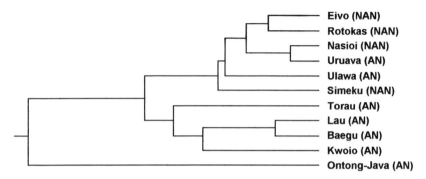

Fig. 6.3   Dendrogram based on tooth crown trait frequencies for
non-Austronesian (NAN) and Austronesian (AN) groups in Melanesia. When
Melanesian populations are grouped by language families, NAN and AN
groups can usually be distinguished by dental characteristics. Redrawn after
Harris (1977).

In Harris' (1977) dendrogram based on 44 tooth crown traits (Fig. 6.3),
the four non-Austronesian (NAN) groupings fall in the same cluster. The
three Austronesian (AN) groups from Malaita are in a single cluster along
with an Austronesian group from Bougainville, the Torau. The Polynesian
outlier, Ontong Java, does not cluster with any of the ten samples so it is
also a 'dental outlier.' Only two of ten samples in the array do not show a
correspondence between language group and dental variation. That is, one
AN group (Uruava) and one Polynesian outlier (Ulawa) fall in the cluster
that includes the NAN groups.

In the case of Uruava, the relationship to NAN groups is also indicated
by genetic data. In a tree diagram that includes 31 Melanesian, Micronesian, and Polynesian populations, Cavalli-Sforza *et al.* (1994) found
that Uruava fell in a Melanesian cluster including mostly NAN groups.
This finding supports the speculation of Harris (1977:284–7) that 'the
Uruava, who speak a Melanesian subdivision of AN and are rather recent
migrants to the island, may have had an appreciable genetic contribution
from the NAN speakers.' Another point of congruence between genetic
and dental data is the finding of Cavalli-Sforza *et al.* (1994) that the
Austronesian-speaking Malaita Islanders and the Torau fall in a single
cluster that also includes three Micronesian populations. Harris
(1977:303–4) concludes 'dental morphology is a moderately useful measure
of inter-population affinity' but he is more enthusiastic with the results
from the analysis of language families or local races. At this level, he notes
'there is rather good agreement with dental morphology and the limited
information available from linguistics, geography, and historical reconstruction.'

Another regional study that demonstrates the potential of tooth morphology for distinguishing closely related populations was conducted by Sofaer *et al.* (1986) among Jewish and non-Jewish populations in the Middle East. Using a battery of 19 crown traits, these authors found that East European, Kurdish, and Moroccan Jews grouped closely together on the basis of tooth morphology. The only two Jewish samples that did not show a close relationship with these groups were Habbanite Jews and a 3000 year old skeletal series from Mt. Zion. The authors felt this disparity was not altogether surprising. The Habbanites, a small endogamous group that only recently immigrated to Israel from southern Arabia, possess genetic markers that indicate a significant amount of gene flow from Africa. African admixture, in conjunction with the effects of drift due to small population size, pulls the Habbanites away from other Jewish dental samples. As an overview, the average distances among the Jewish groups, both with and without the Habbanite sample, are smaller than the average distances between the Jewish and non-Jewish populations and among non-Jewish populations. The overall homogeneity of the Jewish samples and the distinctiveness of Samaritans in particular are supported by evidence from simple genetic markers. The authors conclude 'the broad similarity of the group affinities based on tooth morphology with those based on gene frequency data adds further weight to the use of dental morphological variables in the study of biological relationships between populations, especially when the opportunity for including skeletal groups occurs' (Sofaer *et al.* 1986:273–4).

Many other studies demonstrate the utility of crown and root morphology in assessing regional differentiation. Broken down by major geographic area, those that focus on dental morphological comparisons among two or more populations (or local races) for two or more dental traits include: Europe (Brabant, 1964, 1971; Kochiev, 1973, 1979; Brabant *et al.*, 1975; Berry, 1976; Aksianova, 1978, 1979, Aksianova *et al.*, 1977, 1979; Gadzhiev, 1979; Gravere *et al.*, 1979; Tegako and Salivon, 1979; Kaczmarek and Piontek, 1982; Kajajoja and Zubov, 1986; Kaczmarek, 1992; Scott and Alexandersen, 1992); India and Pakistan (Lukacs, 1983, 1987, 1988); Central Asia (Babakov *et al.*, 1979; Ismagulov and Sikhimbaeva 1989; Haeussler and Turner, 1992); East Asia (K. Hanihara *et al.*, 1974; Khaldeeva, 1979; Vorinina and Vashchaeva, 1979; T. Hanihara, 1989a, 1989b, 1991a,b, 1992a); Pacific (Barksdale, 1972; Richards and Telfer, 1979; T. Hanihara 1990a, 1992c, 1993); North Africa (Greene, 1967, 1972, 1982; Greene *et al.*, 1967; Irish and Turner, 1990; Turner and Markowitz, 1990); and North America (Turner, 1969, 1991, 1993; Sciulli *et al.*, 1984; Scott, 1991b, 1994). Some of these studies are primarily descriptive in nature and do not involve an assessment of biological distance. All help illustrate, however, the magnitude of dental differentiation within circumscribed geographic areas and provide the foundation for the following five generalizations:

(1) Patterns of population relationships indicated by dental morphology show fair to good correspondence with distance values based on other types of biological variables (i.e., supporting, in part, the hypothesis of nonspecificity).

(2) Despite, a moderate level of congruence, there remain certain unresolved factors that must account for the disparity in dentally and genetically based distance values. It appears at this time that dental traits do not exhibit as much variance as simple genetic markers at the local and regional levels. This may support the contention of many authors that traits with complex modes of inheritance diverge more slowly than alleles at a single locus.

(3) If dental divergence is more conservative than divergence in simple genetic systems, tooth morphology may have limited utility in resolving problems at the local level (i.e., intragroup differentiation). Dental variation does, however, provide insights into population history at the regional level where larger taxonomic units such as local races and/or language families are the scale of comparison.

(4) Dental and linguistic differences covary to some extent, perhaps not at the level of dialectical difference but among higher order linguistic groupings such as language families and phyla.

(5) An advantage afforded by teeth in the study of regional population variation is that they can provide both synchronic and diachronic perspectives. In the studies listed above, some authors focus on a synchronic analysis of living populations while others concentrate on earlier human populations represented by skeletal samples. In some cases, workers incorporate dental data on both extinct and extant groups allowing concurrent synchronic and diachronic analyses.

From the foregoing, it is evident that crown and root traits are useful indicators of population relationship if the scale of comparison is sufficiently high. As most historical problems have considerable time depth, it is not a major issue that dental traits evolve too slowly to discern fine scale within-population differences. In the next chapter, we review studies that illustrate how crown and root traits have contributed to the resolution of several specific problems of long-standing interest to anthropologists and prehistorians.

7 Tooth morphology and population history

Introduction

As in other areas of evolutionary biology, human skeletal remains provide the only direct link to other past and living populations. Some skeletal biologists study bones and teeth to discern prehistoric population relationships while others are primarily concerned with temporal changes that can be interpreted in processual terms. These avenues of research provide insights into population history and adaptation, but, in terms of precedence, historical relationships must always be established before temporal trends in size, morphology, and other skeletal indicators are interpreted in processual terms. To illustrate this point, assume that collections of human skeletal remains from one region date to (a) 10,000, (b) 5,000, and (c) 1,000 years BP. Assume further that the three skeletal series show linear trends for a decrease in tooth size and an increase in stature from (a) to (b) to (c). One cannot attribute these trends to genetic changes (e.g., relaxed selection pressure), environmental factors (e.g., improved nutrition), or behavioral/cultural modifications (e.g., new food storage or cooking techniques) until it is shown there is an ancestral-descendant relationship between (a) and (b) and between (b) and (c). Changes perceived as temporal trends might instead reflect new populations moving into our hypothetical region at times (b) and/or (c). It should not be assumed that two groups occupying the same space but at different times are somehow related. With the well-documented human penchant for moving around the landscape, this is not the case in many instances.

Tooth morphology has contributed to the resolution of a number of historical problems that have long attracted anthropological interest. For various reasons, dental morphologists have concentrated much of their attention on Asian and Asian-derived populations so our examples focus on Asia, the Pacific, and the New World. Fewer analytical studies have been directed at historical problems in Europe and Africa. Unfortunately for this review, but fortunately for future dental morphologists, there remain a myriad of historical questions beyond those addressed that await a systematic analysis of crown and root trait variation.

269

The Asian dichotomy: Sinodonty and Sundadonty

Asia, the world's largest continent with the largest living population, includes a wide diversity of habitats, from the world's wettest tropical rainforests (Assam), to deserts, steppes, mountains, and the coldest regions on the planet (Yakutia). Despite Asia's scale in size, numbers of people, and environmental extremes, anthropologists have traditionally included all groups contained therein in a broadly defined Mongoloid race (excluding most peoples of the Indian subcontinent). Even authors who hesitate to subsume all Asian physical variation under one heading simply add an adjective to describe Southeast Asian and Pacific populations as 'southern', 'tropical', 'Pacific', or 'Oceanic' Mongoloids.

In 1968, K. Hanihara defined a 'Mongoloid dental complex' based on his observations of four crown traits in Japanese, American Indian, and Eskimo populations. Members of this complex were characterized by a high frequency of shovel-shaped incisors, cusp 6, the protostylid, and the deflecting wrinkle. Research in the last decade suggests there are two major dental divisions within this Mongoloid complex that are definable on geographic grounds. This division is supported in many of its details by ancillary data from linguistics, genetics, and skeletal biology. Moreover, the recognition of these two divisions has clarified a number of issues in terms of the timing and direction of migrations from within and outside of Asia during the later stages of the Pleistocene and much of the Holocene.

The fundamental dental division in Asia was recognized by Turner (1983b, 1987, 1989, 1990a) following his observations on thousands of prehistoric and recent skeletal remains and living populations, from North, East, and Southeast Asia, the Pacific islands, and North and South America. With descriptive and heuristic aims in mind, he coined the terms Sinodont and Sundadont to denote the two subdivisions of the Mongoloid complex. Employing a standard battery of 29 crown and root traits, Turner (1990a) found eight traits dichotomized these divisions. Sinodonts have significantly higher frequencies of incisor shoveling and double shoveling, 1-rooted upper first premolars, upper first molar enamel extensions, missing-pegged-reduced upper third molars, lower first molar deflecting wrinkles, and 3-rooted lower first molars, as well as a lower frequency of 4-cusped lower second molars. In general, the Sinodont pattern is characterized by intensified and specialized traits while Sundadonty exhibits a more conservative pattern, typified by trait retention rather than elaboration. From the descriptions in chapter 5, the reader will note that most groups of the Sino-American subdivision show the Sinodont pattern

(Jomon and South Siberia are the exceptions) while groups in the Sunda-Pacific division are all Sundadonts.

As the neologisms imply, the Sinodont pattern is centered in North and East Asia. Today, the major populations of China, Mongolia, Japan, Korea, Northeast Asia, and North and South America are Sinodonts. Living Sundadonts reside primarily in mainland and insular Southeast Asia, on what was once the Sunda Shelf, but also extend into Micronesia and Polynesia. The time depth for this Asian dental division is thought to go back perhaps 20,000 to 30,000 years as fossils from Upper Cave Zhoukoudien show the Sinodont pattern (Turner, 1985b) while Sundadonty is evident in the 17,000 year old Minatogawa skeletons and perhaps much earlier fossils from China and Southeast Asia (Turner, 1992b). As the Sundadont pattern is the more generalized of the two, it is thought that Sinodonty developed in Northeast Asia during late Pleistocene times out of an earlier Sundadont population base (Turner, 1989, 1990a). Australians, New Guineans, and some Melanesians are excluded from either dental designation (more on that later) although they show more dental similarities to Sundadonts than to Sinodonts.

The existence of a dual Asian dental division has been corroborated by Japanese dental anthropologists studying dental casts and skeletons from a variety of modern and prehistoric Asian and Pacific populations (cf. T. Hanihara, 1990a, 1990b, 1991a, 1992a; Takei, 1990; Manabe *et al.*, 1991, 1992). Analysis of craniometric variation among Asian and Pacific populations shows the same division (Pietrusewsky *et al.*, 1992; T. Hanihara, 1994). Thus, the recognition of Sinodonty and Sundadonty has been confirmed repeatedly by independent researchers. The question is what this concept tells us about past population history? In that regard, we take up four specific issues, from more recent to very ancient times, regarding (1) the later peopling of the Japanese archipelago, (2) the origins of Polynesians and Micronesians, (3) the settlement of the Americas, and (4) the dental status of Australians and other Sahul-Pacific groups.

The later settlement of Japan

During much of the Pleistocene, islands of the Japanese archipelago were connected by a land bridge to the Asian mainland. Until the later stages of this epoch, around 12,000 to 18,000 years ago, human populations could have reached Japan without resorting to watercraft. It remains unclear as to when humans first inhabited Japan, but unequivocal stone tools have been dated to around 30,000 BP. Around 12,000–15,000 BP, the Japanese archaeological record shows a shift toward the production of microlithic tools (e.g., microblades), not unlike those found earlier in nearby Korea. It was in this terminal Pleistocene context that some of the world's oldest pottery was discovered, for example at Fukui Cave on Kyushu, ushering in the Jomon tradition (Koyama, 1992).

According to Japanese archaeologists, Jomon was a distinctive, albeit conservative, hunting, gathering, and fishing tradition that spanned some ten millennia from 12,500 to 2200 BP. Numerous Jomon sites have been excavated and carbon-14 dated, allowing archaeologists to divide this tradition into various stylistic periods. The end of the Jomon tradition starts in southern Japan around 2200 BP. This date marks the beginning of the Yayoi culture that involved an entirely new economic complex centered on wet-rice agriculture.

Archaeologists are in agreement that there were major cultural influences from mainland Asia during the final stages of the Jomon period that became even more striking at the beginning of the Yayoi period. There is less agreement on the biological make-up of the populations responsible for the Jomon and Yayoi cultural traditions. One view holds that the biological contribution from the mainland was minimal during the transition from Jomon to Yayoi, so changes observed in the archaeological record reflect culture change through diffusion rather than significant population replacement. This local evolution scenario has been challenged by other authors who feel such revolutionary culture change was accompanied by a significant infusion of people who first entered southwestern Japan from China and/or Korea, spreading rapidly north into Honshu and eventually Hokkaido.

The anthropological controversy surrounding continental influences during Jomon and Yayoi times must take into account the recognition of two distinct living populations of the Japanese archipelago, the Japanese proper and the Ainu. For many centuries, the Japanese population has dominated the southern Japanese islands while the Ainu have resided primarily in the northern islands of Hokkaido and Sakhalin. Only three decades ago, Coon (1962:150) wrote 'No single population in the world has had more written about its origins, which are unknown, and its racial classification, which is undetermined, than the Ainu.' The question is, how are these two living populations of Japan related to the bearers of the Jomon and Yayoi traditions?

Levin (1963) reviews three primary theories on how the Neolithic Jomon are related to the extant populations of Japan: (1) there is no ancestral-descendant relationship between the Jomon and either the Ainu or Japanese; (2) the Jomon are directly ancestral to the Ainu; and (3) there is continuity between the Jomon population and modern Japanese. Dental

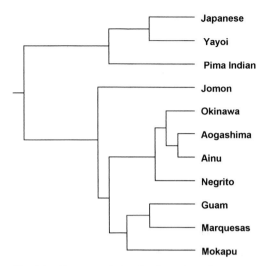

Fig. 7.1   Cluster analysis of prehistoric and recent populations of Japan and the Pacific based on tooth crown traits. Redrawn after T. Hanihara (1990a). Used with permission from the *Journal of the Anthropological Society of Nippon*.

morphological evidence supports the second theory. Preceding his distinction of Sinodont and Sundadont dentitions in Asia, Turner (1976) compared 3000-year-old Chinese and Jomonese skeletal series with recent and living Ainu and Japanese samples. For nine dental traits, he found several significant differences when either China or recent Japan were compared to the Jomon and Ainu, but there were no differences between China and Japan or between Jomon and Ainu. Turner (1976:911) says 'the obvious conclusion to draw from this dichotomy is that the Ainu are descended from the prehistoric Jomon people, and the recent Japanese are probably not. Instead, the Japanese may be descended from the post-Jomon Yayoi agriculturalists.' This observation has been refined in subsequent studies by Turner (1990a, 1992a) and supported by the cranial and dental studies of K. Hanihara (1984, 1991, 1992) and T. Hanihara (1990a,b, 1991a,b, 1992a).

Using a suite of seven crown traits, Hanihara (1990a) found that the Yayoi clustered with modern Japanese and Pima Indians (i.e., Sinodonts). His Jomon sample is distinctly removed from these three groups and clusters with the Ainu, Okinawans, Aogashima Islanders, Oceanic Negritos, Polynesians, and Micronesians (Fig. 7.1). Matsumura (1990) made dental comparisons between Kofun period (AD 300–700) burials in Japan and Jomon, Ainu, and recent Japanese samples and also found a linkage between Ainu and Jomon and between Kofun period remains and the recent Japanese population. Additional support for an ancestral relationship between Jomon and Ainu, and the distinction of both from modern Japanese, comes from studies of anthropometry (Cheboksarov, 1966), craniometry (Brace and Tracer, 1992; T. Hanihara, 1994), nonmetric cranial traits (Dodo *et al.*, 1992; Ossenberg, 1992), and odontometry (Brace and Nagai, 1982). Some geneticists take issue with the 'dual structure' model of Japanese prehistory (e.g. Nei, 1992), although a reliance on synchronic genetic markers in modern populations is suspect given the high level of admixture between Japanese and Ainu populations (Omoto, 1992).

Dental evidence links Jomon to the living Ainu and Yayoi and Kofun period skeletons to the recent population of Japan. Still, questions remain on the origins of the Jomon and the broader biological affinities of the Jomon-Ainu lineage. At various times, the Ainu were thought to be 'archaic Caucasoids,' Australoids, or some subtype of Mongoloid. Coon (1962) favored the view that Ainu were 'archaic Caucasoids,' citing such Ainu-Caucasoid parallels as a high frequency of loops on the fingerprints, a high frequency of sticky-wet ear wax, and a reduced incidence of shovel-shaped incisors. For these traits, the Ainu and Europeans differ markedly from modern Japanese populations. While it is possible to select traits that seemingly link the Ainu with an older hypothetical stratum of Caucasoids in East Asia, many early workers were not swayed by such parallels. As early as 1872, Vivien de Saint-Martin proposed the Ainu were representatives of a race of people who inhabited island Southeast Asia, citing groups such as the Bataks of Sumatra and the Dayaks of Borneo as their closest living relatives. Early in the twentieth century, the Russian anthropologist Shternberg assessed the ethnogenesis of the Ainu using ethnographic, linguistic, archaeological, and physical anthropological data and concluded this group reached Japan from the south as their closest resemblances were to the populations of insular and mainland Southeast Asia (Levin, 1963).

While Coon (1962) used one dental trait, the reduced frequency of incisor shoveling, to bolster his position that the Ainu were related to Europeans, the dental characteristics of both the Jomon and unadmixed Ainu show the time has come to abandon the notion of 'archaic Caucasoids' in the Far East. A less romantic and more empirically-based view places the ultimate ancestors of the Jomon somewhere in Southeast Asia in late Pleistocene times. They could have reached the Japanese archipelago by moving north along the exposed East Asian continental shelf or by island hopping from the Philippines through the Nansei Island chain to southern Japan. Although workers over the past century have concentrated mostly on the 'Ainu question,' many have overlooked other populations in islands south of Japan (e.g., Aogashima Islands, Nansei Islands). These populations also exhibit Sundadonty, linking them with the Jomon and Ainu rather than with recent Japanese (T. Hanihara, 1991b). These groups also show close dental similarities to Polynesian and Micronesian populations that have recently peopled some of the smallest and most remote islands of the Pacific Ocean.

Polynesians and Micronesians

When European voyagers reached the vastly scattered Pacific islands in the 1700s, observers were impressed by the superficial similarities of Polynesians to Europeans. At the time Heyerdahl (1952) made his famous Kon Tiki voyage, he had prepared a lengthy treatise that attempted to show American Indians made a substantial cultural and biological contribution to the populations of the Pacific, with a special focus on Easter and the Hawaiian islands. Such ideas about the colonization of the Pacific are no longer tenable in light of research in genetics, skeletal biology, archaeology and linguistics, all of which have established beyond any reasonable doubt, the ultimate origins of Polynesians and, in many details, the routes and probable timing of dispersal events (Hill and Serjeantson, 1989).

Polynesians are members of the far-flung Austronesian language family, one of two divisions of the Austro-Tai phylum centered in mainland and insular Southeast Asia. Linguistic research suggests the earliest split of Austronesian divided Formosan (the languages spoken by Taiwan aborigines) and Malayo-Polynesian. The western branch of Malayo-Polynesian is spoken by groups scattered throughout the major islands of the East Indian archipelago and parts of mainland Southeast Asia as well as the Malagasay of Madagascar. Eastern Malayo-Polynesian includes finer subdivisions of the Oceanic and Polynesian branches, with speakers extending from the coastal regions of New Guinea and Melanesia to the whole of Micronesia and Polynesia (Bellwood, 1989).

The expansion of Austronesian populations from insular Southeast Asia began about 5000 years ago. These groups carried with them their language, agricultural traditions, and boating skills, settling along the already inhabited coasts of Papua New Guinea, the Admiralty Islands, and parts of the Bismarck archipelago. Between 3500 and 3000 years ago, the Lapita culture, an archaeological tradition defined by a distinctive pottery style, appears without a developmental sequence in western Melanesia. Following this appearance, seafaring populations bearing the Lapita culture settled almost all the Pacific islands east of the Solomons

(Bellwood, 1989). While all lines of evidence suggest this was the point of origins for Polynesian populations, Micronesians appear to have been derived from a different source. The small coral atolls of Micronesia have no Lapita pottery, suggesting the Polynesian expansion into the eastern Pacific did not involve a route through Micronesia. On linguistic grounds, some Micronesian groups show ties to Indonesia and the Philippines while others are related to Oceanic groups in Melanesia. Skeletal and genetic reconstructions support this dual origin model with some Micronesian groups allied with Melanesians and others with Polynesians (Pietrusewsky, 1990; Cavalli-Sforza *et al.*, 1994). Turner (1990b) found that Micronesians, including prehistoric Guamanians, were dentally most similar to Borneo peoples, followed closely by samples from Fiji-Rotuma, Marquesas, Hawaii, and Java-Sumatra. In summary, dental morphology supports the hypothesis of a Southeast Asian origin for Polynesians and Micronesians, since all three groups possess the Sundadont dental pattern.

Within Melanesia, which includes a heterogeneous mix of Austronesian and non-Austronesian speakers, it is generally possible on dental grounds to sort out the recent Holocene migrants from the earlier Indo-Pacific inhabitants. Working with crown casts, Harris (1977) found that living non-Austronesian (NAN) and Austronesian (AN) populations in the Solomon Islands exhibited both trait gradients and boundaries between northern (NAN) and southern (AN) groups. Like mainland Southeast Asians, AN groups show more winging, incisor shoveling, Carabelli's trait, and cusp 6 and considerably less mandibular molar cusp reduction than NAN groups. In his cluster analysis, four NAN groups clustered together along with two AN groups, suggesting either or both admixture and sampling error. The southern AN groups, especially those from the Malaita islands, show a clear separation from the northern AN groups. A similar result was obtained by Doran (1977) who studied four living Papua New Guinea groups. He found the Pari, a coastal population in the vicinity of Port Moresby, had a distinctly higher frequency of shovel-shaped incisors than two highland samples from Lufa and Goroka and another coastal group, the Wewak. Barksdale's (1972) data on shoveling among East New Guinea highland groups correspond closely to the data of Doran. Although northern New Guinea and some of the larger islands of Melanesia have both Austronesian and non-Austronesian speaking populations, dental morphology distinguishes clearly the early NAN inhabitants and later AN migrants.

Teeth have been brought to bear on several other specific questions of Pacific basin prehistory, for example, the origins of Easter Islanders. Using archaeology, folklore, ethnobotany, and the spirit of adventure, Thor

Heyerdahl (1952) attempted to show that Easter Island, the most remote Polynesian island in the Pacific, had been strongly influenced and partly peopled from western South America. Although archaeologists acknowledge that architectural and botanical clues suggest some contact between Easter Islanders and South American Indians, the dental relationships of Easter Islanders are clearly with other eastern Polynesians rather than with American Indians or East Asians (Turner and Scott, 1977). This finding has been recently confirmed by other biological evidence (Cavalli-Sforza *et al.*, 1994).

In the 1970s, little was known about Southeast Asian dental variation and the recognition of the Sinodont and Sundadont dental patterns had not taken place. While living Easter Islanders are clearly allied with other Polynesians, observing the teeth of prehistoric Easter Islanders allows workers to evaluate the question of external relationships for earlier time horizons. For example, if American Indians contributed directly to the population of Easter Island, a Sinodont dental pattern should characterize at least some portion of the earlier skeletal remains, as is suggested by cranial data (Turner, 1968). In a recent study of pre-contact Easter Island dental remains, Swindler *et al.* (1995) found molar crown trait frequencies indicative of Polynesian/Asian affiliations. Still, to sort out any potential genetic contribution from the Americas to the early inhabitants of Easter Island, it would be even more useful to assess traits of the anterior teeth (e.g., shoveling, winging) and roots (e.g., 3RM1) as these show the sharpest contrasts between Sinodont and Sundadont populations.

Peopling of the Americas

The origin of Native Americans has been a subject of intense speculation and analysis ever since the European discovery of the New World in 1492 (the Norse in Greenland aside). Before systematic data were obtained on physical characteristics, language, and archaeology, natural historians of the late eighteenth and early nineteenth centuries inferred that Native Americans originally entered the New World from Asia across the Bering Strait, primarily on the basis of geographic proximity and the many physical resemblances between Asians and Native Americans. As with all questions of origins, the Asian roots of American Indians and Eskimos have been challenged by theories ranging from 'not likely but possible' to the bizarre. Attempts to derive some Indian tribes from Phoenicians, Welshmen, the lost tribes of Israel, sunken continents (e.g., Atlantis, Mu), and, somewhat more plausibly, Southeast Asians, have generated a fascinating if fictive literature. Some Native American groups deny any origin other than in the New World (Echo-Hawk, 1994). They are, in fact, strongly opposed to the Bering Land Bridge migration concept. Still, there are too many threads of evidence linking Native Americans to North and East Asians. Dental morphology provides one of the strongest threads in a broader tapestry of physical evidence.

The best known dental trait that indicates a close relationship between Native Americans and North Asians is shovel-shaped incisors. This was noted by Hrdlička (1911, 1920) and verified by dozens of later workers. High frequencies of incisor winging, enamel extensions on the molars, and single rooted upper first premolars were additional traits shared in common by North Asians and American Indians. However, early attempts to show the dental relationship between Asia and the Americas ordinarily stopped at the point of noting that Native Americans were derived from members of the Mongoloid race. Studies of crown and root morphology in Asian and American living and skeletal populations over the past 25 years allow us to add more details to this generalization.

In 1971, Turner surveyed the frequency of 3-rooted lower first molars in a large sample of North American natives. This study revealed a distinct dichotomy between Eskimo-Aleuts of the American Arctic, who had trait frequencies of 25–40%, whereas most North American Indian groups had frequencies around 6%. Although Athapaskan skeletal remains were not examined, Turner reported that X-ray studies on one Athapaskan group, the Navajo, revealed a 3RM1 frequency of 27%. From this, Turner (1971:239) concluded 'Three migrations into the New World seem to best explain 3RM1 variation in this hemisphere. Pre-Indians, pre-Na-Dene Indians, and pre-Aleut-Eskimos are the three suggested ancestral groups. Importantly, these coincide with major New World linguistic divisions recognized by Sapir and later Greenberg.' That is, the widely scattered tribes of American Indians with a low frequency of 3RM1 were thought to represent the descendants of the original Paleoindian colonists in the New World. The second wave from Asia, with intermediate 3RM1 frequencies, involved the ancestors of Na-Dene speakers who inhabited the western half of the North American subarctic, including much of the Northwest Coast. Migrants of the third and final wave to the New World, with the highest frequencies of 3RM1, were the ultimate ancestors of all Eskimo-Aleut populations distributed from western Alaska to eastern Greenland. The sequence and timing of the second two groups is less well understood than the initial colonization of Amerinds who are thought to be associated with the rapid spread of the Clovis tradition in the Americas.

Focusing on 29 key crown and root traits in dozens of North and South American Indian, Northwest Coast Indian, and Eskimo-Aleut skeletal samples, Turner (1983a, 1984, 1985a, 1985b, 1986a) was unable to discover any patterned dental variation that contradicted his initial three-wave model for the settlement of the Americas. His phenetic analyses consistently showed all Native Americans were more closely related to Northeast Asians than to Southeast Asians. Additional findings included: (1) a greater similarity between Northeast Asians and Eskimo-Aleuts/Greater Northwest Indians than with North and South American Indians; (2) relatively minor levels of diversity between North and South American Indians, supporting the notion of their relatively recent common origins; (3) a significant difference between a small 'Paleoindian' sample and all Native Americans, suggesting considerable microevolution following entry into the New World; and (4) a consistently close relationship between Eskimos and Aleuts. One major unresolved question is the relative relationships among the three major founding groups. In all instances, there is a dental (and genetic) gulf separating American Indians from Eskimo-Aleuts, but the phylogenetic placement of the Greater Northwest Coast group, including Na-Dene populations, remains equivocal. In one analysis, Turner (1985a) produced a dendrogram linking Greater Northwest populations and all other American Indians while Eskimos and Aleuts were outliers. Another tree diagram (Turner, 1986a), however, linked Greater Northwest groups with Eskimo-Aleuts rather than American Indians. In a reanalysis of Turner's data, Powell (1993) encountered the same difficulty in discerning the placement of the intermediate Greater Northwest populations. In a dendrogram based on the unweighted pair-groups method, Northwest Coast clustered with Eskimos, Aleuts, and Northeast Siberians. By contrast, a maximum parsimony consensus tree showed the Northwest Coast grouping clustered most closely with North and South American Indians (Fig. 7.2).

Genetic studies are in even less agreement about numbers of migrations, dating, sequencing, sourcing, and phylogenetic relationships between Eskimo-Aleuts, Northwest Coast, and North and South American Indian populations. All such studies are, of course, synchronic in nature, depending on phylogenetic verification or rejection by diachronic means, which can only come from archaeological and physical anthropological sources. Thus, at one extreme, Merriwether *et al.* (1995) propose a single migration based on their study of mtDNA. Williams *et al.* (1985) found that Gm haplotypes supported the three wave model of Native American origins. Schanfield's (1992) genetic analysis favors four migrations. In many publications, Szathmary (1984, 1985, 1993; Szathmary and Ossenberg, 1978) opts for a pre-Clovis Indian migration and a later colonization by ancestral Eskimo-Aleut/Na-Dene who were thought to have diverged in

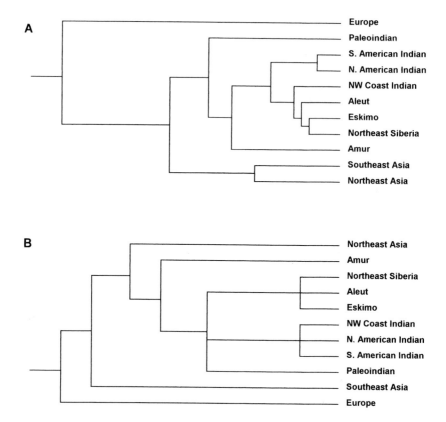

Fig. 7.2   Powell's reanalysis of Turner's dental data on New World and Asian populations. (A) Unweighted pair-group method using arithmetic averages (UPGMA tree). (B) Maximum parsimony tree. Redrawn after Powell (1993). Used with permission from Wayne State University Press.

Beringia. Other geneticists similarly lean towards a pre-Clovis entry for the first Americans, even though there is no archaeological evidence in northeastern Siberia or Alaska for such a parent population (Turner, 1985c). As for Na-Dene Indians, Torroni *et al.* (1992) find them distinctive from other Indians, and Shields *et al.* (1993) see them as closely similar to Eskimos, viewpoints not easily supportable by linguistic, archaeological, or dental data. In summary, there is not much agreement among geneticists on any aspect of the peopling of the Americas.

Although geneticists have reached disparate conclusions on the peopling of the Americas using mtDNA, immunoglobulin allotypes, serum proteins, and blood group antigens, there is some broad consilience between dental and genetic data on this issue. Cavalli-Sforza *et al.* (1994:340), for example, conclude that 'the genetic patterns in the Americas fully confirm the three waves of migrations suggested by dental and linguistic evidence: Amerinds, Na-Dene, and Eskimo.' As for the linkage of Na-Dene, in their chapter on the Americas, Cavalli-Sforza *et al.* (1994) provide a dendrogram that suggests Na-Dene populations are more closely related to Eskimos than to other American Indians, a position long advocated by Szathmary (1984, 1985) and more recently by Shields *et al.* (1993). However, in their global analysis, Northwest Amerinds (= Na-Dene) are linked with other American Indians while Eskimos cluster with Chukchi and North Turkic groups (Cavalli-Sforza *et al.*, 1988, 1994). This discrepancy, also noted by Szathmary (1993), simply highlights again the equivocal relationship of Na-Dene groups of Northwest North America. Resolving this impasse, reflected in both teeth and gene frequencies, remains an interesting problem for archaeologists, physical anthropologists, geneticists, and linguists alike.

Australian aboriginals and proto-Sundadonty

Prehistorians are in general agreement that Sahulland (Australia, New Guinea, Tasmania) was peopled as early as 40,000 to 60,000 years ago (Green, 1994). There is less agreement on whether the modern inhabitants of these areas are all descended from a single basic stock or if their biological make-up is a product of two or even three fundamentally different founding populations. The archaeological record provides few clues to this mystery. Based on the antiquity of sites in Australia, it has been inferred that the original colonists arrived in some form of watercraft from Southeast Asia before the advent of the European Upper Paleolithic (Macintosh and Larnach, 1976; Birdsell, 1977). Following initial settlement, the Australian archaeological record is remarkably conservative, with no significant additions or changes that might bespeak subsequent colonizations. The earliest sites in New Guinea and some of the larger Melanesian islands fall in the 30,000–40,000 year range, but it may be premature to conclude that Australia was settled significantly in advance of the northern reaches of Sahulland, especially until technical issues of dating have been resolved. The Holocene advance of Austronesian populations into the Pacific, beginning about 5000 BP, is evident in northern coastal New Guinea and Melanesia, but this major colonization event did not touch Australia.

Linguists have not found any links between the languages spoken in Australia and those of any other region, including New Guinea. The diverse languages of Australia are a world unto themselves (Ruhlen, 1987).

New Guinea languages have long been known for their diversity and complexity, but megalinguistic comparisons include the non-Austronesian speakers of New Guinea in an older Indo-Pacific phylum that stretched from Timor (and possibly the Andaman Islands) in the west to the Solomon Islands in the east. In other words, New Guinea, unlike Australia, is not linguistically isolated from other island populations of the South Pacific.

Genetically, Australian aboriginals and the non-Austronesian populations of New Guinea are more closely allied to one another than to any other group (Nei and Roychoudhury, 1982, 1993; Cavalli-Sforza *et al.*, 1994), but this relationship is still distant and presumably deeply rooted in time. The land bridge connecting Australia and New Guinea was cut off by rising sea levels in the late Pleistocene and all evidence indicates the populations of these areas were mostly, if not entirely, isolated from one another during the subsequent 12,000 years. It is likely that 12 millennia is not a sufficient period of time to allow for the genetic and phenotypic differentiation that now separates these populations. Other isolating mechanisms were surely involved.

Birdsell (1951, 1977) has long advocated a trihybrid theory of Australian aboriginal origins. In reconstructing the late Pleistocene populations of Asia, he posited the presence of three distinct racial elements: Carpentarians in India and Burma, Oceanic Negritos in insular and mainland Southeast Asia (as far north as southern China) and 'Caucasoids', or Murrayians, in China, Mongolia, and Japan. Birdsell (1951) accounted for the variation within and between Australian and New Guinea populations by positing differential contributions from these three purported racial groups. For Australia, he felt the Negrito element introduced by the initial migrants to Sahulland represents only a minor component of recent populations. He characterized populations in south Australia as mostly Murrayian in origin while those of the north were considered primarily of Carpentarian descent. By contrast, for New Guinea and island Melanesia, living populations were thought to be descended primarily from the earliest Negritoid migrants to Sahulland, with some admixture from later Carpentarians. He did not, however, believe that these groups migrated from Southeast Asia in their present form. He stresses that 'These groups were formed *in situ* in New Guinea and Melanesia as a result of hybridization of late Pleistocene racial elements' (Birdsell, 1951:6).

Birdsell's scenario for deriving Australian and New Guinea populations from mixtures of either two or three primary racial elements during the late Pleistocene has been dismissed by the majority of recent anthropologists and all Australian archaeologists. A popular view is that Australian biological variation can be accounted for by one founding population and subsequent internal differentiation over a period of 40,000+ years (Habgood, 1985). The question remains, however, as to just 'who' differentiated once Australia was settled. Thorne and Wolpoff (1981) contend the ancestral population of modern 'Australoids' evolved *in situ* on Sundaland from a middle Pleistocene *Homo erectus* foundation. These proto-Australoids ultimately developed watercraft that allowed them to reach Australia some 50,000 years ago. Unfortunately, the fossil record in Southeast Asia is too spotty and poorly dated to provide a definitive answer on the issue of long-term local evolution. We do know, however, that the population base that gave rise to Australoids was quite distinct from that of Europe and North Asia. Another issue to resolve is how Oceanic Negritos fit into the schema of Australoid evolution. They are often thought to represent an ancient and once widely distributed substratum throughout mainland and insular Southeast Asia (Howells, 1976; Bellwood, 1989). Did both Negritos and Australoids develop from an early proto-Australoid base on Sundaland?

The dental evidence provides tantalizing clues on the history of Pleistocene populations in Southeast Asia and Australia. Sunda-Pacific and Sahul-Pacific groups show both similarities and differences in crown and root traits. Although the neologism of 'Sahullodont' has never been applied to Australo-Melanesian dentitions, Townsend *et al.* (1990) and T. Hanihara (1992b) feel the Australian dentition is sufficiently distinctive to warrant the designation of an Australian dental pattern, parallel to Sinodonty and Sundadonty. In 1990, Turner proposed, at a major international conference on human origins in Tokyo, that because Sundadonty and the Australian dental pattern were relatively similar, the term proto-Sundadonty or early Australasian be applied to this common ancestral stock. He extended the proto-Sundadont theme in a subsequent Circum-Pacific analysis (Turner, 1992b), which is diagrammed in Fig. 7.3. This suggestion has been followed upon by T. Hanihara (1992b) who developed a model to propose how the three primary Asian dental patterns might have developed from an early common Southeast Asian ancestor. In this model, stimulated by genetic (Omoto, 1984) and dental (Turner, 1992b) research, peoples with a proto-Sundadont dental pattern inhabited Sundaland around 90,000–100,000 BP. From this founding population, Australia was colonized early in the Upper Pleistocene. The Australian dental pattern, with a low frequency of incisor winging and shoveling, a high frequency of cusp 5 (UM1) and cusp 6 (LM1), and retention of the hypocone and hypoconulid on the second molars, developed in isolation from its ancestral population in Sundaland. The Sundadont dental pattern,

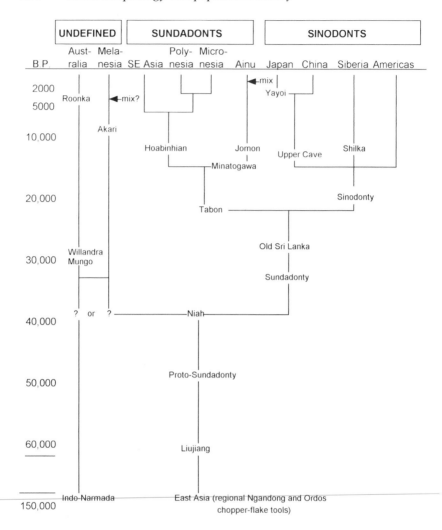

Fig. 7.3   Turner's working model of Circum-Pacific phylogenetic relationships.
(Redrawn after Turner, 1992b. (Used with permission from *Perspectives in
Human Biology 2/Archaeology in Oceania*.)

with a higher frequency of derived traits, is thought to have evolved *in situ*
on Sundaland during the course of the Upper Pleistocene. Late in the
Pleistocene, Sundadont populations spread north from Southeast Asia and
eventually reached the Japanese archipelago. Although the timing is
unclear, Sinodonts, with the most highly derived dentitions in the world,
developed either out of proto-Sundadonts in the early stages of the Upper
Pleistocene or later, from a Sundadont population base. In any case, the

dating of the Upper Cave human remains that possess Sinodonty at 30,000 BP push the probable time of the first modern humans in eastern Asia back far beyond what was imaginable in the 1970s and 1980s.

Dentally, Negritos show closer affinity to other Southeast Asian populations than to Australians and North Asians. The proto-Sundadont model (Turner, 1992b; Hanihara, 1992b), suggesting a common thread underlying all Asian biological variation, is supported in its fundamental details by the genetic analysis of Nei and Roychoudhury (1993). By contrast, Cavalli-Sforza *et al.* (1994) support a genetic linkage between Southeast Asian and Australo-Melanesian populations, but tie North Asians more closely to Caucasoids. Such a relationship is contraindicated by the dental evidence.

Global analysis

During the first half of the twentieth century, dental morphologists emphasized differences in crown and root trait frequencies among the major human races, more often out of necessity than choice. There was not enough information to discuss the differences between groups within a continent, and certainly not within regions or within populations. In Hellman's (1928) classic paper on lower molar crown trait variation, he compared American whites, modern Europeans, ancient Europeans, West Africans, American blacks, Buriats, Chinese, Eskimos, American Indians, and Australians. This illustrates the level of comparison in most early studies.

Although Hellman's study predates by three decades the development of distance statistics for qualitative traits, he would not be surprised by our analysis of his data for four lower molar crown traits observed on six teeth (4-cusped LM1, LM2; Y groove pattern LM1, LM2; cusp 6 LM1; cusp 7 LM1). The tree diagram produced on the basis of his small data set (Fig. 7.4) is largely in accord with modern expectations. His ten geographic samples fall within three major clusters: Europeans, Africans, and Asians. The only anomaly is the branch linking Australians and Eskimos, which given the absence of other Sahul-Pacific or even Sunda-Pacific groups in the matrix, causes no concern.

Hellman personally observed samples from all these groups, but it was more common for early workers to compare their morphological observations on one specific group with data on samples selected from the literature. If someone observed shoveling in a sample, it was standard practice to make comparisons with the data assembled by Hrdlička (1920)

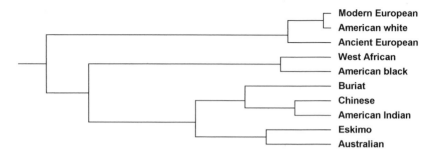

Fig. 7.4   UPGMA tree based on analysis of Hellman's (1928) data on lower molar crown traits.

on American whites, American blacks, Hawaiians, Japanese, Chinese, American Indians, Eskimos, Mongolians, and Melanesians. For other variables, workers commonly enlisted data from the limited sources available on Australians, Bantus, and a few other groups to put their sample into a broader perspective. In his monograph on Greenlandic Eskimo teeth, Pedersen (1949:214) could only conclude that '*There is no definite trend in the dentition of the East Greenland Eskimo to lend decisive support to the view that he be either very primitive or very far advanced. What is definitely borne out, however, is the substantiation of his Mongoloid ancestry*' (author emphasis). This was the level of generalization possible in the middle of the twentieth century. In an article appearing shortly after Pedersen's monograph, Moorrees (1951:820) remarked 'We are always dealing with subraces and composite subraces, whereas the primary races are abstractions or generalizations and a pure Mongoloid does not exist. Therefore, the determination of these variations in racial subgroups by future investigators would be of great value. That will ultimately determine the value of the dentition in racial studies.' Basically, Moorrees was encouraging workers to look below such broad classificatory levels as Mongoloid, Caucasoid, and American Indian, and examine the dental variability of populations subsumed under these headings. Some 40 years later, we are in a position to follow this advice.

Global analysis of dental morphological variation

Few researchers have attempted a worldwide analysis of dental trait variation. A modest effort toward this end was undertaken by Sofaer *et al.* (1972a) who estimated biological distances on the basis of crown trait variation among eight high-order groups: (1) Zuni-Pima-Papago; (2)

American Indian; (3) Aleut and Eskimo; (4) Asia; (5) Pacific and Australia; (6) Caucasian; (7) Negro; and (8) Semitic. This analysis was based on data from 30 sources and focused on five traits: UI1 shoveling, UM1 Carabelli's, UM2 hypocone, LM1 groove pattern, and LM2 cusp number. In their graphical portrayal of relative relationships, Zuni-Pima-Papago, Eskimo-Aleuts, Asians, and American Indians are at one pole and Caucasian and Semitic groups are at the other. Pacific and Negro groupings, which are relatively close together, fall between the two major clusters. These findings, while reasonable, remain at the coarse level of ahistoric and nonevolutionary resolution characteristic of Hellman (1928) and other early nineteenth-century biological anthropologists. Moreover, they were without theoretical orientation, being linked neither to population history nor microevolutionary concerns.

Comparing the data employed in the global synthesis of Sofaer *et al.* (1972a) to that now available on tooth morphology indicates the rapid pace of research in this area during the past two decades. To illustrate, our summary tables in chapter 5 are based on more than 140 sources, many of which contain data on multiple populations (Appendix B). These citations do not include the many publications of C.G. Turner II on Native American, Asian, and Pacific groups nor the dissertation of J. Irish on African dentitions, because the data from these sources were used to derive the independent characterizations of 23 crown and root trait variants shown in Figs. 5.2–5.24.

For global analysis, we are in a much better position to assess crown and root trait variation than was possible 20 years ago. We make no pretense that the subsequent analysis is the final word on dental morphological variation as some geographic deficiencies in knowledge remain and more traits should be added to the analysis. However, as will be shown, it is a quantum jump beyond all previous efforts to assess worldwide dental variation.

To accomplish a global analysis, we calculated biological distances for three sets of dental data that involved, in order: (1) 23 crown and root traits in 21 high-order groups from the five principal subdivisions of humankind (from Figs. 5.2–5.24 and Appendix A); (2) 9 crown traits in 22 high-order groups and language families (from Tables 5.1–5.8); and (3) 9 crown traits in 43 high-order groups, obtained through a combination of data from (1) and (2). A few points regarding these analyses should be noted. Analysis 1 is based almost exclusively on skeletal groups while analysis 2 involves mostly data on living samples. The nine traits used in analysis 3 are dictated by available data from analysis 2. Analysis 3 includes some groups with the same name (e.g., Polynesia) but provides two interesting contrasts: between mostly prehistoric and mostly living populations and between the data of C.G. Turner II and the summarized observations of other workers. We did not include the distal trigonid crest in analyses (2) and (3) because several of the 22 groups have not been surveyed for this trait.

For the three analyses, we used a common distance metric, clustering algorithm, and ordination procedure. The distance statistic employed in all instances is that of Nei (1972), devised originally for the analysis of gene frequency data. Although this statistic has its detractors, it has been widely used in genetic distance studies and it yields high cophenetic correlations with other distance methods. Nei's distance is a measure of dissimilarity so groups with similar frequencies have small pairwise distance values while groups with divergent frequencies have the largest associated distances. Each trait was treated as a single 'locus' and sample size was standardized (conservatively) to UI1 shoveling samples for analysis 1 and to n = 100 for analysis 2. The common hierarchical clustering algorithm is the 'unweighted pair-group method using arithmetic averages,' abbreviated as UPGMA (Romesburg, 1990). For ordinations, we derived either two or three coordinates through multidimensional scaling. The authors are aware of the diversity of approaches and opinions associated with estimates of biological affinity. The data were analyzed through other distance statistics and subjected to a variety of clustering algorithms but, in the main details, all gave comparable results. As noted before, there is no SINGLE best method for affinity assessment. Readers are encouraged to use their favorite distance statistic and clustering or ordination procedure to analyze the data in Appendix A1 and the tables from chapter 5.

Analysis 1

The tree diagram shown in Fig. 7.5 is in general accord with our five geographic subdivisions. A primary split separates six Sino-American groups, or Sinodonts (Turner, 1983b, 1987, 1989), from other world populations. Within Sino-Americans, distance values (Appendix A3) indicate Northeast Siberians and all Native Americans are the most extreme outliers. China-Mongolia and Recent Japan cluster with Northeast Siberians and Native Americans, but their pairwise distances from European, African, and Pacific populations are not as extreme. The three Western Eurasian groups have small pairwise distance values with one another and cluster together at the top of the dendrogram. For Sub-Saharan Africans, West and South Africans cluster together but not with the San. The smallest pairwise distances for the San are, however, between West and South Africa. Minor inconsistencies involving the relationships

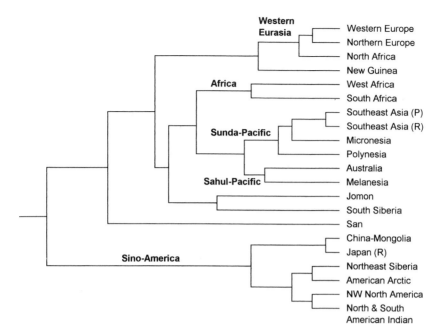

Fig. 7.5 Analysis 1: UPGMA tree based on distance analysis of 23 crown and root traits in 21 regional groups (samples – see Appendix 1; frequency data and sample sizes – see Appendix A2; Nei distance matrix – see Appendix A3).

of the three African groups may be partly attributed to the fact that Irish (1993) sometimes used slightly different breakpoints than those employed in the remainder of our data set. Interestingly, Sunda-Pacific and Sahul-Pacific groups fall primarily in one large cluster. There is some internal division with Australians and Melanesians in one cluster and Southeast Asians, Polynesians, and Micronesians in another, but the distinction is not great nor are the associated distance values large. Two groups that are not easily classified within a world framework are Jomon and South Siberia. Despite their geographic placement, they do not fall within the Sino-American province, a point also apparent in the characterizations of single traits (chapter 5). The smallest distance values for Jomon are with Sunda-Pacific groups. For South Siberians, the smallest distances are with Sunda-Pacific and Western Eurasian groups. In both South Siberian and Jomon groups, hybridization has probably occurred, producing novel populations.

The oddity of Fig. 7.5 is that New Guinea links with Western Eurasians rather than Australians, Melanesians, or Southeast Asians. The distance values between New Guinea and the three Western Eurasian samples are as small or smaller than the pairwise distance with Melanesia, so this is not an artifact of the clustering algorithm. Different distance metrics and clustering methods all gave the same result. Given this methodological agreement, it is reasonable to conclude that one or both of two other explanations underlies the New Guinea similarity with Western Eurasia. These are sampling error and evolutionary history. We tend to discount the former because of the strong similarities between our eastern lowland New Guinea crown trait frequencies and the morphological observations made by J.T. Barksdale (1972) on eastern highlanders. His study was based on dental casts collected by R.A. Littlewood in 1962–63. With the close parallels in the two independent data sets (e.g., very low UI1 shoveling and LM1 cusp 6 and very high 4-cusped LM1 and LM2), sampling error is not the entire explanation. Evolutionary history seems more likely to be the primary reason for the pattern shown by New Guinea crown trait frequencies. Given that the prehistory of Melanesia is sketchy at best, it is beyond our current ability to do much more than suggest that the eastern New Guinea dental pattern could represent genetic drift or an evolutionary random walk towards the Western Eurasian pattern, and away from the pattern first introduced in New Guinea more than 40,000 years ago. However, the modern New Guinea pattern could be the retention of what has been called proto-Sundadonty. This hypothetical dental pattern, and its implications for understanding the origin of modern humans, will be discussed at the end of the chapter.

The ordination presented in Fig. 7.6 adds one more dimension to the group relationships indicated by cluster analysis. This three-dimensional plot shows the close ties between the six Sino-American groups. For Africans, the San are again an outlier, but their placement is closest to South and West Africans. While European and North African samples fall closely together, New Guinea maintains its proximity to the Western Eurasian grouping. Most Sunda-Pacific and Sahul-Pacific groups fall in the middle of the diagram. Interestingly, Melanesia is closer to the Western Eurasian cluster than any other Southeast Asian or Pacific population. Australia is not close to any other specific group, but its smallest distance values are with Melanesia and West Africa. The intermediate status of Jomon and South Siberia is also evident in the ordination, with South Siberia pulled slightly toward Western Eurasia and Jomon toward Sino-America – conditions expected with admixture.

Analysis 2

A cluster analysis based on the summarized observations of other researchers is shown in Fig. 7.7. The 22 composite samples are broken

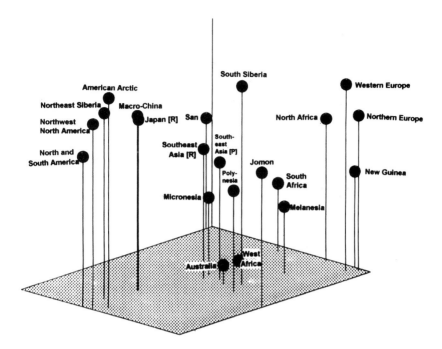

Fig. 7.6   Analysis 1: three-dimensional ordination based on multidimensional scaling of distance matrix for 23 crown and root traits in 21 regional groups.

down into five primary clusters: (1) Sub-Saharan Africans; (2) East Asians and Native Americans; (3) Australia and Pacific groups; (4) Central and North Asians, plus the Ainu and Southeast Asia; and (5) Western Eurasians, plus East Africans. A complete linkage analysis gave exactly the same five clusters although some high-order divisions differed in some respects (e.g., South Africa and Khoisan grouped with Australia and the Pacific rather than standing alone).

The combination of language, geography, and prehistory would predict the homogeneity of Western Eurasians and Sino-Americans, as well as the linkage of Australia and some Pacific populations. The placement of East Africans is perhaps the most surprising as they are more closely tied to Western Eurasians than Sub-Saharan Africans. We noted in chapter 5 that East Africans show a few striking dental differences from South Africans (e.g., in lower molar cusp number). Even so, East Africa shows its smallest pairwise distance value to South Africa. However, South Africa shows consistently higher distance values than East Africans when compared to other groups, especially Western Eurasians. As few populations from East Africa have been observed for crown morphology, these results are suggestive but tentative. The dental ties between Central Asians (Turkic-

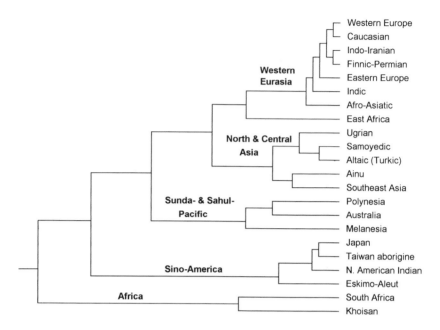

Fig. 7.7   Analysis 2: UPGMA tree based on distance analysis of 9 crown traits in 22 regional groups/language families.

speakers) and North Eurasians (Ugrians and Samoyeds) are very interesting. These groups exhibit neither a Western Eurasian nor Sino-American dental pattern but, instead, fall between these dental extremes. At a higher level, they cluster with the Ainu and Southeast Asians, two other groups characterized by intermediate rather than extreme crown trait frequencies. Hybridization must be the primary reason for this intermediacy.

A two-dimensional ordination of the 22 groups (Fig. 7.8) adds subtle details to the cluster analysis. Sino-Americans and Western Eurasians are the primary outliers in the context of world dental variation. Sub-Saharan Africans also appear distinctive although only two samples are available for comparison. On the one hand, Central Asians, North Eurasians, Southeast Asians, Polynesians, and the Ainu fall between Sino-Americans and Western Eurasians but are clearly closer to Sino-Americans. East Africans, Melanesians, and Australians, on the other hand, are closer to Western Eurasians. Of all the world's subdivisions, Western Eurasians are the most homogeneous—this may reflect in part their shared pattern of dental morphological simplification, augmented to some extent by the large number of samples used to characterize groups from this region.

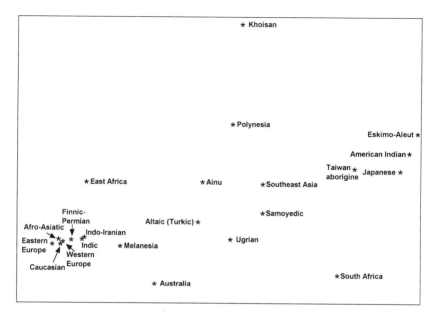

Fig. 7.8   Analysis 2: Two-dimensional ordination based on multidimensional scaling of distance matrix for 9 crown traits in 22 regional groups/language families.

Analysis 3

To consider jointly the entire array of samples from the first two analyses, it is unfortunate that we are restricted to the small trait set (9) of analysis 2 rather than the large trait set (23) of analysis 1 for this final global assessment. Despite this limitation, the combined analysis of 43 samples shows clear geographic patterning.

The dendrogram that contains the entire array of samples from the Figures and Tables of chapter 5 is presented in Fig. 7.9. To distinguish the origin of the samples, those in capital letters are from analysis 1 while those in small letters are from analysis 2. As noted, this analysis contrasts data from two independent sources and, to a large extent, data on skeletal versus living populations. A complete linkage analysis gave precisely the same primary clusters as the UPGMA shown, with only minor rearrangements within low-order clusters.

Cluster I. Western Eurasia

At the top of the dendrogram are 12 groups that cluster closely together, indicating little differentiation. Of these groups, 10 fall within the biological

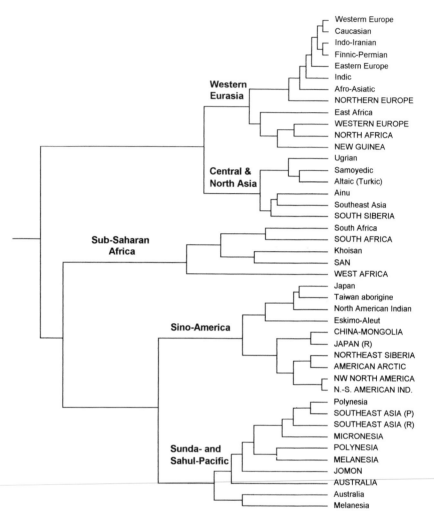

Fig. 7.9   Analysis 3: UPGMA tree based on distance analysis of 9 crown traits in 43 regional groups/language families. (Small letters indicate living populations.)

and linguistic community defined for Western Eurasia. Indo-European, Indo-Iranian, Indic, Caucasian, and Afro-Asiatic speakers are all found in this cluster along with prehistoric and protohistoric populations from Europe and North Africa. The two unexpected groups that link to the Western Eurasian cluster are East Africa and New Guinea. We have already discussed these anomalies in the context of analyses 1 and 2. This may reflect an accident of sampling (too few traits and samples), dental convergence through gene flow and/or selection, chance convergence through genetic drift, or a reflection of population history. Given the overall consistency of the cluster analysis, re-sampling East African and New Guinea populations should be a high priority for dental researchers.

Cluster 2. Central and North Asia

In some respects, this is the most interesting grouping of all – the peoples contained in this cluster have trait frequencies that fall betwixt and between Western Eurasian and Sino-American populations. Two of the groups, Ugrian and Samoyed, are classified in the two major branches of the Uralic language family, Samoyedic and Finno-Ugric (Ruhlen, 1987). Excluding Ugrians and Ob Ugrians, the Finno-Ugric division is dominated by peoples, such as Finns, Hungarians (also Ugric speakers), and the Saami (Lapps), who are biologically akin to Europeans. The Turkic-speakers, however, fall within the Altaic family, which includes Mongolian and Tungusic groups that are dentally East Asian Sinodonts. The geographically intermediate prehistoric sample from South Siberia also falls in this cluster, indicating they were neither European nor Asian but broadly Eurasian. Each of these groups – Ugrian, Samoyed, Altaic, and South Siberian – has some degree of admixture between Europeans and Sinodonts (Levin and Potapov, 1964). Their location on the dendrogram suggests more of the former than the latter. The inclusion of Southeast Asia in this cluster is believed to be due to the fact that Sundadonty was, or was similar to, the ancestral dental pattern for both Europeans and East Asian Sinodonts before each differentiated by drift in their respective and isolated ends of sub-Arctic Eurasia. Recombining the highly differentiated Europeans and Sinodonts returns the resulting hybrids to, or near to, the intermediate trait frequencies that characterize the generalized Sundadont dental pattern. The Ainu cluster with Southeast Asia because both are Sundadont populations. Numerous lines of evidence, including dental, indicate that the ancient progenitors of the Ainu came from Southeast Asia by human groups expanding northward more than 30,000 years ago along the shores of the now-submerged East Asian continental shelf.

Cluster 3 Sub-Saharan Africa

The five groupings that fall within this cluster all represent African populations. The two independently derived Bushmen samples (Khoisan and San) cluster together as do the two South African samples. West Africa appears to be the most divergent group within this cluster, but the distance values suggest the San groups are the most dissimilar from other African populations. This is due to the UPGMA algorithm because complete linkage analysis shows the two Bushmen groups separate at the highest level within the African domain.

Cluster 4. Sino-America

All ten groups within this large cluster come from East and Northeast Asia and the Americas. Interestingly, the primary cluster is divided into two subclusters that contain either groups observed by other workers (summarized data) or those in our characterizations of chapter 5 based on Turner's data. This distinction may reflect subtle inter-observer differences, but a more likely possibility is that there are indeed differences between the living populations (small letters – Fig. 7.9) and the earlier skeletal populations, caused by admixture in historic times. The overall integrity of this grouping is its most salient feature, so we need not disentangle minor distinctions. However, it is important to note that Northwest North America links with North and South American Indians rather than Eskimo-Aleuts, contrary to some of the genetic studies discussed earlier.

Cluster 5. Southeast Asia and the Pacific

The final major cluster includes the remaining ten populations. All of these groups fall within the Sunda-Pacific and Sahul-Pacific subdivisions, that is, Sundadonty and undefined dental Australmelanesians. Southeast Asians, Polynesians, and Micronesians cluster most closely together, with Australia and Melanesia grouping at a slightly higher level. The differentiation of these subdivisions is not as pronounced as that of Western Eurasians, Sub-Saharan Africans, and Sino-Americans who fall within their own major clusters. This suggests the distinction we made between Sunda-Pacific and Sahul-Pacific groups is not as sharp as originally imagined, providing some support for the model deriving Sundadonts and Australians from a common proto-Sundadont ancestor (Turner, 1992b; Hanihara, 1992b). The placement of Jomon in this grouping rather than with Sino-Americans also supports the idea that the Japanese archipelago was settled during earlier times by peoples from this larger Southeast Asia/Pacific grouping rather than from Sino-American populations to the west and north who later colonized Japan and gave rise to its extant population.

Until recently, there has been almost no objection to the long-standing view that 'southern Mongoloids' arose as the result of mixing due to Neolithic gene flow from China into an indigenous Australmelanesian type of people then inhabiting Southeast Asia. This view originated with J. Huxley's Australoid identification of early cranial finds from Gua Kepah, mainland Malaya, followed by Mijsberg's analysis of a single and non-representative mandible from the same site (Turner, 1987). In 1983, Sangvician proposed, on the basis of measurements of modern and archaeologically-derived crania, that modern Thai originated in Thailand, not China, as then generally accepted. Turner reached the same conclusion based on comparisons of recent and prehistoric Southeast Asian teeth. Both workers proposed that the proto-Mongoloid Thai phenotype arose through local evolution, rather than being the result of admixture between indigenous Australmelanesians and southward-migrating Mongoloid farmers. Archaeologists also began to question the orthodox view of north China as having been the only hearth for technological advancement in eastern Asia (Solheim, 1968), especially for metallurgy, ceramics, and agriculture. In the course of time, Turner (1990a) turned the argument around completely, hypothesizing instead that the specialized and intensified Northeast Asian Sinodont dental pattern had its origin in the more generalized and retained Sundadont dental pattern of Southeast Asia. However, Turner (1989) envisioned much more time was involved, minimally 20,000 years.

The major challenge to the local evolution hypothesis was made by Bellwood (1985) who asked why Southeast Asians are not as deeply pigmented as their latitudinal relatives in Melanesia if there had not been a migration of lesser-pigmented Neolithic farmers from the north. The answer, of course, is that degree of skin pigmentation is not due to latitude alone. There are other factors, including sexual selection, ultraviolet light blockage with cloud cover, clothing rules, activity-patterned sunlight exposure, forestation, humidity, and other considerations. Given the wide range of micro-environmental and climatic variation in Southeast Asia, it should be no surprise that skin color is also variable. Even if genes for lighter pigmentation were introduced from the north, they are not necessarily tied to Neolithic farmers. These genes could have been introduced in the last few centuries. Such is the ever-existing problem of attempting phylogenetic reconstructions with synchronic data.

A two-dimensional ordination of the 43 groups in analysis 3 is shown in Fig. 7.10. Most of the details noted in reference to the dendrogram are also reflected in Fig. 7.10. To reiterate briefly, Western Eurasians, Sino-Americans, and Sub-Saharan Africans exhibit the world's three extreme dental patterns. As for other major subdivisions, Sunda-Pacific groups fall closest to Sino-Americans while Sahul-Pacific groups fall between Western Eurasians, Africans, and Southeast Asians. Central Asians and North Eurasians are positioned between Western Eurasians and Sino-Americans.

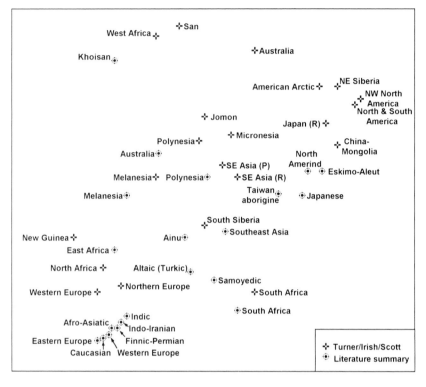

Fig. 7.10   Analysis 3: Two-dimensional ordination based on multidimensional scaling of distance matrix for 9 crown traits in 43 regional groups/language families.

Craniometric and genetic data in light of the dental evidence for human population relationships

The rapid development of computers since the 1950s has made possible the analysis of very large data sets. Computational ease and statistical advances, in conjunction with the efforts of countless workers to amass data on genetic and phenotypic variation within and among human populations (cf. Mourant, 1954; Howells, 1973b, 1989; Mourant *et al.*, 1976; Steinberg and Cook, 1981; Roychoudhury and Nei, 1988), has allowed researchers to assess cladistic and phenetic historical relationships on a global scale. Taking somewhat different approaches, anthropologists and geneticists have both made attempts toward this end since the 1970s. The major anthropological syntheses have been accomplished through the analysis of craniometric data where workers have traveled the world over

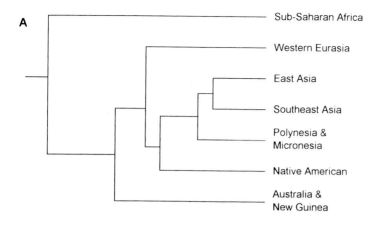

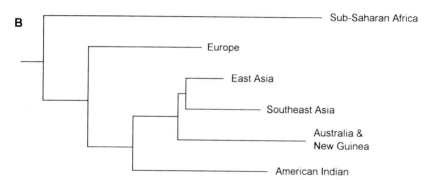

Fig. 7.11   Trees based on distance analysis of genetic markers (121 alleles) in representative world populations. (A) UPGMA tree. (B) Neighbor-joining tree. Modified and redrawn after Nei and Roychoudhury (1993).

to amass large data sets of their own design without recourse to using information compiled by other researchers (Howells, 1973b, 1989; Brace and Hunt, 1990; Brace and Tracer, 1992). Geneticists have employed more than 100 allele and haplotype frequencies from an enormous literature on serological variation to estimate the relationships among representatives of the major living groups of humankind (Nei and Roychoudhury, 1982, 1993; Cavalli-Sforza *et al.*, 1988, 1994; Chen *et al.*, 1995). At this point, there are noteworthy commonalities in the findings of these different approaches, but there is no overall consensus. Given this lack of agreement, we cannot compare our dental-based findings with some agreed upon phylogenetic scenario for the world pattern of human differentiation. However, dental data do provide an additional perspective on the course of

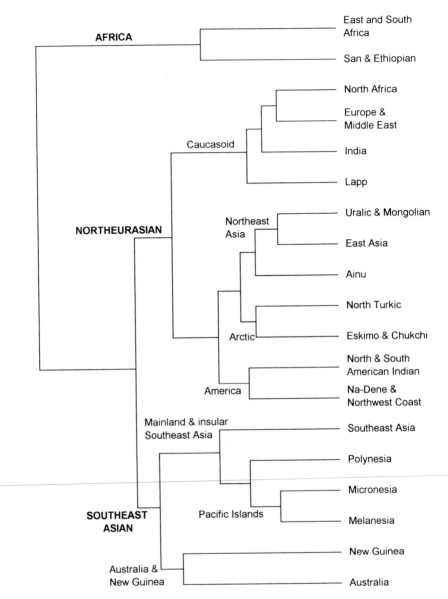

Fig. 7.12   Tree based on distance analysis of genetic markers (120 alleles) in representative world populations. Modified and redrawn after Cavalli-Sforza *et al.* (1994).

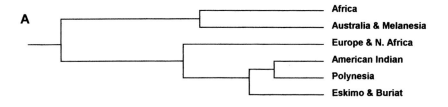

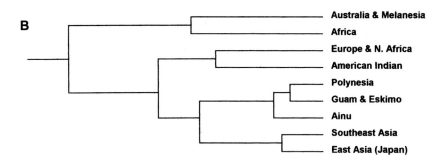

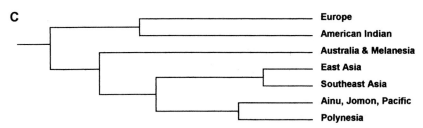

Fig. 7.13 Trees based on analysis of craniometric and craniofacial variables in representative world populations. (A) Craniometric analysis, modified and redrawn after Howells (1973b). (B) Craniometric analysis, modified and redrawn after Howells (1989). (C) Craniofacial analysis, modified and redrawn after Brace and Hunt (1990).

human history that includes both points of agreement and disagreement with craniometric and genetic analyses.

As this discussion is at a global scale, the salient features of several tree diagrams based on either craniometric or genetic data are presented in Figs. 7.11–7.13. From these analyses, we can address several fundamental issues regarding the relationships among major geographic populations.

Are Sub-Saharan Africans the most widely divergent group in a worldwide biological context?

This question is particularly important in the debate over the 'Out of Africa' model of modern human origins. If the ancestral population of all human groups was initially African, the initial dispersal event into the Middle East and beyond would have involved a generalized proto-Eurasian population. With subsequent dispersal to Europe and Asia, differences would eventually develop between the migrants to these two major regions. If this scenario did occur, then Europeans and Asians should be more similar to one another than either is to Africans because of their more recent common ancestry.

Geneticists believe that Africans are the most highly differentiated of all human populations (Nei and Roychoudhury, 1982, 1993; Cavalli-Sforza *et al.*, 1988, 1994). This can be seen in their many tree diagrams where the first-order split is between Africans and non-Africans (Figs. 7.11 and 7.12). This distinctiveness of Africans is evident in mtDNA variation as well as in classical genetic markers (Cann *et al.*, 1987; Vigilant *et al.*, 1989, 1991; Horai, 1995). These reconstructions focus exclusively on genetic variation, holding demographic parameters constant. Relethford and Harpending (1995) have pointed out, however, that differences in population size between Africa, Europe, and Asia in the late Middle and early Upper Pleistocene could be partly responsible for this pattern of 'African distinctiveness' in a world context.

Based on an analysis of craniometric data (Fig. 7.13), Howells (1989:70) concluded 'There is no sign that, cranially, Africans are set off as most distant from other populations – quite the reverse.' In Howells' analyses, Africans typically cluster with Australians and Melanesians, the same result reported by T. Hanihara (1996) for craniofacial measurements. Dentally, we do not find that Africans are the most highly differentiated group. Africans, while distinctive in some morphological features, cannot match Sino-Americans as the most dentally distinctive human grouping. In terms of overall dental differences, the two world extremes are Western Eurasians and Sino-Americans, not Africans.

What are the primary Native American affiliations with Old World populations?

Geneticists find a linkage between Native Americans and the populations of East and Northeast Asia, although distance values and tree diagrams indicate a deep rooting for this relationship. Nei and Roychoudhury (1993) consider Amerindians one of the five major human groups along with Africans, Caucasians, Greater Asians, and Australopapuans (Fig. 7.11). Cavalli-Sforza *et al.* (1988, 1994) have American Indians clustered with Northeast Asians who together form a branch, along with Caucasoids, of a Northeurasian supercluster (7.12). Craniometric analysis does not show any close linkage between American Indians and either North, East, or Southeast Asians but instead groups them with Europeans (Howells, 1973b, 1989; Brace and Hunt, 1990). In other words, serological data indicate a distinct but distant relationship between American Indians and Asians while craniometric data suggests Native Americans are phenotypically most similar to Europeans. In this case, the dental evidence falls more in line with genetic than craniometric evidence. Morphologically, American Indians are Sinodonts, with their greatest resemblance to East and Northeast Asian populations. Dentally, the greatly simplified dentitions of Europeans (and all Western Eurasians) could hardly be farther removed from the hyper-Sinodonts of the Americas.

Is Mongoloid a valid term for all Asians or should East and North Asians be distinguished from Southeast Asians?

In one analysis, Nei and Roychoudhury (1982) found hardly any distinction between East (China, Japan) and Southeast (Malay) Asian populations. In a more recent article, their tree diagram distinguishes East Asians and Southeast Asians but at a relatively low order so both are combined in a 'Greater Asian' grouping (Fig. 7.11; Nei and Roychoudhury, 1993). In contrast to these findings, Cavalli-Sforza *et al.* (1988, 1994) define three superclusters: Africa, Northeurasia and Southeast Asia. If their phylogeny is taken literally in terms of high-order divisions, this would mean that Northeast Asians are more closely related to Europeans than they are to Southeast Asians. Craniometrically, Howells (1989) and Brace and Hunt (1990) arrive at tree diagrams that clearly distinguish East and Southeast Asian populations, but this division is at a low rather than high level of differentiation (Fig. 7.13). A somewhat different result was obtained by Hanihara (1996) who found East and Southeast Asians to be more distantly linked, at least in terms of a 2-dimensional analysis. For the most part, the dental evidence is in closest accord with craniometric analyses and the more recent of the two genetic analyses of Nei and Roychoudhury (1982, 1993). We feel there is a clear dental distinction between North and East Asians (Sinodonts) and Southeast Asians (Sundadonts), but they are more similar to one another than to either Africans or Western Eurasians. As for the validity of the term Mongoloid, or the 'Greater Asians' group of Nei and Roychoudhury (1993), this may well denote an Asian supergroup of considerable time depth and extensive geographic range. However, it does obscure a considerable amount of internal genetic and dental variability. Given the importance of Asian populations in the peopling of the New World and the Pacific, we feel that some distinction should be maintained between the two primary population systems of this area.

Do the populations of the Sahul-Pacific subdivision (Australia, New Guinea, Melanesia) show any special affinity to African populations?

Genetic analyses ordinarily fail to show any relationship between African and Australian/New Guinea populations. In one analysis, Nei and Roychoudhury (1982) found that Australia and New Guinea groups were the second most highly differentiated of all human groups, after Africans. Their more recent reconstruction places these populations more squarely in the province of Southeast Asia, although they still maintain the distinctiveness of Australopapuans as they consider them one of their five major human groups (Nei and Roychoudhury, 1993). Cavalli-Sforza *et al.* (1988, 1994) also found no strong genetic ties between Australopapuans (Australia, New Guinea) and any other human populations, but on a world scale, they fell within the Southeast Asian supercluster. By contrast, in the craniometric analyses of Howells (1973b, 1989) and Hanihara (1996), African and Australomelanesian populations cluster together on one major branch and all European, Asian, and Asian-derived groups cluster on a second major branch.

The dental evidence for Sahul-Pacific groups is not easy to decipher, but there are remarkable possibilities for an ancient African connection. In a cladistic analysis of Turner's published dental morphological data, Stringer (1993) found two equally parsimonious trees—one with Africa as an outgroup, in accord with the 'Out of Africa' model, and one with Southeast Asia as an outgroup, with Africa linking with Melanesia and Australia. New Guinea is clearly the most enigmatic group as they show their closest dental affinities to Western Eurasia and Melanesia but not Australia. Australia and Melanesia show the greatest resemblance to Southeast Asian populations and thus cluster with these groups. Australia shows large pairwise distance values compared to Western Eurasian and Sino-American populations, with the smallest distances to Southeast Asia and West and South Africa. Melanesians in some respects fall between Australian and New Guinea groups in showing ties to both Western Eurasia, like New

Guinea, and ties to Africa, like Australia. Given the inconstancy of distance values involving Sahul-Pacific groups, any conclusion regarding deep African ties are held in abeyance. However, from the overall pattern of relationships, a novel biohistoric hypothesis is emerging that envisions Sundadonty or more likely proto-Sundadonty, as the ancestral pattern for all modern humans (Turner, 1992b,c, 1995). This hypothesis, called 'shifting continuity', has been proposed on at least three bases:

First, of all world populations, average dental divergence is least in Sundaland, a condition that parallels anthropometric and anthroposcopic considerations. Southeast Asians are quite generalized in the sense that they possess various external physical features of many geographic races, although usually in relatively low frequencies. Because Southeast Asians have many dental and non-dental traits in intermediate frequencies, they constitute the best type of population from which all other modern populations could most easily have evolved.

Second, when South Siberian teeth are compared with those of Sundadonts, they show remarkable similarities. Because South Siberians are a mixture of European and Northeast Asian stocks, this hybrid condition retrodicts the probable dental pattern of the common ancestors of Europeans and Asians before these derived groups drifted to their distinctive patterns by late Pleistocene times. The ancestral pattern must have been much like whatever Sundadonty was 50,000 or more years ago, and that condition has been called proto-Sundadonty. The characteristics of proto-Sundadonty can be estimated by averaging the frequencies of dental traits of Australians and Southeast Asians.

Third, the unexpected dental link between Africans and part of the Sahul-Pacific community (Turner, 1992c, 1995; Fig. 7.14) suggests that because both of these regional populations have been strongly isolated, their initial dental pattern drift was preserved when vast and rich territories permitted large and rapid human population growth, fixing the dental patterns as they were more-or-less at 50,000 years ago. It is now well accepted that Australia was colonized from Sundaland at least 50,000 years ago with the aid of watercraft. Sundaland must have been a cultural hearth for boating invention and innovation since this technology had also extended northward to Japan no later than 30,000 years ago. The same sorts of Sundaland watercraft could have just as easily carried parties of Sundadonts or proto-Sundadonts westward along the coast of India, and eventually into the Middle East and North Africa. Refinements to this model will be possible when dental characterizations are obtained for the Pleistocene population of the Indian subcontinent (i.e., pre-Dravidian, pre-Indic).

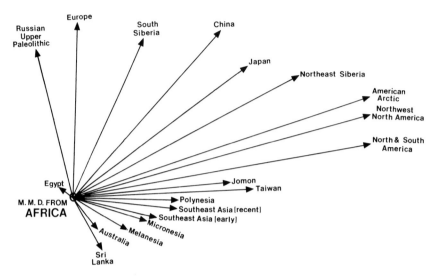

Fig. 7.14　Starburst of Mean Measures of Divergence (MMD) from Africa, based on 29 crown and root trait frequencies. Each line is scaled to the pairwise MMD between Africa and 19 world populations; the arrows indicate geographic direction of population from Africa. Note that shortest lines are to Australia and Pacific populations. Redrawn after Turner (1995).

A potential trigger for a human dispersal event could have been the Toba volcanic eruption in Sumatra 73,500 years ago, an eruption more explosive than any known for the last few 100,000 years and one that Rampino and Self (1993) have suggested could have related to a late Pleistocene bottleneck in human evolution caused by several years of volcanic winter, especially outside the tropics. With the worldwide flooding of all continental shelves in terminal Pleistocene times, very little trace of this dispersal event along the coast of India would be preserved, but derived populations might be expected to show increasing divergence caused by drift the further away they moved from Southeast Asia. As dental analyses show, this is exactly the case. These data also challenge the genetic hypothesis that claims Africans are equally unlike all other populations. This is clearly not the case with teeth nor with crania. There is strong bonding between southern Asian and Pacific populations and those of Africa – a condition that favors an origin in Southeast Asia rather than Africa since boating technology apparently developed earlier in Asia than Africa, and the Toba volcanic eruption provides a physical mechanism for driving people out of Southeast Asia.

The ongoing and sometimes vituperative debate on the origins of anatomically modern humans has occupied the attention of dozens of paleoanthropologists, geneticists, physical anthropologists, archaeologists, and linguists over the past decade. Protagonists generally adhere to one of two major models: 'multiregional evolution' or 'rapid replacement' (also known as 'Noah's Ark' or 'Garden of Eden'). The first model contends that human racial diversity is rooted deeply in time, with all modern humans last sharing a common ancestor more than one million years ago. The second model posits that all humans share a much more recent common ancestor, somewhere in the range of 50,000 to 250,000 years ago. In almost all instances, advocates of the replacement model assume that this recent common ancestor (in some cases ancestors) originated in Africa. The dental evidence is generally supportive of the replacement model, but diverges on the homeland of the hypothetical common ancestor, proposing Southeast Asia rather than Africa. Testing the shifting continuity hypothesis, and other possible variants of the rapid replacement model, as well as testing the contending multi-regional model, will make for exciting future research possibilities. We believe that the best explanation for the origin of modern humans will benefit greatly by including diachronic and synchronic dental morphological information along with the traditional archaeological, osteological, genetic, and natural historical observations.

Epilogue

We have focused on the description, classification, genetics, and variation of crown and root traits of the permanent dentition in recent and modern human populations. This emphasis was dictated first, by our own interests and second, by the available literature. Dental morphologists have concentrated traditionally on just those issues presented here. Other promising avenues of research in dental morphology were barely touched upon, if at all. The intent of this epilogue is to review briefly areas where the potential of tooth morphology has been partially but not fully exploited.

The morphology of deciduous teeth

For reasons noted in chapter 1, dental anthropologists have concentrated on permanent teeth to the near exclusion of deciduous teeth. This seems to indicate general agreement with Lasker's (1957:415) remark that 'A systematic analysis of racial traits of the deciduous dentition does not seem warranted... in general the characteristics are similar to those of the permanent dentition.' Many crown traits expressed in the primary dentition foreshadow the expression of the 'same trait' in the permanent dentition. Populations with high trait frequencies on deciduous teeth typically have high, albeit reduced, frequencies on the permanent teeth. It is not infrequent, however, to find dentitions where traits expressed on the primary second molars are not expressed on the permanent molars. An individual may exhibit, for example, a well-developed protostylid on dm_2 and no manifestation of this trait on LM1, LM2, or LM3. If only permanent teeth are observed, such an individual would be scored as absent or unaffected for the protostylid, even though they seemingly have the genes for trait expression. What are the ramifications of this inconsistency? The assumption is that genes regulating trait development on deciduous teeth are the same genes acting on permanent teeth. A thorough study of the genetics of deciduous crown morphology is needed to resolve this problem, but such an effort would require a long-term project. Dental casts

308

would have to be obtained from children and then, 15–25 years later, from their children.

It has often been observed that deciduous teeth are more primitive, or evolutionarily conservative, than permanent teeth. Brabant (1967:897) states 'The working hypothesis here is that, as the permanent dentition succeeds the deciduous teeth, the pathologic characteristics of the permanent teeth must be more frequent, and evolution must be more rapid than in the deciduous dentition.' In comparing several thousand deciduous and permanent teeth of Europeans, dating from the Upper Paleolithic to the present, Brabant found deciduous teeth showed less crown size reduction through time than permanent teeth (12–16% vs. 33%). Deciduous teeth were also more stable in terms of numerical variations, with significantly lower frequencies of missing and extra teeth than the permanent dentition. Regarding morphology, the deciduous teeth of Europeans showed more incisor shoveling than permanent teeth but fewer enamel extensions on the molars. Some characteristics, like Carabelli's trait, cusp 6, and lower molar groove pattern had similar frequencies in the primary and permanent dentitions of Europeans. We have made some reference to trait intensification (derived), simplification (derived), and retention (primitive) for permanent crown and root traits, but the deciduous dentition might provide additional insights on 'primitive' and 'derived' traits in the hominid dentition.

In 1961, K. Hanihara developed three-dimensional standards for several crown traits of the deciduous dentition, including shoveling, Carabelli's trait, the protostylid, molar cusp number and pattern, and others. Workers using these standards have found crown and root traits of the deciduous dentition effectively discriminate populations from the major subdivisions of humankind. Sciulli's (1990) affinity assessment using deciduous crown traits linked South Africans with Afroamericans and Archaic Ohio Valley Indians with recent Japanese populations. Euroamericans were clearly set apart from these two groupings while a Chalcolithic sample from India fell between Euroamericans on the one hand and American Indians and Japanese on the other. Given the findings from the permanent dentition, these results conform to expectations. Kitagawa *et al.* (1995), focusing on Jomon deciduous dental variation, compared six nonmetric dental traits among ten populations. Through correspondence analysis, the authors showed that Japanese, Pima Indians, and Eskimos grouped closely together as did the Kalahari San and South African Blacks. American Blacks were pulled slightly in the direction of Asiatic Indians and Euroamericans, an expected finding given gene flow from the Euroamerican to Afroamerican population of the United States. Jomon linked most closely with Japanese and Native Americans, but there were noteworthy differences between Jomon and the Sino-American groups. Shoveling, for example, was less frequent in Jomon deciduous upper incisors, a difference already noted for permanent incisor shoveling. Jomon deciduous lower second molars had a higher frequency of cusp 6, more closely corresponding to Polynesian and Australian frequencies than to those of Sino-Americans. Although there are tantalizing hints that the deciduous dentition of the Jomon is Sundadont or 'Proto-Sundadont' in nature, the authors were reluctant to make this inference until more comparative data become available on the deciduous teeth of Southeast Asian and Pacific populations.

At this time, standardized morphological data on deciduous dental traits are but a 'drop in the bucket' compared to data on permanent teeth. The level of population comparison for deciduous dental variation is about where permanent crown morphology stood some 50 years ago, i.e., between major geographic subdivisions. For additional information on variation in morphological traits of the deciduous dentition, the reader is referred to K. Hanihara (1954, 1955, 1956a, 1956b, 1957, 1963, 1965, 1966, 1968b, 1970), Jørgensen (1956), Sciulli (1977), Kaul and Prakash (1981), Lukacs and Walimbe (1984), Grine (1986, 1990), and Sekikawa *et al.* (1990).

Fossil hominid tooth morphology

Although teeth and jaws make up a substantial portion of the fossil record, there have been few systematic studies of nonmetric crown and root traits in Pleistocene hominids. Researchers have measured most, if not all, fossil hominid teeth for basic crown dimensions (length, breadth, height) and frequently cuspal components, making it possible to summarize tooth size for the majority of Lower, Middle, and Upper Pleistocene hominids (cf. Wolpoff, 1971; Frayer, 1978). It would not, however, be possible to summarize the frequency of, say, cusp 6 of the lower molars, for the same sample of fossils.

In an excellent series of articles on Plio-Pleistocene hominid teeth, B.A. Wood and his collaborators (Wood and Abbott, 1983; Wood *et al.*, 1983, 1988; Wood and Uytterschaut, 1987; Wood and Engleman, 1988) reported frequencies for a number of key premolar and molar crown and root traits in East and South African robust and gracile Australopithecines and early *Homo*. Through their efforts, we know the frequencies for groove pattern, cusp number, cusp 6, cusp 7, the protostylid, lower premolar cusp number,

Carabelli's trait, and upper and lower premolar root number in these early hominid lineages. Their observations are wholly comparable to those on recent and living human populations. Bermúdez de Castro (1988, 1993) also provides dental morphological data for late Middle Pleistocene hominid fossils from Atapuerca, Spain. Studies of this sort will hopefully stimulate workers to systematically observe and tabulate the frequencies of crown and root traits in early hominid fossils dating to all stages of the Pleistocene.

If detailed morphological descriptions were available for hominid fossil teeth, an independent source of data, little used to date, could be applied to a number of problems in paleoanthropology. For example, one major argument in hominid evolution centers on the origins of anatomically modern humans. Early twentieth century paleontologists, including Gorjanovic-Kramberger (1906) who excavated the remains at Krapina, observed that Neanderthal teeth exhibited a number of peculiarities, including the common occurrence of a pronounced basal eminence and associated spines and ridges on the upper anterior teeth (i.e., *tuberculum dentale*). Neanderthal incisors also commonly express well-developed shoveling (Carbonell, 1963). Neanderthal cheek teeth exhibit unusual accessory ridges and tubercles as well, but the frequencies of these variants have not been quantified. A thorough comparative analysis of Neanderthal and Upper Paleolithic *Homo sapiens* crown and root morphology could make a significant contribution to the debate over Middle–Upper Paleolithic continuity in Europe and the Middle East. This would, of course, require observations on a large suite of dental characters, not just a select few like shoveling and Carabelli's trait.

Many of the traits described in chapter 2 are present in polymorphic frequencies in fossil hominid teeth – from the earliest Australopithecines to Upper Paleolithic *H. sapiens*. Some crown and root traits observable in modern humans have a long evolutionary history in not only the hominid but also the primate lineage (e.g., hypocone, hypoconulid, cusp 6, cusp 7). Dental morphological trends we see in long-term human evolution, including hypocone and hypoconulid loss, are possibly tied to trends in tooth and jaw size reduction, increased frequencies of hypodontia, basi-cranial flexion, and other dental and skeletal indicators of ontogenetic change. Given the temporal and spatial variation in recent human populations, it is still not clear why some groups show more cusp and root number reduction than others.

Some crown and root traits are derived features that make their first appearance in the hominid dentition during the Pleistocene. Establishing the location and timing of derived traits could shed light on early human population relationships and movements. For example, 3-rooted lower first molars have not been observed in Plio-Pleistocene hominids. This trait does, however, appear in a *H. erectus* mandible excavated at Zhoukoudien in 1959 (Wu and Xianglong, 1995). Prior to this finding, the Tabon mandible from the Philippines, at 22,000 BP, provided the earliest evidence for the appearance of 3RM1. If 3RM1 was present in Asia for more than 200,000 years, what are the ramifications of this finding to the debate over regional continuity vs. replacement? Is it coincidental that this trait, now most common in Asia and the Americas, makes its first appearance in Asia? This same question might apply to any number of crown and root variants that, in recent times, show a regional focus, including incisor winging, premolar odontomes, and 2-rooted lower canines.

The only dental trait brought to bear in the debate over regional evolution vs. rapid replacement is incisor shoveling. As we and earlier workers have shown, North and East Asians and their derivative populations in the Americas are distinguished, in both frequency and degree of shoveling expression, from all other human groups. The pronounced expressions of shoveling in hominid fossils from Zhoukoudien provided one line of evidence for Weidenreich (1937) and Coon (1962) to link, in an ancestral-descendent relationship, Sinanthropus and living East Asians. For this trait, it is not just presence but degree of expression that is critical. We know, for example, that Australopithecines exhibit moderate shoveling while Neanderthals show moderate to pronounced shoveling. Although recent Western Eurasians characteristically show no or trace shoveling, earlier populations in Europe exhibit shoveling in higher frequencies and to more pronounced degrees. The shoveled incisors of Bronze Age skeletons from Spain, illustrated by Du Souich (1974), would rarely be found in a recent or modern European population. While the potential is there, establishing evolutionary linkages between Middle and Upper Pleistocene and Holocene populations will require a broader comparative baseline on shoveling variation among earlier hominid groups than is currently available.

Dentochronology

'When' questions abound in paleoanthropology and prehistory – e.g., when did hominoids first appear in the fossil record?; When did hominids diverge from a common ancestor with chimpanzees and gorillas?; When did hominids first reach Asia?; When did Indo-Europeans make their entry into Europe?; When was the New World colonized?; When did populations first reach the far-flung islands of the Pacific basin? All of these 'when' questions, and many more, have generated an enormous amount of controversy in the anthropological literature. Stimulated by the importance of putting historic events into chronological perspective, geochronologists, paleontologists, and archaeologists have utilized a wide variety of relative and absolute dating methods to pin down the elusive quantity of time. The use of biological data to measure 'time' adds an interesting dimension to this quest.

Evolutionists have long used phylogenetic trees to illustrate proposed historic relationships among species and higher categories. Only within the past few decades have molecular biologists attempted to estimate branching times for divergence events separating two or more taxa. Molecular dating has been shown to be a valuable tool in delineating the pattern and timing of species divergence within many mammalian orders, including primates. Although greeted at first with great skepticism by paleoanthropologists, one of the most revolutionary developments from 'molecular clocks' is the remarkably recent estimate of four to six million years for the divergence of hominids from chimpanzees and gorillas.

Genetic differences among polytypic populations of a single species are of much lower magnitude than between-species differences. There have been, however, some efforts to estimate divergence dates among major geographic races based on genetic distance values (Nei and Roychoudhury, 1982; Cavalli-Sforza *et al.*, 1988). These estimates of branching time assume: (a) differentiation is primarily the result of genetic drift and founder effect; (b) gene flow between the diverging groups has been nonexistent or minimal; and (3) the rate of genetic change is constant, resulting in a linear relationship between genetic distance and time.

Along a similar vein, but using teeth rather than genes, Turner (1986b) developed 'dentochronology' to estimate divergence dates among a wide range of human populations. Employing Mean Measures of Divergence (MMD) calculated from 28 crown and root trait frequencies, Turner's premise was the same as that of Cavalli-Sforza *et al.* (1988). That is, there is a strong positive correlation between distance values and time – large paired distances indicate temporally deep divergence dates while small distances indicate recent common ancestry. If a single pairwise distance can be anchored to a point in time by independent lines of evidence, a 'distance clock' can be established to estimate divergence dates for other paired populations. Cavalli-Sforza *et al.* (1988) set their 'distance clock' by assuming the anatomically modern human fossils found at Jebel Qafza in Israel, dated at 92,000 BP by thermoluminescence and electron spin resonance, marks the point of divergence between African and non-African populations. Turner (1986b) sets his dental clock by assuming some New World populations, including Aleuts, colonized the Americas about 12,000 years ago. Mean measures of divergence between New World and Northeast Asian populations indicated an average dental microevolution rate of about 0.01 ± 0.033 MMD/thousand years. With a large database on populations from Asia, the Pacific, and the Americas, and smaller samples from Europe and Africa, Turner correlated dentally-based divergence dates with independent data from archaeology, linguistics, and geomorphology. In 45 pairwise comparisons, he found good overall agreement between dentochronology and other lines of evidence.

A few select divergence estimates derived by dentochronology that cover the range from very high to very low levels of divergence include: New World/Africa, 60,000 years; New World/Europe, 50,000 years; Malay-Java/Polynesia, 4964 years; Ceylon/Europe, 4757 years; Japan/China, 1805 years. The first two dates support the replacement or Noah's ark model that posits a recent common ancestry for all modern humans. The final three dates correspond to peopling events discussed in chapter 7: (1) the expansion of Austronesian populations into the Pacific; (2) the mid-Holocene invasion of India by Indo-Aryan 'Caucasoid' peoples of the middle East; and (3) the recent movement of mainland Asian populations to the Japanese archipelago.

While many workers remain sceptical of efforts to estimate times of human population divergence based on genetic and/or dental distance values, the importance of chronology dictates the need for more research in this area. In labs around the world, workers are using mtDNA nucleotide differences to estimate times of divergence among various human populations. In the near future, it should be possible to assess more comprehensively the differences and similarities in divergence estimates based on mtDNA, nuclear genetic markers, and dental traits.

Dental morphology and forensic science

Forensic anthropologists utilize a wide array of methods to help law enforcement officials with problems of individual identification. In some instances, when only skeletal-dental remains are available, this involves determining: (1) age; (2) sex; (3) stature/body build; (4) ethnicity; (5) elapsed time since death; (6) distinctive anatomical or pathological features; and (7) perimortem damage or changes. When medical or dental records are available, especially in the form of radiographs, dental and skeletal data are used to determine the exact identity of an individual, a process known as 'individuation.' The use of dental X-rays to identify individuals who were victims of a criminal act or mass disaster is widely known and accepted by legal authorities and the lay public.

Forensic odontology is a division of forensic science that utilizes dental data of different kinds to help resolve legal disputes. The use of dental X-rays for individual identification is only one way teeth are used in forensic cases. Bite-marks left on the skin of victims and/or assailants provide a form of physical evidence that is being increasingly exploited by forensic odontologists. Albert A. Dahlberg once recalled a case where a thief unwittingly took a bite out of a piece of cheese while robbing a grocery store. The distinctive imprint of the thief's teeth in the cheese proved to be damning physical evidence that allowed Dahlberg and the police to place the individual at the scene of the crime. Bitemarks and peculiarities of tooth alignment can be extremely useful for individuation in criminal cases and are interesting topics in their own right. However, our discussion is herein limited to describing how tooth morphology can help law enforcement officials when confronted with isolated skeletal-dental remains.

Tooth morphology provides few clues as to age, sex, body size, elapsed time since death, etc., so its primary usage is in discerning the ethnic affiliation or race of an individual. Are there specific crown or root traits that aid the forensic anthropologist in identifying racial affiliation? Lasker (1957:404) noted that 'To speak of a dental trait as 'racial' merely implies that it occurs in higher frequency in one race than in another; it is very unlikely that a certain feature will be found exclusively in a given race.' The extensive dental surveys undertaken since 1957 corroborate this conclusion. To our knowledge, only one trait – the disto-sagittal ridge or Uto-Aztecan premolar (UP1) – has been found almost exclusively in one of our major subdivisions (i.e., Sino-American). However, the forensic utility of this trait is limited by its extreme rarity (< 2%), even in American Indian groups where it occurs most frequently.

An individual from any one of our five subdivisions could theoretically exhibit all or none of the crown and root traits reviewed in chapter 5. However, these traits do show significant differences in frequency and grade of expression between the subdivisions that can provide clues about group affiliation. When viewed as constellations of characters rather than as isolated traits, they can be used to estimate the relative probability of ethnic or racial identity. For example, an individual with no or trace incisor shoveling, a Carabelli's tubercle or cusp, and a 4-cusped lower second molar would most likely be of European or Western Eurasian descent. However, moderate incisor shoveling, molar enamel extensions, 1-rooted upper first premolars, and cusp 6 on the lower first molar would form a set

Table E.1. *Tooth crown and root traits with distinct patterns of geographic variation that are the most useful for identifying the ethnic affiliation of isolated human remains in a forensic context; used singly or in combination, these characteristics could help establish the least and most likely group affiliation of an individual.*

Trait (grade of expression)	Least likely affiliation	Most likely affiliation
Shoveling (grade 0)	Sino-American	European, African
Shoveling (grades 4-7)	European, African	Sino-American
Winging (bilateral)	European, African	Sino-American
Carabelli's trait (grades 5-7)	Native American	All others
Cusp 5 (grades 2-5)	European, Sino-American	Australian, African
UM1 enamel extensions	European, African	Sino-American
UP1 root number (2-rooted)	Sino-American	African, Australian, European
Premolar odontomes	European, African	Sino-American, Sunda-Pacific
LC root number (2-rooted)	All others	European
4-cusped LM1 and/or LM2	All others	European
Cusp 6 (grades 1-5)	European	All others
Cusp 7 (grades 1-4)	All others	African
3-rooted LM1	European, African	Eskimo-Aleut, North Asian

UM1: upper first molar; UP1: upper first premolar; LC: lower canine; LM1: lower first molar; LM2: lower second molar.

of characters rarely found in a Western Eurasian. This suite of traits would occur much more commonly in someone of North Asian or Native American descent. Preliminary ethnic assessments could be made using the major distinguishing dental features listed in Table E.1 under the dichotomous headings of least and most likely group affiliation. These traits would be most useful in North American forensic cases where the primary concern is to determine if an individual was Euroamerican, Afroamerican, or Native American.

Assuming excellent preservation, discriminant function analysis of craniometric traits is often the preferred method for determining ethnicity in forensic cases. In some instances, however, crania are too damaged by perimortem (e.g., trauma, burning) and/or postmortem (e.g., weathering, soil acidity) factors to allow for a full suite of cranial measurements. When a cranium is well preserved, dental morphology is a useful adjunct to craniometric analysis in the assessment of ethnic affiliation. In severely damaged human remains, tooth morphology often provides the most useful clues to identification.

Discriminant function solutions have been used sparingly for dental data. Matis and Zwemer (1971), using a combination of dental metric and morphological traits, were very successful in classifying individuals to the correct group in their discriminant analysis of five Native American samples (Eskimo, Papago, Pima, Navajo, Apache). They were able to classify Eskimo vs. Indian about 90% of the time and could assign individuals to their correct group about 50% of the time. While the latter estimate does not appear highly accurate, the groups involved are all Sinodonts, exhibiting the same overall dental morphological pattern. Rarely in forensic cases is a worker expected to classify individuals down to the tribal level. More often than not, the question is whether the remains belong to a Native American or are of European or African descent.

We are optimistic that the ever-expanding world database on crown and root trait variation will one day allow dental researchers to determine the ethnicity of isolated human remains with more precision. The geographic differences in dental trait frequency and expression are often pronounced, as shown in chapter 5. When these differences are assessed through advanced methods of classification (e.g., discriminant function analysis, Bayes' theorem, neural networks), it will be possible to transcend educated guesses and calculate the probability that an individual belonged to a particular ethnic group. We intended to pursue this avenue of research for the current volume but ran out of both time and space.

Appendix A: Information base for global descriptions and analysis

A1 Samples used in regional characterization and map
(Samples are in geographic subdivisions; see associated map Fig. A1.1)

WESTERN EURASIA

Western Europe (C.G. Turner II)
 Lapps, Reindeer Is., Karil Peninsula, England (Poundbury), Netherlands (Dorestad de Heul), Lent, Danish Neolithic
Northern Europe (G.R. Scott)
 Medieval Norway, Greenland, Iceland
North Africa (J.D. Irish)
 Algeria, Bedouin, Canary Islands, Carthage, Chad, Christian (Sudan), El Hesa, Kabyle, Kharga, Lisht, Meroitic, Mesolithic Nubians, Soleb, X-Group, North Africa (Algeria, Chad)

SUB-SAHARAN AFRICA

West Africa (C.G. Turner II)
 West Africa, Nubia #117, Nubia #67/80
South Africa (J.D. Irish)
 Congo, Gabon, Ghana, Nigeria/Cameroon, Pygmy, South Africa, Senegambia, Sotho, Tanzania, Togo/Benin, Tukulor
San (J.D. Irish)
 San, Khoikhoi

SINO-AMERICAS (C.G. Turner II)

China-Mongolia
 Urga, An-Yang, South China #1,2, Buriat #1,2, Mongol #2,3, Tibet, North China, China, Hong Kong (living, recent, prehistoric)

318

Japan (recent)
Japan, Hiogo, Kamakura Japanese, Japanese (recent), Kanto Japanese
Japan (Jomon)
Jomon (Japan), Jomon (Ota), Jomon (Tsukumo), S.W. Jomon, Jomon (Yosekura), Jomon (Hokkaido), Jomon (Yoshiko)
Northeast Siberia
Chukchi, Uelen Eskimo, Ekven Eskimo, Ulchi, Goldi, Orochi, Negedal, Tungus, Gilyak
South Siberia
Ket, Tuva, Tuvinci, Krasnoyarsk, Altai, Sopka2 #2
American Arctic
Eastern Aleut #1,2,3,4, Western Aleut #1,2, St. Lawrence Island, Point Hope, Point Barrow #1,2, MacKenzie Eskimo, Smith Sound, Southampton Is., East Greenland, Greenland, S.W. Greenland, N.N.W. Greenland, N.W. Greenland, West Greenland, N.E. Greenland
Northwest North America
Kachemak, Alaska Peninsula, Kodiak Is., Northern Maritime #1,2,3,4, Central Maritime #1,2,3, Lower Columbia, Gulf of Georgia #1,2,3, Intermountain Fraser, Apache, Yukon Athapaskans
North and South America
Canada (Archaic Saskatchewan, Archaic Quebec, Roebuck Iroquois, Toronto Iroquois), Maryland (Nanjemoy/Juhle ossuaries #2,4), Alabama (Shellmound, Kroger Island), Arkansas (Quapaw, Togo, Nodena, Golightly, Wapanoca, Vernon Paul), Archaic California, Southern California, Northern California (Humboldt, Sacramento, Alameda), North Dakota, American Southwest (Point of Pines, Mogollon, Chavez Pass, Ciudad, Cottonwood, Grand Gulch, San Cristobal, Pecos, Grasshopper, Canyons de Chelly & Del Muerto, Kayenta Anasazi), Mexico (Coahuila, Tlatelolco, Cuicuilco, Chichen Itza, Tehuacan), Panama #1,2, Ecuador (Ayalan, Santa Elena, Valdivia, Chanduy, Cotocoallao), Peru (#1,2, Paloma, Preceramic), Chile (Herradura), Patagonia, Bolivia, Brazil (Sambaqui North and South, Lagoa Santa #1,2,3, Minas Gerais, Corondo), Paleoindian

SUNDA-PACIFIC (C.G. Turner II)

Southeast Asia (prehistoric)
Thailand (Central, Non Nak Tha, Ban Chiang early & late, Don Klang/Ban Tong, Ban Kao, Ban Na Di), Kelantan (Gua Cha), Laos (Tam Hang), Tonkin (Mesolithic/Neolithic), Malay Peninsula (Gua Kepah), Celebes (Leang Tjadang), Java (Sampung)

Southeast Asia (recent)
Thailand (living Mrabri, Yao, & Meo, Bangkok, recent Thai), Cambodia, Annam, Recent Tonkin, Laos, Malaya (Singapore, recent), Philippines (#1,2, living Batak, Banton Is., Calatagan, Penablanca), Sumatra, Java, Borneo, Niah Cave, Sarawak, Celebes, Moluccas, Lesser Sunda, Timor, Andaman Is., Nicobar Is.

Polynesia
Marquesas #1,2, Tahiti, Easter Island #2, Fiji #1,2, Society Is., Ellice Is., Tonga, Samoa, Cook Is., New Zealand, Chatham Is., Rotuma, Raiatea, Mokapu, Wallis Is., Gambier Is., Tuamotu archipelago

Micronesia
Guam #1,2,3,4, Gilbert Is., Caroline Is., Tinian, Mariana Is., Saipan

SAHUL-PACIFIC (C.G. Turner II)

Australia
Australia (south, north), Australia, Tasmania

New Guinea
New Guinea

Melanesia
New Britain #1,2,3, New Hebrides, Solomon Is., New Caledonia, Torres Strait

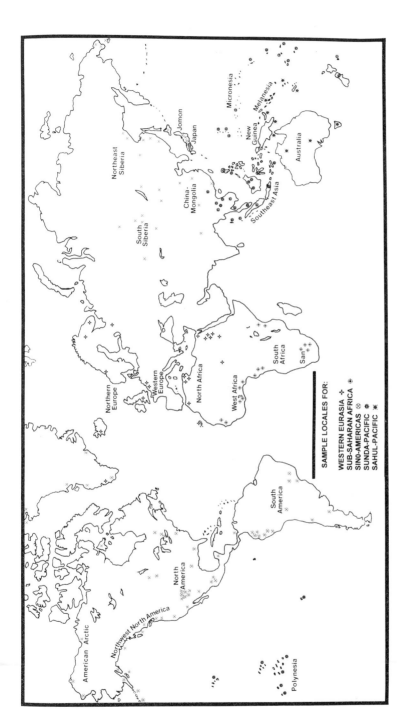

Fig. A1.1   World map to show geographic subdivisions for samples listed in Appendix A1.

A2. *Regional crown and root trait frequencies (with sample sizes)*

Traits	n	WE	n	NE	n	NA	n	WA	n	SA	n	KH	n	CM	n	JO	n	RJ	n	NES
Winging	180	.072	150	.047	460	.075	52	.192	496	.042	90	.167	591	.245	166	.199	265	.219	112	.339
Shoveling UI1	186	.027	46	.022	194	.075	41	.073	220	.093	155	.133	542	.720	117	.257	276	.660	61	.620
Double Shoveling	184	.038	100	.050	175	.086	39	.026	282	.018	79	.000	545	.288	138	.014	267	.195	43	.325
Interr. Grooves	224	.420	100	.300	241	.324	48	.104	301	.120	83	.157	537	.430	189	.646	301	.445	95	.463
Bushman canine	230	.043	125	.000	261	.061	55	.291	398	.126	77	.351	615	.028	136	.022	365	.030	119	.025
Odontomes	246	.008	111	.000	545	.002	56	.000	531	.004	86	.000	639	.055	260	.004	462	.050	95	.021
3-cusped UM2	308	.247	239	.192	446	.106	83	.036	531	.072	86	.062	798	.108	206	.180	482	.135	192	.218
Carabelli's cusp	249	.273	138	.181	200	.200	61	.213	246	.114	155	.168	774	.162	181	.023	458	.149	172	.053
Cusp 5 UM1	238	.118	140	.264	357	.185	48	.625	439	.216	66	.348	633	.242	146	.315	390	.197	106	.104
Enamel Exten.	371	.038	229	.022	503	.068	99	.000	387	.000	15	.000	718	.532	278	.097	522	.546	328	.497
4-cusped LM1	217	.078	170	.100	250	.100	47	.000	346	.000	133	.008	538	.002	214	.000	314	.003	90	.000
4-cusped LM2	284	.711	225	.844	381	.664	75	.120	370	.300	88	.068	639	.208	244	.287	345	.136	138	.065
Y pattern LM2	257	.272	319	.210	402	.306	67	.328	392	.457	89	.719	646	.076	290	.321	352	.131	145	.186
Cusp 6 LM1	217	.083	130	.169	352	.077	47	.447	362	.188	85	.047	538	.359	214	.467	314	.427	90	.500
Cusp 7 LM1	291	.045	179	.050	414	.094	71	.437	385	.265	87	.264	721	.079	285	.031	382	.057	151	.060
Deflecting wrinkle	154	.052	75	.160	267	.082	30	.167	298	.181	60	.167	343	.157	162	.049	262	.149	81	.395
2-rooted UP1	317	.407	194	.459	468	.571	87	.667	386	.611	15	.200	645	.272	241	.245	506	.249	375	.069
3-rooted UM2	265	.574	227	.612	364	.786	82	.829	341	.845	16	.750	591	.650	254	.469	495	.689	260	.508
2-rooted LC	314	.057	214	.061	347	.023	33	.000	192	.000	14	.000	401	.000	203	.010	335	.012	206	.000
Tomes' root	270	.059	168	.066	372	.086	49	.387	217	.230	15	.000	248	.141	282	.032	200	.100	99	.101
3-rooted LM1	357	.006	198	.000	337	.012	92	.076	240	.004	15	.000	604	.283	377	.034	429	.242	238	.223
1-rooted LM2	318	.280	269	.208	333	.117	82	.085	225	.036	15	.286	548	.398	336	.098	407	.329	220	.355
Dist. trig. crest	219	.064	200	.024	276	.033	45	.000	321	.016	46	.043	448	.058	292	.069	334	.180	132	.121

WE: Western Europe; NE: Northern Europe; NA: North Africa; WA: West Africa; SA: South Africa; KH: Khoisan; CM: China-Mongolia; JO: Jomon; RJ: Recent Japan; NES: Northeast Siberia.
UI1: upper central incisor; UM1: upper first molar; UM2: upper second molar; LM1: lower first molar; LM2: lower second molar; UP1: upper first premolar; LC: lower canine; Dist. trig. crest: distal trigonid crest.

A2 (cont.)

Traits	n	SS	n	AA	n	NWA	n	NSAI	n	SEE	n	SER	n	PO	n	MI	n	AU	n	NG	n	ML
Winging	109	.183	220	.232	226	.358	1177	.500	131	.275	270	.226	274	.204	78	.397	508	.094	170	.076	209	.187
Shoveling UI1	98	.367	172	.692	172	.831	1368	.919	184	.305	261	.349	275	.207	83	.313	274	.201	30	.000	135	.089
Double Shoveling	92	.152	155	.349	158	.567	1231	.705	182	.159	199	.120	287	.045	85	.082	261	.042	32	.000	134	.045
Interr. Grooves	145	.545	275	.596	223	.650	1405	.510	200	.385	272	.290	329	.353	102	.245	357	.182	56	.161	165	.188
Bushman canine	155	.084	339	.000	268	.004	1402	.016	235	.060	366	.060	382	.029	121	.051	391	.020	54	.019	174	.029
Odontomes	155	.006	372	.062	371	.065	1787	.044	213	.028	564	.025	572	.023	170	.012	336	.030	119	.000	218	.028
3-cusped UM2	233	.142	569	.306	459	.142	2381	.115	368	.082	730	.115	632	.077	186	.151	643	.033	191	.047	295	.075
Carabelli's cusp	186	.140	477	.019	388	.055	2054	.056	262	.187	701	.208	617	.217	160	.225	332	.214	197	.187	291	.203
Cusp 5 UM1	191	.251	418	.167	378	.214	1780	.167	328	.324	581	.310	565	.427	163	.276	449	.615	151	.457	234	.444
Enamel Exten.	289	.249	936	.459	699	.509	3016	.437	581	.225	746	.361	741	.201	193	.078	797	.092	240	.050	289	.035
4-cusped LM1	195	.031	355	.000	332	.000	1847	.000	248	.004	418	.007	417	.012	148	.000	235	.004	66	.045	210	.019
4-cusped LM2	225	.542	484	.052	447	.044	2462	.086	314	.322	555	.303	461	.332	161	.205	413	.097	93	.591	234	.500
Y pattern LM2	270	.222	529	.200	498	.118	2473	.098	348	.184	587	.175	501	.188	160	.212	465	.127	102	.392	254	.268
Cusp 6 LM1	195	.205	355	.504	322	.503	1847	.551	248	.403	418	.325	417	.535	148	.453	235	.617	166	.152	210	.495
Cusp 7 LM1	272	.099	565	.085	473	.068	2756	.085	370	.075	588	.073	495	.071	175	.058	294	.053	100	.070	267	.124
Deflecting wrinkle	142	.169	230	.300	192	.365	1311	.381	150	.220	290	.159	322	.140	149	.228	35	.171	52	.038	184	.179
2-rooted UP1	278	.313	1022	.049	693	.067	2849	.143	299	.432	845	.386	808	.336	196	.556	642	.424	278	.302	299	.462
3-rooted UM2	247	.470	836	.374	523	.415	2054	.559	196	.730	752	.774	739	.495	184	.734	644	.809	260	.554	297	.751
2-rooted LC	260	.030	733	.003	500	.000	2404	.007	204	.005	568	.011	528	.004	196	.010	409	.000	100	.000	200	.000
Tomes' root	196	.173	493	.034	494	.093	1833	.199	172	.216	307	.222	372	.162	125	.175	383	.273	126	.072	160	.169
3-rooted LM1	242	.025	871	.311	741	.165	3276	.065	400	.082	652	.141	628	.086	204	.029	612	.049	157	.000	251	.032
1-rooted LM2	242	.463	772	.312	659	.387	2703	.328	282	.234	630	.267	617	.313	211	.166	523	.065	142	.162	242	.062
Dist. trig. crest	210	.176	391	.187	294	.078	1990	.042	224	.058	415	.099	453	.046	177	.040	291	.041	80	.000	209	.014

SS: South Siberia; AA: American Arctic; NWA: Northwest North America; NSAI: N. & S. American Indian; SEE: Southeast Asia (Early); SER: Southeast Asia (Recent); PO: Polynesia; MI: Micronesia; AU: Australia; NG: New Guinea; ML: Melanesia. UI1: upper central incisor; UM1: upper first molar; UM2: upper second molar; LM1: lower first molar; LM2: lower second molar; UP1: upper first premolar; LC: lower canine; Dist. trig. crest: distal trigonid crest.

A3. Distance matrix among 21 regional groupings (Nei distance values)

	1	2	3	4	5	6	7	8	9	10	11	12	13	14	15	16	17	18	19	20	21
1. Western Europe	0.000																				
2. Northern Europe	0.031	0.000																			
3. North Africa	0.046	0.040	0.000																		
4. West Africa	0.427	0.364	0.262	0.000																	
5. South Africa	0.230	0.205	0.094	0.092	0.000																
6. San	0.400	0.457	0.312	0.220	0.174	0.000															
7. China-Mongolia	0.446	0.480	0.409	0.461	0.444	0.529	0.000														
8. Japan (Jomon)	0.250	0.276	0.261	0.346	0.324	0.387	0.244	0.000													
9. Japan (Recent)	0.451	0.499	0.411	0.449	0.416	0.491	0.013	0.206	0.000												
10. NE Siberia	0.623	0.677	0.626	0.618	0.603	0.584	0.060	0.255	0.051	0.000											
11. South Siberia	0.127	0.159	0.172	0.395	0.310	0.385	0.133	0.141	0.153	0.209	0.000										
12. American Arctic	0.679	0.760	0.727	0.741	0.746	0.692	0.084	0.237	0.072	0.024	0.237	0.000									
13. NW America	0.751	0.811	0.759	0.755	0.795	0.747	0.075	0.294	0.087	0.030	0.234	0.030	0.000								
14. N. & S. Amerind	0.740	0.763	0.685	0.648	0.678	0.716	0.086	0.330	0.105	0.050	0.253	0.079	0.021	0.000							
15. SE Asia (Prehist.)	0.204	0.194	0.146	0.164	0.153	0.302	0.109	0.119	0.106	0.181	0.087	0.251	0.243	0.213	0.000						
16. SE Asia (Recent)	0.224	0.225	0.164	0.193	0.163	0.301	0.077	0.172	0.072	0.170	0.099	0.240	0.248	0.226	0.016	0.000					
17. Polynesia	0.231	0.210	0.224	0.188	0.248	0.363	0.162	0.101	0.156	0.230	0.102	0.276	0.292	0.298	0.041	0.065	0.000				
18. Micronesia	0.267	0.253	0.184	0.133	0.138	0.297	0.196	0.167	0.182	0.258	0.188	0.352	0.348	0.278	0.033	0.062	0.084	0.000			
19. Australia	0.451	0.363	0.306	0.092	0.191	0.374	0.302	0.230	0.274	0.409	0.346	0.491	0.497	0.441	0.101	0.123	0.099	0.096	0.000		
20. New Guinea	0.110	0.071	0.081	0.234	0.158	0.253	0.506	0.263	0.514	0.731	0.214	0.827	0.831	0.876	0.194	0.212	0.175	0.246	0.257	0.000	
21. Melanesia	0.183	0.117	0.105	0.104	0.099	0.299	0.348	0.178	0.335	0.470	0.217	0.572	0.598	0.518	0.074	0.110	0.080	0.076	0.078	0.082	0.000

Appendix B: Sources of comparative data used for compiling tables in chapter 5

(The Sources of comparative data used in chapter 5 are presented by geographic region; when author(s) provided data on trait(s) for more than one region, they are cited under the heading of multiple regions.)

Multiple regions

Carbonell, V.M. (1963). Variations in the frequency of shovel-shaped incisors in different populations. In *Dental Anthropology*, ed. D.R. Brothwell, pp. 211–34. Oxford: Pergamon Press.

Hanihara, K. (1977). Dentition of the Ainu and the Australian aborigines. In *Orofacial Growth and Development*, eds. A.A. Dahlberg and T.M. Graber, pp. 195–200. The Hague: Mouton Publishers.

Hanihara, T. (1989). Affinities of the Philippine Negritos as viewed from dental characters: a preliminary report. *Journal of the Anthropological Society of Nippon* **97**, 327–39.

Harris, E.F. (1977). *Anthropologic and Genetic Aspects of the Dental Morphology of Solomon Islanders, Melanesia.* PhD dissertation, Department of Anthropology, Arizona State University, Tempe.

Hellman, M. (1928). Racial characters in human dentition. *Proceedings of the American Philosophical Society* **67**, 154–74.

Hrdlička, A. (1920). Shovel-shaped teeth. *American Journal of Physical Anthropology* **3**, 429–65.

Kanazawa, E., Sekikawa, M., Akai, J. and Ozaki, T. (1985). Allometric variation on cuspal areas of the lower first molar in three racial populations. *Journal of the Anthropological Society of Nippon* **93**, 425–38.

Kieser, J.A. and Becker, P.J. (1989). Correlations of dimensional and discrete dental traits in the post-canine and anterior dental segments. *Journal of the Dental Association of South Africa* **44**, 101–3.

Lavelle, C.L.B. (1971). Mandibular molar tooth configuration in different racial groups. *Journal of Dental Research* **50**, 1353.

Morris, D.H. (1970). On deflecting wrinkles and the Dryopithecus pattern in human mandibular molars. *American Journal of Physical Anthropology* **32**, 97–104.

Scott, G.R. (1973). *Dental Morphology: A Genetic Study of American White Families and Variation in Living Southwest Indians.* PhD dissertation, Department of Anthropology, Arizona State University, Tempe, Arizona.

de Souza-Freitas, J.A., Lopes, E.S. and Casati-Alvares, L. (1971). Anatomic variations of lower first permanent molar roots in two ethnic groups. *Oral Surgery, Oral Medicine, Oral Pathology* **31**, 274–8.

Riesenfeld, A. (1956). Shovel-shaped incisors and a few other dental features among the native peoples of the Pacific. *American Journal of Physical Anthropology* **14**, 505–21.

Tratman, E.K. (1938). Three-rooted lower molars in man and their racial distribution. *British Dental Journal* **64**, 264–74.

Wissler, C. (1931). Observations on the face and teeth of the North American Indians. *Anthropological Papers of the American Museum of Natural History* **33**, 1–33.

Zubov, A.A. and Khaldeeva, N.I. (1979). *Ethnic Odontology of the USSR.* Moscow: Nauka. (In Russian.)

Western Eurasia

Aksianova, G.A. (1978). Some dental material in connection with the problem of ancient populations of northern Europe. *Journal of Human Evolution* **7**, 525–8.

Aksianova, G.A., Zubov, A.A. and Kochiev, R.S. (1977). Odontological description of the Komi-Zyrians. In *Physical Anthropology of the Komi*, Moscow: *Suomen Anthropologisen Seuran Toimituksia* No. 4, pp. 65–74.

Aksianova, G.A., Zubov, A.A., Segeda, S.P., Peskina, M.Y. and Khaldeeva, N.I. (1979). Slavic peoples of the European part of the USSR. Russians. In *Ethnic Odontology of the USSR*, eds. A.A. Zubov and N.I. Khaldeeva, pp. 9–31. Moscow: Nauka. (In Russian.)

Alexandersen, V. (1963). Double-rooted human lower canine teeth. In *Dental Anthropology*, ed. D.R. Brothwell, pp. 235–44. Oxford: Pergamon Press.

Alexandersen, V. (1978). Sūkās V: a study of teeth and jaws from a Middle Bronze Age collective grave on Tall Sūkās. *Det Kongelige Danske Videnskabernes Selskab Biologiske Skrifter* **22**(2), 1–56.

Axelsson, G. and Kirveskari, P. (1977). The deflecting wrinkle on the teeth of Icelanders and the Mongoloid dental complex. *American Journal of Physical Anthropology* **47**, 321–4.

Axelsson, G. and Kirveskari, P. (1979). Sixth and seventh cusp on lower molar teeth of Icelanders. *American Journal of Physical Anthropology* **51**, 79–82.

Axelsson, G. and Kirveskari, P. (1982). Correlations between lower molar occlusal traits in Icelanders. In *Teeth: Form, Function, and Evolution*, ed. B. Kurtén, pp. 237–44. New York: Columbia University Press.

Berry, A.C. (1976). The anthropological value of minor variants of the dental crown. *American Journal of Physical Anthropology* **45**, 257–68.

Beynon, A.D. (1971). The dentition of the Afghan Tajik. In *Dental Morphology and Evolution*, ed. A.A. Dahlberg, pp. 271–82. Chicago: University of Chicago Press.

Brabant, H.E. (1964). Observations sur l'evolution de la denture permanente humaine en Europe Occidentale. *Bulletin du Groupement International pour la Recherche Scientifique en Stomatologie et Odontologie* **7**, 11–84.

Brabant, H.E. (1971). The human dentition during the Megalithic era. In *Dental Morphology and Evolution*, ed. A.A. Dahlberg, pp. 283–97. Chicago: University of Chicago Press.

Brabant, H.E. and Ketelbant, R. (1975). Observations sur la frequence de certains caracteres Mongoloides dans la denture permanente de la population Belge. *Bulletin du Groupement International pour la Recherche Scientifique en Stomatologie et Odontologie* **18**, 121–34.

Du Souich, P. 1974. Estudio antropológico de los dientes de una población del Bronce I de Gorafe (Granada). *Anales del Desarrollo* **18**, 137–66.

Gadzhiev, Y.M. (1979). Peoples of the Caucasus, Daghestan. In *Ethnic Odontology of the USSR*, eds. A.A. Zubov and N.I. Khaldeeva, pp. 141–63. Moscow: Nauka. (In Russian.)

Goose, D.H. and Roberts, E.E. (1982). Size and morphology of children's teeth in North Wales. In *Teeth: Form, Function, and Evolution*, ed. B. Kurtén, pp. 228–36. New York: Columbia University Press.

Gravere, R.U., Zubov, A.A. and Sarap, G.G. (1979). Baltic peoples. In *Ethnic Odontology of the USSR*, eds. A.A. Zubov and N.I. Khaldeeva, pp. 68–92. Moscow: Nauka. (In Russian.)

Greene, D.L. (1967). Dentition of Meroitic, X-group, and Christian populations from Wadi Halfa, Sudan. *Anthropological Papers, Department of Anthropology, University of Utah*, No. 85. Salt Lake City: University of Utah Press.

Jørgensen, K.D. (1955). The *Dryopithecus* pattern in recent Danes and

Dutchmen. *Journal of Dental Research* **34**, 195–208.

Joshi, M.R. (1975). Carabelli's trait on maxillary second deciduous molars and first permanent molars in Hindus. *Archives of Oral Biology* **20**, 699–700.

Joshi, M.R., Godiawala, R.N., and Dutia, A. (1972). Carabelli's trait in Hindu children from Gujarat. *Journal of Dental Research* **51**, 706–11.

Kaczmarek, M. (1981). Studies on dental morphology of a modern Polish population. *Prznglad Antropologiczny* **47**, 63–82.

Kaczmarek, M. (1992). Dental morphological variation of the Polish people and their eastern neighbors. In *Structure, Function and Evolution of Teeth*, eds. P. Smith and E. Tchernov, pp. 413–23. London: Freund Publishing House.

Kaczmarek, M. and Piontek, J. (1982). Human cremated remains and the diversity of man. *Homo* **33**, 230–6.

Kajajoja, P. and Zubov, A.A. (1986). Somatology and population genetics of the Bashkirs. *Annales Academiae Scientiarum Fennicae, Series A, V. Medica* **175**, 67–72.

Kaul, V. and Prakash, S. (1981). Morphological features of Jat dentition. *American Journal of Physical Anthropology* **54**, 123–7.

Kirveskari, P. (1974). *Morphological Traits in the Permanent Dentition of Living Skolt Lapps.* PhD dissertation, Institute of Dentistry, University of Turku, Turku, Finland.

Kochiev, R.S. (1979). Peoples of the Caucasus, Trans-Caucasus and north Caucasus. In *Ethnic Odontology of the USSR*, eds. A.A. Zubov and N.I. Khaldeeva, pp. 114–41. Moscow: Nauka. (In Russian.)

Koski, K. and Hautala, E. (1952). On the frequency of shovel-shaped incisors in the Finns. *American Journal of Physical Anthropology.* **10**, 127–32.

Kulkarni, V.S., Bhanu, B.V., and Walimbe, S.R. (1985). Some observations on dental morphology of Andh tribe in Maharashtra. In *Dental Anthropology: Applications and Methods*, ed. V. Rami-Reddy, pp. 139–47. New Delhi: Inter-India Publications.

Lukacs, J.R. (1983). Dental anthropology and the origins of two Iron Age populations from northern Pakistan. *Homo* **34**, 1–15.

Lukacs, J.R. (1987). Biological relationships derived from morphology of permanent teeth: recent evidence from prehistoric India. *Anthropologischer Anzeiger* **45**, 97–116.

Lukacs, J.R. (1988). Dental morphology and odontometrics of early agriculturalists from Neolithic Mehrgarh, Pakistan. In *Teeth Revisited*, eds. D.E. Russell, J.-P. Santoro, and D. Sigogneau-Russell, *Mémoires du Muséum Nationale d'Histoire Naturelle, Série C, Sciences de la Terre* **53**, 285–303.

Mayhall, J.T., Saunders, S.R. and Belier, P.L. (1982). The dental morphology of North American whites: a reappraisal. In *Teeth: Form, Function, and Evolution*, ed. B. Kurtén, pp. 245–58. New York: Columbia University Press.

Ohno, N. (1986). The dentition of the Ladakhi, India. *Journal of the Anthropological Society of Nippon* **94**, 137–46.

Pal, A. (1972). Double-rooted human lower canine – a rare anomaly. *Journal of the Indian Anthropological Society* **7**, 171–4.

Risnes, S. (1974). The prevalence and distribution of cervical enamel projections reaching into the bifurcation of human molars. *Scandinavian Journal of Dental Research* **82**, 413–19.

Rosenzweig, K.A. and Zilberman, Y. (1967). Dental morphology of Jews from Yemen and Cochin. *American Journal of Physical Anthropology* **26**, 15–22.

Rosenzweig, K.A. and Zilberman, Y. (1969). Dentition of Bedouin in Israel. II. Morphology. *American Journal of Physical Anthropology* **31**, 199–204.

Sharma, J.C. and Kaul, V. (1977). Dental morphology and odontometry in Panjabis. *Journal of the Indian Anthropological Society* **12**, 213–26.

Smith, P. (1977). Variations in dental traits within populations. In *Orofacial Growth and Development*, eds. A.A. Dahlberg and T.M. Graber, pp. 171–81. The Hague: Mouton Publishers.

Sofaer, J.A., Smith, P. and Kaye, E. (1986). Affinities between contemporary and skeletal Jewish and non-Jewish groups based on tooth morphology. *American Journal of Physical Anthropology* **70**, 265–75.

Tegako, L.I. and Salivon, I.I. (1979). Byelorussians. In *Ethnic Odontology of the USSR*, eds. A.A. Zubov and N.I. Khaldeeva, pp. 48–65. Moscow: Nauka. (In Russian.)

Thomsen, S. (1955). *Dental Morphology and Occlusion in the People of Tristan da Cunha*. Results of the Norwegian Scientific Expedition to Tristan da Cunha 1937–1938, No. 25. Oslo: Det Norske Videnskaps-Akademi.

Toth, T. (1992). On the frequency of shovel-shaped incisors in Hungarians. In *Structure, Function and Evolution of Teeth*, eds. P. Smith and E. Tchernov, pp. 491–9. London: Freund Publishing House.

Townsend, G.C., Richards, L.C., Brown, T., Burgess, V.B., Travan, G.R. and Rogers, J.R. (1992). Genetic studies of dental morphology in south Australian twins. In *Structure, Function and Evolution of Teeth*, eds. P. Smith and E. Tchernov, pp. 501–18. London: Freund Publishing House.

Younes, S.A., Al-Shammery, A.R. and Al-Angbawi, M.F. (1990). Three-rooted permanent mandibular first molars of Asian and black groups in the Middle East. *Oral Surgery, Oral Medicine, Oral Pathology* **69**, 102–5.

Sub-Saharan Africa

Barnes, D.S. (1968). Variations in tooth morphology between male and female in the Iteso. *Journal of Dental Research* **47**, 971–2. (Abstract.)
Barnes, D.S. (1969). Tooth morphology and other aspects of the Teso dentition. *American Journal of Physical Anthropology* **30**, 183–94.
Chagula, W.K. (1960). The cusps on the mandibular molars of East Africans. *American Journal of Physical Anthropology* **18**, 83–90.
Drennan, M.R. (1929). The dentition of a Bushman tribe. *Annals of the South African Museum* **24**, 61–88.
Greene, D.L. (1982). Discrete dental variations and biological distances of Nubian populations. *American Journal of Physical Anthropology* **58**, 75–9.
Grine, F.E. (1981). Occlusal morphology of the mandibular permanent molars of the South African Negro and the Kalahari San (Bushman). *Annals of the South African Museum* **86**, 157–215.
Haeussler, A.M, Irish, J.D., Morris, D.H. and Turner, C.G., II (1989). Morphological and metrical comparison of San and Central Sotho dentitions from southern Africa. *American Journal of Physical Anthropology* **78**, 115–22.
Hassanali, J. (1982). Incidence of Carabelli's trait in Kenyan Africans and Asians. *American Journal of Physical Anthropology* **59**, 317–19.
Hassanali, J. and Amwayi, P. (1988). Report on two aspects of the Maasai dentition. *East African Dental Journal* **65**, 798–803.
Kieser, J.A. (1978). The incidence and expression of Carabelli's trait in two South African ethnic populations. *Journal of the Dental Association of South Africa* **33**, 5–9.
Morris, D.H. (1975). Bushman maxillary canine polymorhphism. *South African Journal of Science* **71**, 333–5.
Reid, C., van Reenan, J.F. and Groeneveld, H.T. (1991). Tooth size and the Carabelli trait. *American Journal of Physical Anthropology* **84**, 427–32.
Sakuma, M. and Ogata, T. (1987). Sixth and seventh cusp on lower molar teeth of Malawians in east-central Africa. *Japanese Journal of Oral Biology* **29**, 738–45.

Sakuma, M., Yamamoto, M. and Ogata, T. (1987). On the deflecting wrinkle in the lower molars of Malawians derived from pureblooded Negroid racial stock. *Japanese Journal of Oral Biology* **29**, 371–7.

Sakuma, M., Mine, K. and Ogata, T. (1988). The incidence and expression of Carabelli complex in Malawians and Japanese. *Japanese Journal of Oral Biology* **30**, 545–9.

Sakuma, M., Irish, J.D. and Morris, D.H. (1991). The Bushman maxillary canine of the Chewa tribe in east-central Africa. *Journal of the Anthropological Society of Nippon* **99**, 411–17.

Shaw, J.C.M. (1927). Cusp development on the second lower molars in the Bantu and Bushmen. *American Journal of Physical Anthropology* **11**, 97–100.

Shaw, J.C.M. (1931). *The Teeth, the Bony Palate and the Mandible in Bantu Races of South Africa*. London: Bale and Danielsson.

Sino-Americas

Babakov, O.B., Dubova, N.A., Zubov, A.A., Rykushina, G.B. and Khodzhiev, T.K. (1979). Peoples of Central Asia and Kazakhstan. In *Ethnic Odontology of the USSR*, eds. A.A. Zubov and N.I. Khaldeeva, pp. 164–86. Moscow: Nauka. (In Russian.)

Brewer-Carias, C.A., LeBlanc, S. and Neel, J.V. (1976). Genetic structure of a tribal population, the Yanomama Indians. XII. Dental microdifferentiation. *American Journal of Physical Anthropology* **44**, 5–14.

Chang, S.Y. and Kim, M.K. (1961). A note on shovel-shaped incisors, instanding lateral incisor and occlusal type of incisors in Koreans. *The Seoul Journal of Medicine* **2**, 277–8.

Dahlberg, A.A. (1951). The dentition of the American Indian. In *The Physical Anthropology of the American Indian*, ed. W.S. Laughlin, pp. 138–76. New York: The Viking Fund.

Dahlberg, A.A. (1963a). Analysis of the American Indian dentition. In *Dental Anthropology*, ed. D.R. Brothwell, pp. 149–77. Oxford: Pergamon Press.

DeVoto, F.C.H. and Perrotto, B.M. (1972). Groove pattern and cusp number of mandibular molars from Tastilian Indians. *Journal of Dental Research* **51**, 205.

Escobar, V., Conneally, P.M. and Lopez, C. (1977). The dentition of the Queckchi Indians: anthropological aspects. *American Journal of Physical Anthropology* **47**, 443–52.

Goaz, P.W. and Miller, M.C., III (1966). A preliminary description of the dental morphology of the Peruvian Indian. *Journal of Dental Research* **45**, 106–19.

Goldstein, M.S. (1931). The cusps in the mandibular molar teeth of the Eskimo. *American Journal of Physical Anthropology* **16**, 215–35.

Goldstein, M.S. (1948). Dentition of Indian crania from Texas. *American Journal of Physical Anthropology* **6**, 63–84.

Goose, D.H. (1977). The dental condition of Chinese living in Liverpool. In *Orofacial Growth and Development*, eds. A.A. Dahlberg and T.M. Graber, pp. 183–94. The Hague: Mouton Publishers.

Hanihara, K., Kuwashima, T. and Sakao, N. (1964). The deflecting wrinkle on the lower molars in recent man. *Journal of the Anthropological Society of Japan* **72**, 1–8.

Hanihara, K., Masuda, T. and Tanaka, T. (1974). Affinities of dental characteristics in the Okinawa Islanders. *Journal of the Anthropological Society of Nippon* **82**, 75–82.

Hanihara, T. (1989b). Comparative studies of dental characteristics in the Aogashima Islanders. *Journal of the Anthropological Society of Nippon* **97**, 9–22.

Hanihara, T. (1991b). Dentition of Nansei Islanders and peopling of the Japanese archipelago: the basic populations in East Asia, IX. *Journal of the Anthropological Society of Nippon* **99**, 399–409.

Ismagulov, O. and Sikhimbaeva, K.B. (1989). *Ethnic Odontology of Kazakhstan*. Alma-Ata: Nauka. (In Russian.)

Jien, S.-S. (1970). The Chinese dentition. II. Shovel incisors, Carabelli's cusps, groove patterns, cusp numbers, and abnormalities in morphology of the permanent teeth. *Journal of the Formosan Medical Association* **69**, 264–71.

Kanazawa, E., Sekikawa, M., Kamiakito, Y. and Ozaki, T. (1989). A quantitative investigation of irregular cusps in lower permanent molars. *Nihon University Journal of Oral Science* **15**, 450–6.

Kieser, J.A. and Preston, C.B. (1981). The dentition of the Lengua Indians of Paraguay. *American Journal of Physical Anthropology* **55**, 485–90.

Kraus, B.S. (1959). Occurrence of the Carabelli trait in Southwest ethnic groups. *American Journal of Physical Anthropology* **17**, 117–23.

Lasker, G.W. (1945). Observations on the teeth of Chinese born and reared in China and America. *American Journal of Physical Anthropology* **3**, 129–50.

Liu, K.-L. (1977). Dental condition of two tribes of Taiwan aborigines – Ami and Atayal. *Journal of Dental Research* **56**, 117–27.

Manabe, Y., Rokutanda, A., Kitagawa, Y. and Oyamada, J. (1991). Genealogical position of native Taiwanese (Bunun tribe) in East Asian populations based on tooth crown morphology. *Journal of the Anthropological Society of Nippon* **99**, 33–47.

Matsumura, H. (1990). Geographical variation of dental characteristics in the Japanese of the protohistoric Kofun period. *Journal of the Anthropological Society of Nippon* **98**, 439–49.

Mayhall, J.T. (1979). The dental morphology of the Inuit of the Canadian central Arctic. *OSSA* **6**, 199–218.

Merrill, R.G. (1964). Occlusal anomalous tubercles on premolars of Alaskan Eskimos and Indians. *Oral Surgery, Oral Medicine, Oral Pathology* **17**, 484–96.

Mizoguchi, Y. (1985). *Shovelling: A Statistical Analysis of its Morphology.* Tokyo: University of Tokyo Press.

Moorrees, C.F.A. (1957). *The Aleut Dentition: A Correlative Study of Dental Characteristics in an Eskimoid People.* Cambridge: Harvard University Press.

Nelson, C.T. (1938). The teeth of the Indians of Pecos Pueblo. *American Journal of Physical Anthropology* **23**, 261–93.

Oschinsky, L. and Smithurst, R. (1960). On certain dental characters of the Eskimo of the eastern Canadian Arctic. *Anthropologica* **11**, 105–12.

Pedersen, P.O. (1949). The East Greenland Eskimo dentition. *Meddelelser om Grønland* **142**, 1–244.

Rothhammer, F., Lasserre, E., Blanco, R., Covarrubias, E. and Dixon, M. (1968). Microevolution in Chilean populations. IV. Shovel-shape, mesial-palatal version and other dental traits in Pewenche Indians. *Zeitschrift für Morphologie und Anthropologie* **60**, 162–9.

Sciulli, P.W., Schneider, K.N. and Mahaney, M.C. (1984). Morphological variation of the permanent dentition in prehistoric Ohio. *Anthropologie* **22**, 211–15.

Scott, G.R. (1991). Continuity or replacement at the Uyak site: a physical anthropological analysis of population relationships. In *The Uyak Site on Kodiak Island: Its Place in Alaskan Prehistory.* University of Oregon Anthropological Papers No. 44, pp. 1–56.

Scott, G.R. (1994). Teeth and prehistory on Kodiak Island. In *Reckoning with the Dead: The Larsen Bay Repatriation and the Smithsonian Institution*, eds. T.L. Bray and T.W. Killion, pp. 67–74. Washington: Smithsonian Institution Press.

Scott, G.R. and Gillispie, T. (1997). The dentition of the prehistoric inhabitants of St. Lawrence Island, Alaska. In *St.- Lorenz Insel Studien*, ed. B. Haupt, Geneva: Publications Interdisciplinaires de l'Academie Suisse des Sciences Humaines et de la Societé Helvetique des Sciences Naturelle. (In press.)

Scott, G.R., Potter, R.H.Y., Noss, J.F., Dahlberg, A.A., and Dahlberg, T. (1983). The dental morphology of Pima Indians. *American Journal of Physical Anthropology* **61**, 13–31.

Snyder, R.G., Dahlberg, A.A., Snow, C.C. and Dahlberg, T. (1969). Trait analysis of the dentition of the Tarahumara Indians and mestizos of the Sierra Madre Occidental, Mexico. *American Journal of Physical Anthropology* **31**, 65–76.

Sofaer, J.A., Niswander, J.D., MacLean, C.J. and Workman, P.L. (1972a). Population studies on Southwestern Indian tribes. V. Tooth morphology as an indicator of biological distance. *American Journal of Physical Anthropology* **37**, 357–66.

Somogyi-Csizmazia, W. and Simons, A.J. (1971). Three-rooted mandibular first permanent molars in Alberta Indian children. *Journal of the Candian Dental Association* **37**, 105–6.

Suzuki, M. and Sakai, T. (1965). Labial surface pattern on permanent upper incisors of the Japanese. *Journal of the Anthropological Society of Japan* **73**, 1–8. (In Japanese, with English summary.)

Takei, T. (1990). An anthropological study on the tooth crown morphology in the Atayal tribe of Taiwan aborigines: comparative analysis between Atayal and some Asian-Pacific populations. *Journal of the Anthropological Society of Japan* **98**, 337–51. (In Japanese, with English summary.)

Turner, C.G., II and Hanihara, K. (1977). Additional features of Ainu dentition. V. Peopling of the Pacific. *American Journal of Physical Anthropology* **46**, 13–24.

Voronina, V.G. and Vaschaeva, V.F. (1979). Maritime territories. In *Ethnic Odontology of the USSR*, eds. A.A. Zubov and N.I. Khaldeeva, pp. 212–28. Moscow: Nauka. (In Russian.)

Wright, H.B. (1941). A frequent variation of the maxillary central incisors with some observations on dental caries among the Jivaro (Shuara) Indians of Ecuador. *American Journal of Orthodontics* **27**, 249–54.

Zubov, A.A. (1969) Odontological analysis of the cranial series from the Ekven and Uelen cemeteries. In *Ancient Cultures of the Asiatic Eskimos*, by S.A. Arutiunov and D.A. Sergeev, pp. 185–94. Moscow: Nauka. (In Russian.)

Sunda-Pacific

Abrahams, L.C. (1949). Shovel-shaped incisors in the Cape Malays. *Journal of the Dental Association of South Africa* **4**, 7–13.

Ganguly, P. (1960). Observations on the teeth of Nicobar Islanders.

Bulletin of the Anthropological Survey of India **9**, 43–50.

Hanihara, T. (1992). Biological relationships among Southeast Asians, Jomonese, and the Pacific populations as viewed from dental characters: the basic population of East Asia, X. *Journal of the Anthropological Society of Nippon* **100**, 53–67.

Hanihara, T. (1993). Dental affinities among Polynesia and circum-Polynesia populations. *Japan Review* **4**, 59–82.

Harris, E.F., Turner, C.G., II and Underwood, J.H. (1975). Dental morphology of living Yap Islanders, Micronesia. *Archaeology and Physical Anthropology in Oceania* **10**, 218–34.

Hochstetter, R.L. (1975). Incidence of trifurcated mandibular first permanent molars in the population of Guam. *Journal of Dental Research* **54**, 1097.

Jacob, T. (1967). Racial identification of the Bronze Age human dentitions from Bali, Indonesia. *Journal of Dental Research* **46** (suppl. to no. 5), 903–10.

Leigh, R.W. (1929). Dental morphology and pathology of prehistoric Guam. *Memoirs of the Bernice P. Bishop Museum* **11**, 451–79.

Sakai, T. (1975). The dentition of the Hawaiians. *Journal of the Anthropological Society of Nippon* **83**, 49–81.

Srisopark, S.S. (1972). A study on the size of permanent teeth, shovel-shaped incisors and paramolar tubercles in Thai skulls. *Journal of the Dental Association of Thailand* **22**, 199–205.

Suzuki, M. and Sakai, T. (1964). Shovel-shaped incisors among the living Polynesians. *American Journal of Physical Anthropology* **22**, 65–72.

Suzuki, M. and Sakai, T. (1973). Occlusal surface pattern of the lower molars and the second deciduous molar among the living Polynesians. *American Journal of Physical Anthropology* **39**, 305–16.

Turner, C.G., II and Scott, G.R. (1977). Dentition of Easter Islanders. In *Orofacial Growth and Development*, eds. A.A. Dahlberg and T.M. Graber, pp. 229–49. The Hague: Mouton Publishers.

Yamada, H. and Kawamoto, K. (1988). The dentition of Cook Islanders. In *People of the Cook Islands – Past and Present*, eds. K. Katayama and A. Tagaya, pp. 143–209. Rarotonga: The Cook Islands Library and Museum Society.

Sahul-Pacific

Bailit, H.L., DeWitt, S.J. and Leigh, R.A. (1968). The size and morphology of the Nasioi dentition. *American Journal of Physical Anthropology* **28**, 271–88.

Barksdale, J.T. (1972) Appendix III. A descriptive and comparative investigation of dental morphology. In *Physical Anthropology of the Eastern Highlands of New Guinea*, by R.A. Littlewood, pp. 113–74. Seattle: University of Washington Press.

Birdsell, J.B. (1993). *Microevolutionary Patterns in Aboriginal Australia.* New York: Oxford University Press.

Campbell, T.D. (1925). *Dentition and Palate of the Australian Aboriginal.* Adelaide: The Hassell Press.

Dahlberg, A.A. (1961). Relationship of tooth size to cusp number and groove conformation of occlusal surface patterns of lower molar teeth. *Journal of Dental Research* **40**, 34–8.

Doran, G.A. (1977). Characteristics of the Papua New Guinean dentition. I. Shovel-shaped incisors and canines associated with lingual tubercles. *Australian Dental Journal* **22**, 389–92.

Ellicott, D.F. (1979). The incidence of Carabelli's cusp in the Australian aboriginal. *Archaeology and Physical Anthropology in Oceania* **14**, 118–22.

Pal, A. (1972). Shovel-shaped incisors among the Negritoes of Andaman Islands. *Man in India* **52**, 239–51.

Richards, L.C. and Telfer, P.J. (1979). The use of dental characters in the assessment of genetic distance in Australia. *Archaeology and Physical Anthropology in Oceania* **14**, 184–94.

Smith, P., Brown, T. and Wood, W.B. (1981). Tooth size and morphology in a recent Australian aboriginal population from Broadbeach, South East Queensland. *American Journal of Physical Anthropology* **55**, 423–32.

Townsend, G.C. and Brown, T. (1981). The Carabelli trait in Australian aboriginal dentitions. *Archives of Oral Biology* **26**, 809–14.

Townsend, G.C., Yamada, H. and Smith, P. (1990). Expression of the entoconulid (sixth cusp) on mandibular molar teeth of an Australian aboriginal population. *American Journal of Physical Anthropology* **82**, 267–74.

References

Aas, I.H.M. and Risnes, S. (1979a). The depth of the lingual fossa in permanent incisors of Norwegians. I. Method of measurement, statistical distribution and sex dimorphism. *American Journal of Physical Anthropology* **50**, 335–40.

Aas, I.H.M. and Risnes, S. (1979b). The depth of the lingual fossa in permanent incisors in Norwegians. II. Differences between central and lateral incisors, correlations, size asymmetry and variability. *American Journal of Physical Anthropology* **50**, 341–8.

Aksianova, G.A. (1978). Some dental material in connection with the problem of the ancient populations of northern Europe. *Journal of Human Evolution* **7**, 525–8.

Aksianova, G.A. (1979). Peoples of the basin of the Pechora and lower Ob. In *Ethnic Odontology of the USSR*, eds. A.A. Zubov and N.I. Khaldeeva, pp. 93–113. Moscow: Nauka. (In Russian.)

Aksianova, G.A., Zubov, A.A. and Kochiev, R.S. (1977). Odontological description of the Komi-Zyrians. In *Physical Anthropology of the Komi*, Moscow: *Suomen Antropologisen Seuran Toimituksia* No. 4, pp. 65–74.

Aksianova, G.A., Zubov, A.A., Segeda, S.P., Peskina, M.Y. and Khaldeeva, N.I. (1979). Slavic peoples of the European part of the USSR. Russians. In *Ethnic Odontology of the USSR*, eds. A.A. Zubov and N.I. Khaldeeva, pp. 9–31. Moscow: Nauka. (In Russian.)

Alexandersen, V. (1962). Root conditions in human lower canines with special regard to double-rooted canines. II. Occurrence of double-rooted lower canines in Homo sapiens and other primates. *Sætryk af Tandlægebladet* **66**, 729–60.

Alexandersen, V. (1963). Double-rooted human lower canine teeth. In *Dental Anthropology*, ed. D.R. Brothwell, pp. 235–44. New York: Pergamon Press.

Alt, K.W., Pichler, S. and Vach, W. (1995). Dental morphology: teeth as key structures for the detection of biological relationships. In *Proceedings of the 10th International Symposium on Dental Morphology*, eds. R.J. Radlanski and H. Renz, pp. 324–331. Berlin: 'M' Marketing Services, C. & M. Brünne GbR.

Alvesalo, L. and Varrela, J. (1991). Taurodontism and the presence of an extra Y chromosome: study of 47,XYY males and analytical review. *Human Biology* **63**, 31–8.

Alvesalo, L., Nuutila, M. and Portin, P. (1975). The cusp of Carabelli. Occurrence in first upper molars and evaluation of its heritability. *Acta Odontologica Scandinavica* **33**, 191–7.

Anderson, D.L., Thompson, G.W. and Popovich, F. (1977). Molar polymorphisms and the timing of dentition mineralization. *Growth* **41**, 191–7.

Aoyagi, F. (1967). Morpho-genetical studies on similarities in the teeth and dental occlusion of twins. *Shikwa Gakuho (The Journal of the Tokyo Dental College Society)* **67**, 606–24.

Axelsson, G. and Kirveskari, P. (1982). Correlations between lower molar occlusal traits in Icelanders. In *Teeth: Form, Function, and Evolution*, ed. B. Kurtén, pp. 237–44. New York: Columbia University Press.

Babakov, O.B., Dubova, N.A., Zubov, A.A., Rykushina, G.B. and Khodzhiev, T.K. (1979). Peoples of Central Asia and Kazakhstan. In *Ethnic Odontology of the USSR*, eds. A.A. Zubov and N.I. Khaldeeva, pp. 164–86. Moscow: Nauka, (In Russian.)

Bader, R.S. (1965). Fluctuating asymmetry in the dentition of the house mouse. *Growth* **29**, 291–300.

Bailey-Schmidt, S.E. (1995). *Population Distribution of the Tuberculum Dentale Complex and Anomalies of the Maxillary Anterior Teeth*. MA thesis, Department of Anthropology, Arizona State University, Tempe.

Bailit, H.L. (1975). Dental variation among populations: an anthropologic view. *Dental Clinics of North America* **19**, 125–39.

Bailit, H.L. and Sung, B. (1968). Maternal effects on the developing dentition. *Archives of Oral Biology* **13**, 155–61.

Bailit, H.L., DeWitt, S.J. and Leigh, R.A. (1968). The size and morphology of the Nasioi dentition. *American Journal of Physical Anthropology* **28**, 271–88.

Bailit, H.L., Workman, P.L., Niswander, J.D. and MacLean, C.J. (1970). Dental asymmetry as an indicator of genetic and environmental conditions in human populations. *Human Biology* **42**, 626–38.

Bailit, H.L., Anderson, S. and Kolakowski, D. (1974). The genetics of tooth morphology. *American Journal of Physical Anthropology* **41**:468 (abstract).

Bailit, H., Brown, R. and Kolakowski, D. (1975). The heritability of non-metric dental traits. *American Journal of Physical Anthropology* **42**, 289 (abstract).

Barden, H.S. (1980). Fluctuating dental asymmetry: a measure of developmental instability in Down syndrome. *American Journal of Physical Anthropology* **52**, 169–73.

Barksdale, J.T. (1972). Appendix III: a descriptive and comparative investigation of dental morphology. In *Physical Anthropology of the Eastern Highlands of New Guinea*, by R.A. Littlewood, pp. 113–74. Seattle: University of Washington Press.

Barnes, D.S. (1969). Tooth morphology and other aspects of the Teso dentition. *American Journal of Physical Anthropology* **30**, 183–94.

Baume, R.M. and Crawford, M.H. (1979). Discrete dental trait asymmetry in Mexico and Belize. *Journal of Dental Research* **58**, 1811.

Baume, R.M. and Crawford, M.H. (1980). Discrete dental trait asymmetry in Mexican and Belizean groups. *American Journal of Physical Anthropology* **52**, 315–21.

Bellwood, P.S. (1985). *Prehistory of the Indo-Malaysian Archipelago*. Sydney: Academic Press.

Bellwood, P.S. (1989). The colonization of the Pacific: some current hypotheses. In *The Colonization of the Pacific: A Genetic Trail*, eds. A.V.S. Hill and S.W. Serjeantson, pp. 1–59. Oxford: Clarendon Press.

Bellwood, P.S. (1991). The Austronesian dispersal and the origin of languages. *Scientific American* **265**, 88–93.

Benedict, P.K. (1975). *Austro-Thai Language and Culture.* New Haven: Human Relations Area Files Press.

Bennett, K.W. (1979). *Fundamentals of Biological Anthropology.* Dubuque: Wm. C. Brown.

Bermúdez de Castro, J.M. (1988). Dental remains from Atapuerca/Ibeas (Spain). II. Morphology. *Journal of Human Evolution* **17**, 279–304.

Bermúdez de Castro, J.M. (1989). The Carabelli trait in human prehistoric populations of the Canary Islands. *Human Biology* **61**, 117–31.

Bermúdez de Castro, J.M. (1993). The Atapuerca dental remains. New evidence (1987–1991 excavations) and interpretations. *Journal of Human Evolution* **24**, 339–71.

Berry, A.C. (1976). The anthropological value of minor variants of the dental crown. *American Journal of Physical Anthropology* **45**, 257–68.

Berry, A.C. (1978). Anthropological and family studies on minor variants of the dental crown. In *Development, Function and Evolution of Teeth*, eds. P.M. Butler and K.A. Joysey, pp. 81–98. London: Academic Press.

Berry, R.J. (1968). The biology of non-metrical variation in mice and men. In *The Skeletal Biology of Earlier Human Populations*, ed. D.R. Brothwell, pp. 103–33. Oxford: Pergamon Press.

Berry, R.J. and Berry, A.C. (1967). Epigenetic variation in the human cranium. *Journal of Anatomy* **101**, 361–79.

Berryman, H.E., Owsley, D.W. and Henderson, A.M. (1979). Noncarious interproximal grooves in Arikara Indians. *American Journal of Physical Anthropology* **50**, 209–12.

Bhaskar, S.N., ed. (1976). *Orban's Oral Histology and Embryology.* Saint Louis: C.V. Mosby.

Bhussry, B.R. (1976). Development and growth of teeth. In *Orban's Oral Histology and Embryology.* ed. S.N. Bhaskar, pp. 23–44. Saint Louis: C.V. Mosby.

Biggerstaff, R. H. (1970). Morphological variations for the permanent mandibular first molars in human monozygotic and dizygotic twins. *Archives of Oral Biology* **15**, 721–30.

Biggerstaff, R. H. (1973). Heritability of the Carabelli cusp in twins. *Journal of Dental Research* **52**, 40–4.

Biggerstaff, R.H. (1979). The biology of dental genetics. *Yearbook of Physical Anthropology* **22**, 215–27.

Birdsell, J.B. (1951). The problem of the early peopling of the Americas as viewed from Asia. In *The Physical Anthropology of the American Indian*, ed. W.S. Laughlin, pp. 1–68. New York: The Viking Fund.

Birdsell, J.B. (1977). The recalibration of a paradigm for the first peopling of Greater Australia. In *Sunda and Sahul*, eds. J. Allen, J. Golson and R. Jones, pp. 113–67. London: Academic Press.

Birdsell, J.B. (1981). *Human Evolution.* 3rd edition. Boston: Houghton Mifflin.

Birdsell, J.B. (1993). *Microevolutionary Patterns in Aboriginal Australia.* New York: Oxford University Press.

Black, G.V. (1902). *Descriptive Anatomy of Human Teeth.* 5th edition, Philadelphia: S.S. White Dental Mfg. Co.

Blanco, R. and Chakraborty, R. (1977). The genetics of shovel shape in maxillary central incisors in man. *American Journal of Physical Anthropology* **44**, 233–6.

Boas, F. (1912). *Changes in the Bodily Form of Descendants of Immigrants.* New York: Columbia University Press.

Bolk, L. (1915). Das Carabellische Höckerchen. *Schweizerische Vierteljahrsschrift für Zahnheilkunde* **25**, 81–104.

Boyd, W.C. (1950). *Genetics and the Races of Man.* Boston: Little, Brown and Company.

Brabant, H. (1964). Observations sur l'evolution de la denture permanente humaine en Europe Occidentale. *Bulletin du Groupement International pour la Recherche Scientifique en Stomatologie et Odontologie* **7**, 11–84.

Brabant, H. (1967). Comparison of the characteristics and anomalies of the deciduous and the permanent dentition. *Journal of Dental Research* **46** (suppl. to no. 5), 897–902.

Brabant, H. E. (1971). The human dentition during the Megalithic era. In *Dental Morphology and Evolution*, ed. A.A. Dahlberg, pp. 283–97. Chicago: University of Chicago Press.

Brabant, H. and Ketelbant, R. (1975). Observations sur la frequence de certains caracteres Mongoloides dans la denture permanente de la population Belge. *Bulletin du Groupement International pour la Recherche Scientifique en Stomatologie et Odontologie* **18**, 121–34.

Brace, C.L. (1980). Australian tooth-size clines and the death of a stereotype. *Current Anthropology* **21**, 141–53.

Brace, C.L. and Hinton, R.J. (1981). Oceanic tooth size variation as a reflection of biological and cultural mixing. *Current Anthropology* **22**, 549–57.

Brace, C.L. and Hunt, K.D. (1990). A nonracial craniofacial perspective on human variation A(ustralia) to Z(uni). *American Journal of Physical Anthropology* **82**, 341–60.

Brace, C.L. and Montagu, M.F.A. (1977). *Human Evolution: An Introduction to Biological Anthropology.* 2nd edition. New York: Macmillan.

Brace, C.L. and Nagai, M. (1982). Japanese tooth size, past and present. *American Journal of Physical Anthropology* **59**, 399–411.

Brace, C.L. and Tracer, D.P. (1992). Craniofacial continuity and change: a comparison of late Pleistocene and recent Europe and Asia. In *The Evolution and Dispersal of Modern Humans in Asia*, eds. T. Akazawa, K. Aoki, and T. Kimura, pp. 439–71. Tokyo: Hokusen-Sha Publishing Co.

Brewer-Carias, C.A., Le Blanc, S. and Neel, J.V. (1976). Genetic structure of a tribal population, the Yanomama Indians. XII. Dental microdifferentiation. *American Journal of Physical Anthropology* **44**, 5–14.

Brook, A.H. (1984). A unifying aetiological explanation for anomalies of human tooth number and size. *Archives of Oral Biology* **29**, 373–8.

Brothwell, D.R. (Ed.) (1963). *Dental Anthropology.* New York: Pergamon Press.

Brues, A.M. (1977). *People and Races.* New York: Macmillan.

Butler, P.M. (1937). Studies of the mammalian dentition. I. The teeth of *Centetes ecaudatus* and its allies. *Proceedings of the Zoological Society of London* **B107**, 103–32.

Butler, P.M. (1939). Studies of the mammalian dentition. Differentiation of the post-canine dentition. *Proceedings of the Zoological Society of London* **B109**, 1–36.

Butler, P.M. (1956). The ontogeny of molar pattern. *Biological Review* **31**, 30–70.

Butler, P.M. (1963). Tooth morphology and primate evolution. In *Dental Anthropology*, ed. D.R. Brothwell, pp. 1–13. New York: Pergamon Press.

Butler, P.M. (1982). Some problems of the ontogeny of tooth patterns. In *Teeth: Form, Function and Evolution*, ed. B. Kurtén, pp. 44–51. New York: Columbia University Press.

Butler, P.M. and Joysey, K.A. (Eds.) (1978). *Development, Function and Evolution of Teeth*. New York: Academic Press.

Cadien, J.D. (1972). Dental variation in man. In *Perspectives on Human Evolution 2*, eds. S.L. Washburn and P. Dolhinow, pp. 199–222. New York: Holt, Rinehart and Winston.

Campbell, T.D. (1925). *The Dentition and Palate of the Australian Aboriginal*. Adelaide: Hassell Press.

Cann, R.L., Stoneking, M. and Wilson, A.C. (1987). Mitochondrial DNA and human evolution. *Nature* **325**, 31–6.

von Carabelli, G. (1842). *Anatomie des Mundes*. Wien: Braumüller und Seidel.

Carbonell, V. M. (1963). Variations in the frequency of shovel-shaped incisors in different populations. In *Dental Anthropology*, ed. D.R. Brothwell, pp. 211–34. New York: Pergamon Press.

Carlsen, O. (1987). *Dental Morphology*. Copenhagen: Munksgaard.

Carlsen, O. and Alexandersen, V. (1990). Radix entomolaris: identification and morphology. *Scandinavian Journal of Dental Research* **98**, 363–73.

Carter, C.O. (1969). Genetics of common disorders. *British Medical Bulletin* **25**, 52–7.

Cavalli-Sforza, L.L., Piazza, A., Menozzi, P. and Mountain, J. (1988). Reconstruction of human evolution: bringing together genetic, archaeological, and linguistic data. *Proceedings of the National Academy of Sciences USA* **85**, 6002–6.

Cavalli-Sforza, L.L., Menozzi, P. and Piazza, A. (1994). *The History and Geography of Human Genes*. Princeton: Princeton University Press.

Cheboksarov, N.N. (1966). *The Ethnic Anthropology of Eastern Asia*. Moscow: Nauka.

Chagnon, N. A. (1972). Tribal social organization and genetic microdifferentiation. In *The Structure of Human Populations*, eds. G.A. Harrison and A.J. Boyce, pp. 252–82. Oxford: Clarendon Press.

Chappel, H.G. (1927). Jaws and teeth of ancient Hawaiians. *Memoirs of the B.P. Bishop Museum* **9**, 249–68.

Chen, H., Sokal, R.R., and Ruhlen, M.R. (1995). Worldwide analysis of genetic and linguistic relationships of human populations. *Human Biology* **67**, 595–612.

Christian, J.C. (1979). Testing twin means and estimating genetic variance: basic methodology for the analysis of quantitative twin data. *Acta Geneticae, Medicae, et Gemellologiae* **28**, 35–40.

Christian, J.C. and Norton, J.A. (1977). A proposed test of the difference between the means of monozygotic and dizygotic twins. *Acta Geneticae, Medicae, et Gemellologiae* **26**, 49–53.

Christian, J.C., Kang, K.W. and Norton, J.A. (1974). Choice of an estimate of genetic variance from twin data. *American Journal of Human Genetics* **26**, 154–61.

Christian, J.C., Feinleib, M. and Norton, J.A. (1975). Statistical analysis of genetic variance in twins. *American Journal of Human Genetics* **27**, 807.

Constandse-Westermman, T.S. (1972). *Coefficients of Biological Distance*. Oosterhout N.B., The Netherlands: Anthropological Publications.

Coon, C.S. (1962). *The Origin of Races*. New York: Alfred A. Knopf.

Coon, C.S. (1965). *The Living Races of Man*. New York: Alfred A. Knopf.

Coon, C.S., Garn, S.M. and Birdsell, J.B. (1950). *Races: a Study of the Problems of Race Formation in Man*. Springfield: C.C. Thomas.

Cope, E.D. (1874). On the homologies and origin of the types of molar teeth in Mammalia educabilia. *Journal of the Academy of Natural Sciences, Philadelphia* **8**, 71–89.

Cope, E.D. (1888). On the tritubercular molar in human dentition. *Journal of Morphology* **2**, 7–26.

Corruccini, R.S. and Potter, R.H.Y. (1981). Developmental correlates of crown component asymmetry and occlusal discrepancy. *American Journal of Physical Anthropology* **55**, 21–31.

Corruccini, R.S., Sharma, K. and Potter, R.H.Y. (1988). Comparative genetic variance of dental size and asymmetry in U.S. and Punjabi twins. In *Teeth Revisited*, eds. D.E. Russell, J.-P. Santoro, and D. Sigogneau-Russell, Mémoires du Muséum Nationale d'Histoire Naturelle, Série C, Sciences de la Terre No. 53, pp. 47–53.

Count, E.W. (1950). *This is Race*. New York: Henry Schuman.

Cox, G.J., Finn, S.B. and Ast, D.B. (1961). Effect of fluoride ingestion on the size of the cusp of Carabelli during tooth formation. *Journal of Dental Research* **40**, 393–5.

Curzon, M.E.J. (1973). Three-rooted mandibular permanent molars in English Caucasians. *Journal of Dental Research* **52**, 181.

Curzon, M.E.J., Curzon, J.A. and Poyton, H.G. (1970). Evaginated odontomes in the Keewatin Eskimo. *British Dental Journal* **129**, 324–8.

Dahlberg, A.A. (1945a). The changing dentition of man. *Journal of the American Dental Association* **32**, 676–90.

Dahlberg, A.A. (1945b). The paramolar tubercle (Bolk). *American Journal of Physical Anthropology* **3**, 97–103.

Dahlberg, A.A. (1950). The evolutionary significance of the protostylid. *American Journal of Physical Anthropology* **8**, 15–25.

Dahlberg, A.A. (1951). The dentition of the American Indian. In *The Physical Anthropology of the American Indian*, ed. W.S. Laughlin, pp. 138–76. New York: The Viking Fund.

Dahlberg, A.A. (1956). Materials for the establishment of standards for classification of tooth characters, attributes, and techniques in morphological studies of the dentition. Zollar Laboratory of Dental Anthropology, University of Chicago (mimeo).

Dahlberg, A.A. (1959). A wing-like appearance of upper central incisors among American Indians. *Journal of Dental Research* **38**, 203–4.

Dahlberg, A.A. (1961). Relationship of tooth size to cusp number and groove conformation of occlusal surface patterns of lower molar teeth. *Journal of Dental Research* **40**, 34–8.

Dahlberg, A.A. (1963a). Analysis of the American Indian dentition. In *Dental*

Anthropology, ed. D.R. Brothwell, pp. 149–78. New York: Pergamon Press.

Dahlberg, A.A. (1963b). Dental evolution and culture. *Human Biology* **35**, 237–49.

Dahlberg, A.A. (1968). On the teeth of early sapiens. In *Evolution and Hominisation*, ed. G. Kurth, pp. 273–80. Stuttgart: Gustav Fischer.

Dahlberg, A.A., (Ed.) (1971a). *Dental Morphology and Evolution*. Chicago: University of Chicago Press.

Dahlberg, A.A. (1971b). Penetrance and expressivity of dental traits. In *Dental Morphology and Evolution*, ed. A.A. Dahlberg, pp. 257–62. Chicago: University of Chicago Press.

Dahlberg, A.A. and Mikkelsen, O. (1947). The shovel-shaped character in the teeth of the Pima Indians. *American Journal of Physical Anthropology* **5**, 234–5. (Abstract.)

Dahlberg, A.A., Kirveskari, P. and Dahlberg, T. (1982). The Pima Indian studies of the inheritance of dental morphological traits. In *Teeth: Form, Function, and Evolution*, ed. B. Kurtén, pp. 292–7. New York: Columbia University Press.

Davies, P.L. (1968). Relationship of cusp reduction in the permanent mandibular first molar to agenesis of teeth. *Journal of Dental Research* **47**, 499.

Dempster, E.R. and Lerner, I.M. (1950). Heritability of threshold characters. *Genetics* **35**, 212–36.

DeVoto, F.C.H. and Perrotto, B.M. (1971). Phenotypes and genotypes of Carabelli's complex in isolated populations of Argentina highlands. *Journal of Dental Research* **50**, 1152–3.

DeVoto, F.C.H., Arias, N.H., Ringuelet, S. and Palma, N.H. (1968). Shovel-shaped incisors in a northwestern Argentine population. *Journal of Dental Research* **47**, 820–3.

Dietz, V.H. (1944). A common dental morphotropic factor: the Carabelli cusp. *Journal of the American Dental Association* **31**, 784–9.

Dixon, G.H. and Stewart. R.E. (1976). Genetic aspects of anomalous occlusal development. In *Oral Facial Genetics*, eds. R.E. Stewart and G.H. Prescott, pp. 124–50. Saint Louis: C. V. Mosby.

Dixon, R.B. (1923). *The Racial History of Man*. New York: Charles Scribner's Sons.

Dobzhansky, T. (1937). *Genetics and the Origin of Species*. New York: Columbia University Press.

Dodo, Y., Ishida, H. and Saitou, N. (1992). Population history of Japan: a cranial nonmetric approach. In *The Evolution and Dispersal of Modern Humans in Asia*, eds. T. Akazawa, K. Aoki and T. Kimura, pp. 479–92. Tokyo: Hokusen-Sha Publishing Co.

Doran, G.A. (1977). Characteristics of the Papua New Guinean dentition. I. Shovel-shaped incisors and canines associated with lingual tubercles. *Australian Dental Journal* **22**, 389–92.

Drennan, M.R. (1929). The dentition of a Bushman tribe. *Annals of the South African Museum* **24**, 61–88.

Du Bois, C. (1944). *The People of Alor*. Minneapolis: The University of Minnesota Press.

Du Souich, P. (1974). Estudio antropológico de los dientes de una población del Bronce I de Gorafe (Granada). *Anales del Desarrollo* **18**, 137–66.

Ebeling, C.F., Ingervall, B., Hedegard, B. and Lewin, R. (1973). Secular changes in tooth size in Swedish men. *Acta Odontologica Scandinavica* **31**, 141–7.

Echo-Hawk, R.C. (1994). *Kara Katit Pakutu: Exploring the Origins of Native America in Anthropology and Oral Traditions.* M.A. Thesis, Department of Anthropology, University of Colorado, Boulder.

Eckhardt, R.B. (1979). *The Study of Human Evolution.* New York: McGraw-Hill.

Edwards, J.H. (1960). The simulation of Mendelism. *Acta Genetica* **10**, 63–70.

Edwards, J.H. (1969). Familial predisposition in man. *British Medical Bulletin* **25**, 58–64.

Enoki, D. and Nakamura, E. (1959). Bilateral rotation (mesiopalatal torsion) of maxillary central incisors. *Journal of Dental Research* **38**, 204.

Enoki, K. and Dahlberg, A.A. (1958). Rotated maxillary central incisors. *Orthodontic Journal of Japan* **17**, 157–69.

Escobar, V., Melnick, M. and Conneally, P.M. (1976). The inheritance of bilateral rotation of maxillary central incisors. *American Journal of Physical Anthropology* **45**, 109–16.

Escobar, V., Conneally, P.M. and Lopez, C. (1977). The dentition of the Queckchi Indians. Anthropological aspects. *American Journal of Physical Anthropology* **47**, 443–52.

Falconer, D. S. (1960). *Introduction to Quantitative Genetics.* New York: The Ronald Press Company.

Frank, H.A. (1926). *East of Siam: Ramblings in the Five Divisions of French Indo-China.* New York: Century.

Frayer, D.W. (1978). *Evolution of the Dentition in Upper Paleolithic and Mesolithic Europe.* Lawrence, Kansas: University of Kansas Publications in Anthropology 10.

Gabriel, A.C. (1948). *Genetic Types in Teeth.* Sydney: Australasian Medical Publishing Company Limited.

Gadzhiev, Y.M. (1979). Peoples of the Caucasus, Daghestan. In *Ethnic Odontology of the USSR*, eds. A.A. Zubov and N.I. Khaldeeva, pp. 141–63. Moscow: Nauka. (In Russian.)

Garn, S.M. (1971). *Human Races.* 3rd edn. Springfield: C.C. Thomas.

Garn, S. M. (1977). Genetics of dental development. In *The Biology of Occlusal Development*, ed. J. A. McNamara, Jr., pp. 61–88. Ann Arbor: Center for Human Growth and Development,.

Garn, S.M. and Bailey, S.M. (1977). The symmetrical nature of bilateral asymmetry (δ) of deciduous and permanent teeth. *Journal of Dental Research* **56**, 1422.

Garn, S.M., Lewis, A.B. and Vicinus, J.H. (1963). Third molar polymorphism and its significance to dental genetics. *Journal of Dental Research* **42**(suppl. to no. 6), 1344–63.

Garn, S.M., Lewis, A.B. and Kerewsky, R.S. (1964). Sex difference in tooth size. *Journal of Dental Research* **43**, 306.

Garn, S.M., Lewis, A.B. and Kerewsky, R.S. (1965). Size interrelationships of the mesial and distal teeth. *Journal of Dental Research* **44**, 350–53.

Garn, S.M., Dahlberg, A.A., Lewis, A.B. and Kerewsky, R.S. (1966a). Groove pattern, cusp number, and tooth size. *Journal of Dental Research* **45**, 970.

Garn, S.M., Lewis, A.B. and Kerewsky, R.S. (1966b). Bilateral asymmetry and concordance in cusp number and crown morphology of the mandibular first molar. *Journal of Dental Research* **45**, 1820.

Garn, S.M., Lewis, A.B. and Kerewsky, R.S. (1966c) Sexual dimorphism in the buccolingual tooth diameter. *Journal of Dental Research* **45**, 1819.

Garn, S.M., Lewis, A.B. and Kerewsky, R.S. (1966d). Extent of sex influence on Carabelli's polymorphism. *Journal of Dental Research* **45**, 1823.

Garn, S.M., Lewis, A.B., Kerewsky, R.S. and Dahlberg, A.A. (1966e). Genetic independence of Carabelli's trait from tooth size or crown morphology. *Archives of Oral Biology* **11**, 745–7.

Garn, S.M., Lewis, A.B. and Kerewsky, R.S. (1967). Shape similarities throughout the dentition. *Journal of Dental Research* **46**, 1481.

Garn, S.M., Lewis, A.B. and Kerewsky, R.S. (1968a). Relationship between buccolingual and mesiodistal tooth diameters. *Journal of Dental Research* **47**, 495.

Garn, S.M., Lewis, A.B. and Walenga, A. (1968b). Evidence for a secular trend in tooth size over two generations. *Journal of Dental Research* **47**, 503.

Garn, S.M., Cole, P.E., Wainwright, R.L. and Guire, K.E. (1977). Sex discriminatory effectiveness using combinations of permanent teeth. *Journal of Dental Research* **56**, 697.

Garn, S.M., Cole, P.E. and Smith, B.H. (1979). The effect of sample size on crown size asymmetry. *Journal of Dental Research* **58**, 2012.

Garrod, A.E. (1909). *Inborn Errors of Metabolism*. London: Henry Frowde, Hodder and Stoughton.

Gaunt, W.A. and Miles, A.E.W. (1967). Fundamental aspects of tooth morphogenesis. In *Structural and Chemical Organization of Teeth*, Vol. 1, ed. A.E.W. Miles, pp. 151–97. New York: Academic Press.

Gilmore, R.W. (1968). Epidemiology and heredity of accessory occlusal ridges on the buccal cusps of human premolar teeth. *Archives of Oral Biology* **13**, 1035–46.

Glasstone, S. (1952). The development of halved tooth germs: a study in experimental embryology. *Journal of Anatomy* **86**, 12–15.

Glasstone, S. (1967). Development of teeth in tissue culture. *Journal of Dental Research* **46**(suppl. to no. 5), 858–61.

Glasstone, S. (1979). Tissue culture of the development of teeth and jaws. *OSSA* **6**, 89–104.

Goaz, P.W. and Miller, M.C., III (1966). A preliminary description of the dental morphology of the Peruvian Indian. *Journal of Dental Research* **45**, 106–19.

Goldstein, M.S. (1948). Dentition of Indian crania from Texas. *American Journal of Physical Anthropology* **6**, 63–84.

Gomes, E.H. (1911). *Seventeen Years Among the Sea Dyaks of Borneo; a Record of Intimate Association With the Natives of the Bornean Jungles*. London: Seeley.

Goodman, H. O. (1965). Genetic parameters of dentofacial development. *Journal of Dental Research* **44**(suppl. to no. 1), 174–84.

Goose, D.H. (1977). The dental condition of Chinese living in Liverpool. In *Orofacial Growth and Development*, eds. A.A. Dahlberg and T.M. Graber, pp. 183–94. The Hague: Mouton Publishers.

Goose, D.H. and Lee, G. T. R. (1971). The mode of inheritance of Carabelli's trait. *Human Biology* **43**, 64–9.

Gorjanovic-Kramberger, K. (1906). *Der Diluviale Mensch von Krapina in Kroatien.* Wiesbaden: Kreidel.

Gould, S.J. (1977). *Ontogeny and Phylogeny.* Cambridge, Mass.: Belknap Press.

Grainger, R.M, Paynter, K.J., Honey, L. and Lewis, D. (1966). Epidemiologic studies of tooth morphology. *Journal of Dental Research* **45** (suppl. to no. 3), 693–702.

Gravere, R.U., Zubov, A.A. and Sarap, G.G. (1979). Baltic peoples. In *Ethnic Odontology of the USSR,* eds. A.A. Zubov and N.I. Khaldeeva, pp. 68–92. Moscow: Nauka. (In Russian.)

Green, R.C. (1994). Changes over time: recent advances in dating human colonisation of the Pacific Basin area. In *The Origin of the First New Zealanders,* ed. D.G. Sutton, pp. 1–33. Auckland: Auckland University Press.

Green, R.F., Suchey, J.M. and Gokhale, D.V. (1979). The statistical treatment of correlated bilateral traits in the analysis of cranial material. *American Journal of Physical Anthropology* **50**, 629–34.

Greenberg, J.H. (1963). The languages of Africa. *International Journal of American Linguistics* **29**, 1–177.

Greenberg, J.H. (1971). The Indo-Pacific hypothesis. In *Linguistics in Oceania,* ed. T.A. Sebeok, Current Trends in Linguistics, **8**, 807–71.

Greenberg, J.H. (1987). *Language in the Americas.* Stanford: Stanford University Press.

Greenberg, J.H., Turner, C.G., II and Zegura, S. (1985). Convergence of evidence for the peopling of the Americas. *Collegium Antropologicum* **9**, 33–42.

Greenberg, J.H., Turner, C.G., II and Zegura, S. (1986). The settlement of the Americas: a comparison of the linguistic, dental, and genetic evidence. *Current Anthropology* **24**, 477–97.

Greene, D.L. (1967). Dentition of Meroitic, X-group, and Christian populations from Wadi Halfa, Sudan. *Anthropological Papers, Department of Anthropology, University of Utah,* No. 85. Salt Lake City: University of Utah Press.

Greene, D.L. (1972). Dental anthropology of early Egypt and Nubia. *Journal of Human Evolution* **1**, 315–24.

Greene, D.L. (1982). Discrete dental variations and biological distances of Nubian populations. *American Journal of Physical Anthropology* **58**, 75–9.

Greene, D.L., Ewing, G.H. and Armelagos, G.J. (1967). Dentition of a Mesolithic population from Wadi Halfa, Sudan. *American Journal of Physical Anthropology* **27**, 41–56.

Gregory, W.K. (1916). Studies on the evolution of the Primates. I. The Cope-Osborn 'theory of trituberculy' and the ancestral molar patterns of the Primates. *Bulletin of the American Museum of Natural History* **35**, 239–57.

Gregory, W.K. (1922). *The Origin and Evolution of the Human Dentition.* Baltimore: Williams and Wilkins.

Gregory, W.K. and Hellman, M. (1926). The dentition of Dryopithecus and the origin of man. *American Museum of Natural History Anthropological Papers* **28**, 1–117.

Grine, F.E. (1981). Occlusal morphology of the mandibular permanent molars of the South African Negro and the Kalahari San (Bushman). *Annals of the South African Museum* **86**, 157–215.

Grine, F.E. (1986). Anthropological aspects of the deciduous teeth of South

African blacks. In *Variation, Culture and Evolution in African Populations*, eds. R. Singer and J.K. Lundy, pp. 47–83. Johannesburg: Witwatersrand University Press.

Grine, F.E. (1990). Deciduous dental features of Kalahari San: comparison of non-metrical traits. In *From Apes to Angels: Essays in Anthropology in Honor of Philip V. Tobias*, ed. G.H. Sperber, pp. 153–69. New York: Wiley-Liss.

Grüneberg, H. (1952). Genetical studies on the skeleton of the mouse. IV. Quasi-continuous variations. *Journal of Genetics* **51**, 95–114.

Grüneberg, H. (1963). *The Pathology of Development*. New York: John Wiley.

Guthrie, R.D. (1996). The mammoth steppe and the origin of Mongoloids and their dispersal. In *Prehistoric Dispersal of Mongoloids*, eds. T. Akazawa and E.J.E. Szathmary, pp. 172–86. Oxford: Oxford University Press.

Habgood, P.J. (1985). The origin of the Australian Aborigines: an alternative approach and view. In *Hominid Evolution: Past, Present and Future*, ed. P. Tobias, pp. 367–80. New York: Alan R. Liss.

Haeussler, A.M. (1996). *Biological Relationships of Late Pleistocene and Holocene Eurasian and American Peoples: The Dental Anthropological Evidence*. PhD dissertation, Department of Anthropology, Arizona State University, Tempe.

Haeussler, A.M. and Turner, C.G., II (1992). The dentition of Soviet Central Asians and the quest for New World ancestors. In *Culture, Ecology and Dental Anthropology*, ed. J.R. Lukacs, *Journal of Human Ecology* (Special Issue) **2**, 273–97.

Haeussler, A.M., Irish, J.D., Morris, D.H. and Turner, C.G., II (1989). Morphological and metrical comparison of San and Central Sotho dentitions from southern Africa. *American Journal of Physical Anthropology* **78**, 115–22.

Halffman, C.M, Scott, G.R. and Pedersen, P.O. (1992). Palatine torus in the Greenlandic Norse. *American Journal of Physical Anthropology* **88**, 145–61.

Hanihara, K. (1954). Studies on the deciduous dentition of the Japanese and the Japanese-American hybrids. I. Deciduous incisors. *Journal of the Anthropological Society of Nippon* **63**, 168–85.

Hanihara, K. (1955). Studies on the deciduous dentition of the Japanese and the Japanese-American hybrids. II. Deciduous canines. *Journal of the Anthropological Society of Nippon* **64**, 63–82.

Hanihara, K. (1956a). Studies on the deciduous dentition of the Japanese and the Japanese-American hybrids. III. Deciduous lower molars. *Journal of the Anthropological Society of Nippon* **64**, 95–116.

Hanihara, K. (1956b). Studies on the deciduous dentition of the Japanese and the Japanese-American hybrids. IV. Deciduous upper molars. *Journal of the Anthropological Society of Nippon* **65**, 67–87.

Hanihara, K. (1957). Studies on the deciduous dentition of the Japanese and the Japanese-American hybrids. V. General conclusion. *Journal of the Anthropological Society of Nippon* **65**, 151–64.

Hanihara, K (1961). Criteria for classification of crown characters of the human deciduous dentition. *Journal of the Anthropological Society of Nippon* **69**, 27–45.

Hanihara, K. (1963). Crown characters of the deciduous dentition of the Japanese-American hybrids. In *Dental Anthropology*, ed. D.R. Brothwell, pp. 105–24. New York: Pergamon Press.

Hanihara, K. (1965). Some crown characters of the deciduous incisors and canines in Japanese–American hybrids. *Journal of the Anthropological Society of Nippon* **72**, 135–45.

Hanihara, K. (1966). Mongoloid dental complex in the deciduous dentition. *Journal of the Anthropological Society of Nippon* **74**, 61–71.

Hanihara, K. (1968a). Mongoloid dental complex in the permanent dentition. *Proceedings of the VIIIth International Congress of Anthropological and Ethnological Sciences*, Vol. 1, pp. 298–300. Tokyo: Science Council of Japan.

Hanihara, K. (1968b). Morphological pattern of the deciduous dentition in the Japanese-American hybrids. *Journal of the Anthropological Society of Nippon* **76**, 114–21.

Hanihara, K. (1970). Mongoloid dental complex in the deciduous dentition, with special reference to the dentition of Ainu. *Journal of the Anthropological Society of Nippon* **78**, 3–17.

Hanihara, K. (1977). Dentition of the Ainu and the Australian aborigines. In *Orofacial Growth and Development*, eds. A.A. Dahlberg and T.M. Graber, pp. 195–200. The Hague: Mouton Publishers.

Hanihara, K. (1984). Origins and affinities of Japanese viewed from cranial measurements. *Acta Anthropogenetica* **8**, 149–58.

Hanihara, K. (1991). Dual structure model for the population history of the Japanese. *Japan Review* **2**, 1–33.

Hanihara, K. (1992). Dual structure model for the formation of the Japanese population. In *International Symposium on Japanese as a Member of the Asian and Pacific Populations*, ed. K. Hanihara, pp. 244–51. Kyoto: International Research Center for Japanese Studies.

Hanihara, K., Tamada, M. and Tanaka, T. (1970). Quantitative analysis of the hypocone in the human upper molar. *Journal of the Anthropological Society of Nippon* **78**, 200–7.

Hanihara, K., Masuda, T. and Tanaka, T. (1974). Affinities of dental characteristics in the Okinawa Islanders. *Journal of the Anthropological Society of Nippon* **82**, 75–82.

Hanihara, K., Masuda, T. and Tanaka, T. (1975) Family studies of the shovel trait in the maxillary central incisor. *Journal of the Anthropological Society of Nippon* **83**, 107–12.

Hanihara, T. (1989a). Affinities of the Philippine Negritos as viewed from dental characters: a preliminary report. *Journal of the Anthropological Society of Nippon* **97**, 327–39.

Hanihara, T. (1989b). Comparative studies of dental characteristics in the Aogashima Islanders. *Journal of the Anthropological Society of Nippon* **97**, 9–22.

Hanihara, T. (1990a). Dental anthropological evidence of affinities among the Oceania and the Pan-Pacific populations: the basic populations of East Asia, II. *Journal of the Anthropological Society of Nippon* **98**, 233–46.

Hanihara, T. (1990b). Studies on the affinities of Sakhalin Ainu based on dental characters: the basic population of East Asia, III. *Journal of the Anthropological Society of Nippon* **98**, 425–37.

Hanihara, T. (1991a). The origin and microevolution of Ainu as viewed from dentition: the basic populations in East Asia, VIII. *Journal of the Anthropological Society of Nippon* **99**, 345–61.

Hanihara, T. (1991b). Dentition of Nansei Islanders and peopling of the Japanese archipelago: the basic populations in East Asia, IX. *Journal of the Anthropological Society of Nippon* **99**, 399–409.

Hanihara, T. (1991c). Dental and cranial evidence on the affinities of the East Asian and Pacific populations. In *Japanese as a Member of the Asian and Pacific Populations: International Symposium 4*, ed. K. Hanihara, pp. 119–37. Kyoto: International Research Center for Japanese Studies.

Hanihara, T. (1992a). Biological relationships among Southeast Asians, Jomonese, and the Pacific populations as viewed from dental characters: the basic populations in East Asia, X. *Journal of the Anthropological Society of Nippon* **100**, 53–67.

Hanihara, T. (1992b). Negritos, Australian aborigines, and the 'Proto-Sundadont' dental pattern: the basic populations in East Asia, V. *American Journal of Physical Anthropology* **88**, 183–96.

Hanihara, T. (1992c). Dental variation of the Polynesian populations. *Journal of the Anthropological Society of Nippon* **100**, 291–301.

Hanihara, T. (1993). Dental affinities among Polynesian and circum-Polynesian populations. *Japan Review* **4**, 59–82.

Hanihara, T. (1994). Craniofacial continuity and discontinuity of Far Easterners in the late Pleistocene and Holocene. *Journal of Human Evolution* **27**, 417–41.

Hanihara, T. (1996). Comparison of craniofacial features of major human groups. *American Journal of Physical Anthropology* **99**, 389–412.

Harris, E.F. (1977). *Anthropologic and Genetic Aspects of the Dental Morphology of Solomon Islanders, Melanesia*. PhD dissertation, Department of Anthropology, Arizona State University, Tempe.

Harris, E.F. (1980). Sex differences in lingual marginal ridging on the human maxillary central incisor. *American Journal of Physical Anthropology* **52**, 541–8.

Harris, E.F. and Bailit, H.L. (1980). The metaconule: a morphologic and familial analysis of a molar cusp in humans. *American Journal of Physical Anthropology* **53**, 349–58.

Harris, E.F. and Bailit, H.L. (1988). A principal components analysis of human odontometrics. *American Journal of Physical Anthropology* **75**, 87–99.

Harris, E.F. and Nweeia, M.T. (1980). Dental asymmetry as a measure of environmental stress in the Ticuna Indians of Columbia. *American Journal of Physical Anthropology* **53**, 133–42.

Harris, E.F., Turner, C.G., II and Underwood, J.H. (1975). Dental morphology of living Yap Islanders, Micronesia. *Archaeology and Physical Anthropology in Oceania* **10**, 218–34.

Harrison, G.A., Tanner, J.M., Pilbeam, D.R. and Baker, P.T. (1988). *Human Biology*. 3rd edn. Oxford: Oxford University Press.

Hellman, M. (1928). Racial characters in human dentition. *Proceedings of the American Philosophical Society* **67**, 157–74.

Hershkovitz, P. (1971). Basic crown patterns and cusp homologies of mammalian teeth. In *Dental Morphology and Evolution*, ed. A.A. Dahlberg, pp. 95–150. Chicago: University of Chicago Press.

Heyerdahl, T. (1952). *American Indians in the Pacific*. Chicago: Rand McNally.

Hill, A.V.S. and Serjeantson, S.W. (Eds.) (1989). *The Colonization of the Pacific: A Genetic Trail*. Oxford: Clarendon Press.

Hill, O. (1963). The soft anatomy of a North American Indian. *American Journal of Physical Anthropology* **21**, 245–70.

Hillson, S. (1986). *Teeth*. Cambridge: Cambridge University Press.

Hinton, R.J. (1981). Form and patterning of anterior tooth wear among aboriginal human groups. *American Journal of Physical Anthropology* **54**, 555–64.

Hjelmman, G. (1929). Morphologische Beobachtungen an den Zähnen der Finnen. *Acta Societas Medicorum Fennicae 'Duodecim'* **11**, 1–132.

Hochstetter, R.L. (1975). Incidence of trifurcated mandibular first permanent molars in the population of Guam. *Journal of Dental Research* **54**, 1097.

Holzinger, K.J. (1929). The relative effect of nature and nurture influences on twin differences. *Journal of Educational Psychology* **20**, 241–8.

Hooton, E.A. (1918). On certain Eskimoid characters in Icelandic skulls. *American Journal of Physical Anthropology* **1**, 53–76.

Horai, S. (1995). Origin of *Homo sapiens* inferred from the age of the common ancestral human mitochondrial DNA. In *The Origin and Past of Modern Humans as Viewed from DNA*, eds. S. Brenner and K. Hanihara, pp. 171–85. Singapore: World Scientific Publishing Company.

Howells, W.W. (1973a). *The Pacific Islanders*. New York: Charles Scribner's Sons.

Howells, W.W. (1973b). *Cranial Variation in Man: A Study by Multivariate Analysis of Patterns of Difference Among Recent Human Populations*. Papers of the Peabody Museum of Archaeology and Ethnology, Harvard University, Vol. 67. Cambridge, Mass.: Harvard University.

Howells, W.W. (1976). Physical variation and history in Melanesia and Australia. *American Journal of Physical Anthropology* **45**, 641–9.

Howells, W.W. (1977). The sources of human variation in Melanesia and Australia. In *Sunda and Sahul*, eds. J. Allen, J. Golson and R. Jones, pp. 169–86. London: Academic Press.

Howells, W.W. (1989). *Skull Shapes and the Map: Craniometric Analyses in the Dispersion of Modern Homo*. Papers of the Peabody Museum of Archaeology and Ethnology, Harvard University, Vol. 79. Cambridge, Mass.: Harvard University.

Hrdlička, A. (1911). Human dentition and teeth from the evolutionary and racial standpoint. *Dominion Dental Journal* **23**, 403–17.

Hrdlička, A. (1920). Shovel-shaped teeth. *American Journal of Physical Anthropology* **3**, 429–65.

Hrdlička, A. (1921). Further studies of tooth morphology. *American Journal of Physical Anthropology* **4**, 141–76.

Hrdlička, A. (1924). New data on the teeth of early man and certain fossil European apes. *American Journal of Physical Anthropology* **7**, 109–32.

Hylander, W.L. (1977). The adaptive significance of Eskimo craniofacial morphology. In *Orofacial Growth and Development*, eds. A.A. Dahlberg and T.M. Graber, pp. 129–69. The Hague: Mouton Publishers.

Irish, J.D. (1993). *Biological Affinities of Late Pleistocene Through Modern African Aboriginal Populations: The Dental Evidence*. PhD dissertation, Department of Anthropology, Arizona State University, Tempe.

Irish, J.D. and Morris, D.H. (1996). Technical note: canine mesial ridge (Bushman canine) dental trait definition. *American Journal of Physical Anthropology* **99**, 357–9.

Irish, J.D. and Turner, C.G., II (1990). West African dental affinity of late Pleistocene Nubians: peopling of the Eurafrican-South Asian triangle II. *Homo* **41**, 42–53.

Ismagulov, O. and Sikhimbaeva, K.B. (1989). *Ethnic Odontology of Kazakhstan.* Alma-Ata: Nauka. (In Russian.)

Jacob, T. (1967). Racial identification of the Bronze Age human dentitions from Bali, Indonesia. *Journal of Dental Research* **46** (suppl. to no. 5), 903–10.

Jones, D. (1995). Sexual selection, physical attractiveness, and facial neoteny. *Current Anthropology* **36**, 723–48.

Jørgensen, K.D. (1955). The *Dryopithecus* pattern in recent Danes and Dutchmen. *Journal of Dental Research* **34**, 195–208.

Jørgensen, K.D. (1956). The deciduous dentition: a descriptive and comparative anatomical study. *Acta Odontologica Scandinavica* **14**, 1–202.

Kaczmarek, M. (1992). Dental morphological variation of the Polish people and their eastern neighbors. In *Structure, Function and Evolution of Teeth*, eds. P. Smith and E. Tchernov, pp. 413–23. London: Freund.

Kaczmarek, M. and Piontek, J. (1982). Human cremated remains and the diversity of man. *Homo* **33**, 230–6.

Kajajoja, P. and Zubov, A.A. (1986). Somatology and population genetics of the Bashkirs. *Annales Academiae Scientiarum Fennicae*, Series A, V. Medica **175**, 67–72.

Kanazawa, E., Sekikawa, M., Kamiakito, Y. and Ozaki, T. (1989). A quantitative investigation of irregular cusps in lower permanent molars. *Nihon University Journal of Oral Science* **15**, 450–6.

Kanazawa, E., Sekikawa, M, and Ozaki, T. (1990). A quantitative investigation of irregular cuspules in human maxillary permanent molars. *American Journal of Physical Anthropology* **83**, 173–80.

Kanazawa, E., Natori, M. and Ozaki, T. (1992). Anomalous tubercles on the occlusal table of upper first molars in nine populations including Pacific populations. In *Craniofacial Variation in Pacific Populations*, eds. T. Brown and S. Molnar, pp. 53–9. Adelaide: Anthropology and Genetics Lab, Department of Dentistry, University of Adelaide.

Kang, K.W., Christian, J.C. and Norton, J.A. (1978). Heritability estimates from twin studies. I. Formulae of heritability estimates. *Acta Geneticae, Medicae, et Gemellologiae* **27**, 39–44.

Kanner, L. (1928). *Folklore of the Teeth.* New York: The Macmillan Company.

Kato, K. (1937). Contribution to the knowledge concerning the cone shaped supernumerary cusp in the center of the occlusal surface on premolars of Japanese. *Nihon Shika Gakukai Zassi* **30**, 28–49.

Kaul, S.S., Sharma, K., Sharma, J.C. and Corruccini, R.S. (1985). Non-metric variants of the permanent dental crown in human monozygous and dizygous twins. In *Dental Anthropology: Application and Methods*, ed. V. Rami Reddy, pp. 187–93. New Delhi: Inter-India Publications.

Kaul, V. and Prakash, S. (1981). Morphological features of the Jat dentition. *American Journal of Physical Anthropology* **54**, 123–7.

Keene, H.J. (1965). The relationship between third molar agenesis and the morphologic variability of the molar teeth. *Angle Orthodontist* **35**, 289–98.

Keene, H.J. (1968). The relationship between Carabelli's trait and the size, number and morphology of the maxillary molars. *Archives of Oral Biology* **13**, 1023–5.

Keene, H.J. (1982). The morphogenetic triangle: a new conceptual tool for application to problems in dental morphogenesis. *American Journal of Physical Anthropology* **59**, 281–7.

Keene, H.J. (1991). On heterochrony in heterodonty: a review of some problems in tooth morphogenesis and evolution. *Yearbook of Physical Anthropology* **34**, 251–82.

Keith, A. (1931). Forward to *The Teeth, the Bony Palate, and the Mandible in the Bantu Races of South Africa*, by J.C.M. Shaw, London: Bale and Danielsson.

Kelso, A.J. (1974). *Physical Anthropology*. 2nd edn. Philadelphia: J.B. Lippincott.

Kempthorne, O. (1957). *An Introduction to Genetic Statistics*. New York: John Wiley and Sons.

Khaldeeva, N.I. (1979). Peoples of Siberia and the Far East. In *Ethnic Odontology of the USSR*, eds. A.A. Zubov and N.I. Khaldeeva, pp. 187–211. Moscow: Nauka. (In Russian.)

Kieser, J.A. (1991). *Human Adult Odontometrics*. Cambridge: Cambridge University Press.

Kieser, J.A. and Becker, P.J. (1989). Correlations of dimensional and discrete dental traits in the post-canine and anterior dental segments. *Journal of the Dental Association of South Africa* **44**, 101–3.

Kieser, J.A. and Groeneveld, H.T. (1986). Fluctuating odontometric asymmetry in a South African caucasoid population. *Journal of the Dental Association of South Africa* **41**, 185–9.

Kieser, J.A. and Preston, C.B. (1981). The dentition of the Lengua Indians of Paraguay. *American Journal of Physical Anthropology* **55**, 485–90.

Kirveskari, P. and Alvesalo, L. (1979). Quantification of the shovel shape of incisor teeth. *OSSA* **6**, 151–6.

Kirveskari, P. and Alvesalo, L. (1981). Shovel shape of maxillary incisors in 47,XYY males. *Proceedings of the Finnish Dental Society* **77**, 79–81.

Kirveskari, P. and Alvesalo, L. (1982). Dental morphology in Turner's syndrome (45,X females). In *Teeth: Form, Function and Evolution*, ed. B. Kurtén, pp. 298–303. New York: Columbia University Press.

Kitagawa, Y., Manabe, Y., Oyamada, J. and Rokutanda, A. (1995). Deciduous dental morphology of the prehistoric Jomon people of Japan: comparison of nonmetric characters. *American Journal of Physical Anthropology* **97**, 101–11.

Kochiev, R.S. (1973). Odontological characteristic of Caucasian ethnic groups. Paper presented at IXth International Congress of Anthropological and Ethnological Sciences, Chicago, Illinois.

Kochiev, R.S. (1979). Peoples of the Caucasus, Trans-Caucasus and north Caucasus. In *Ethnic Odontology of the USSR*, eds. A.A. Zubov and N.I. Khaldeeva, pp. 114–41. Moscow: Nauka. (In Russian.)

Kolakowski, D., Harris, E.F. and Bailit, H.L. (1980). Complex segregation analysis of Carabelli's trait in a Melanesian population. *American Journal of Physical Anthropology* **53**, 301–8.

Kollar, E.J. (1972). The development of the integument: spatial, temporal, and phylogenetic factors. *American Zoologist* **12**, 125–35.

Kollar, E.J. and Baird, G.R. (1971). Tissue interactions in developing mouse tooth

References

353

germs. In *Dental Morphology and Evolution*, ed. A.A. Dahlberg, pp. 15–29. Chicago: University of Chicago Press.

Kollar, E.J. and Kerley, M.A. (1979). Odontogenesis: interaction between isolated enamel organ epithelium and dental papilla cells. *OSSA* **6**, 163–70.

Korenhof, C.A.W. (1960). *Morphogenetical aspects of the human upper molar.* Utrecht: Uitgeversmaatschappij Neerlandia.

Korenhof, C.A.W. (1961). The enamel-dentin border: a new morphological factor in the study of the (human) molar pattern. *Koninklijke Nederlandse Akademie van Wetenschappen–Amsterdam.*, series B, **64**, 639–64.

Korkhaus, G. (1930). Anthropologic and odontologic studies of twins. *International Journal of Orthodontia* **16**, 640–7.

Koyama, S. (1992). Prehistoric Japanese populations: a subsistence-demographic approach. In *Japanese as a Member of the Asian and Pacific Populations*, ed. K. Hanihara, pp. 187–97. Kyoto: International Research Center for Japanese Studies.

Kraus, B. S. (1951). Carabelli's anomaly of the maxillary molar teeth. *American Journal of Human Genetics* **3**, 348–55.

Kraus, B.S. (1957). The genetics of the human dentition. *Journal of Forensic Sciences* **2**, 419–27.

Kraus, B.S. (1959). Occurrence of the Carabelli trait in Southwest ethnic groups. *American Journal of Physical Anthropology* **17**, 117–23.

Kraus, B.S. (1963). Morphogenesis of deciduous molar pattern in man. In *Dental Anthropology*, ed. D.R. Brothwell, pp. 87–104. New York: Pergamon Press.

Kraus, B.S. and Furr, M.L. (1953). Lower first premolars. Part I. A definition and classification of discrete morphologic traits. *Journal of Dental Research* **32**, 554–64.

Kraus, B.S. and Jordan, R.E. (1965). *The Human Dentition Before Birth.* Philadelphia: Lea and Febiger.

Kraus, B.S., Wise, W.J. and Frei, R.H. (1959). Heredity and the craniofacial complex. *American Journal of Orthodontics* **45**, 172–217.

Kroeber, A.L. (1939). *Cultural and Natural Areas of Native North America.* Berkeley: University of California Press.

Krogman, W.M. (1927). Anthropological aspects of the human teeth and dentition. *Journal of Dental Research* **7**, 1–108.

Krogman, W. M. (1960). Oral structures genetically and anthropologically considered. *Annals of the New York Academy of Sciences* **85**, 17–41.

Kronmiller, J.E., Uphold, W.B. and Kollar, E.J. (1991). EGF antisense oligodeoxynucleotides block murine odontogenesis *in vitro. Developmental Biology* **147**, 485–8.

Kronmiller, J.E., Upholt, W.B. and Kollar, E.J. (1992). Alteration of murine odontogenic patterning and prolongation of expression of epidermal growth factor mRNA by retinol *in vitro. Archives of Oral Biology* **37**, 129–38.

Kruger, B.J. (1962). Influence of boron, fluorine, and molybdenum on the morphology of the rat molar. *Journal of Dental Research* **41**, 215.

Kruger, B.J. (1966). Interaction of fluoride and molybdenum on dental morphology in the rat. *Journal of Dental Research* **45**, 714–25.

Kurtén, B., ed. (1982). *Teeth: Form, Function, and Evolution.* New York: Columbia University Press.

Kustaloglu, O.A. (1962). Paramolar structure of the upper dentition. *Journal of Dental Research* 41, 75–83.

Larsen, C.S. (1985). Dental modifications and tool use in the western Great Basin. *American Journal of Physical Anthropology* 67, 393–402.

Lasker, G.W. (1945). Observations on the teeth of Chinese born and reared in China and America. *American Journal of Physical Anthropology* 3, 129–50.

Lasker, G.W. (1950). Genetic analysis of racial traits of the teeth. *Cold Spring Harbor Symposia on Quantitative Biology* 15, 191–203.

Lasker, G.W. (1957). Racial traits in the human teeth. *Journal of Forensic Sciences* 2, 401–19.

Lasker, G.W. (1976). *Physical Anthropology*. 2nd edn. New York: Holt, Rinehart, and Winston.

Lau, E.C., Mohandas, T.K., Shapiro, L.F., Slavkin, H.C. and Snead, M.L. (1989). Human and mouse amelogenin gene loci are on the sex chromosomes. *Genomics* 4, 162–8.

Lau, T.C. (1955). Odontomes of the axial core type. *British Dental Journal* 99, 219–25.

Lavelle, C.L.B. (1972). Secular trends in different racial groups. *Angle Orthodontist* 42, 19–25.

Lee, G.T.R. and Goose, D.H. (1972). The inheritance of dental traits in a Chinese population in the United Kingdom. *Journal of Medical Genetics* 9, 336–9.

Leigh, R.W. (1928). Dental pathology of aboriginal California. *University of California Publications in American Archaeology and Ethnology* 23, 399–440.

Leigh, R.W. (1929). Dental morphology and pathology of prehistoric Guam. *Memoirs of the Bernice P. Bishop Museum* 11, 451–79.

Lerner, I.M. (1954). *Genetic Homeostasis*. Edinburgh: Oliver and Boyd.

Levin, M.G. (1963). *Ethnic Origins of the Peoples of Northeastern Asia*. Arctic Institute of North America Anthropology of the North Translations from Russian Sources, Number 3, ed. H.N. Michael, Toronto: University of Toronto Press.

Levin, M.G. and Potapov, L.P. (1964). *The Peoples of Siberia*. Translated from the original Russian (Narody Sibiri) by Stephen Dunn. Chicago: University of Chicago Press.

Livingstone, F.B. (1991). Phylogenies and the forces of evolution. *American Journal of Human Biology* 3, 83–9.

Lombardi, A.V. (1975). Tooth size associations of three morphologic dental traits in a Melanesian population. *Journal of Dental Research* 54, 239–43.

Ludwig, F.J. (1957). The mandibular second premolars: morphologic variation and inheritance. *Journal of Dental Research* 36, 263–73.

Lukacs, J.R. (1983). Dental anthropology and the origins of two Iron Age populations from northern Pakistan. *Homo* 34, 1–15.

Lukacs, J.R. (1987). Biological relationships derived from morphology of permanent teeth: recent evidence from prehistoric India. *Anthropologischer Anzeiger* 45, 97–116.

Lukacs, J.R. (1988). Dental morphology and odontometrics of early agriculturalists from Neolithic Mehrgarh, Pakistan. In *Teeth Revisited*, eds. D.E. Russell, J.-P. Santoro and D. Sigogneau-Russell, *Mémoires du Muséum Nationale d'Histoire Naturelle, Série C, Sciences de la Terre* 53, 245–58.

Lukacs, J.R. and Walimbe, S.R. (1984). Deciduous dental morphology and the biological affinities of a late Chalcolithic skeletal series from western India. *American Journal of Physical Anthropology* **65**, 23–30.

Lumholtz, C. (1902). *Unknown Mexico. A Record of Five Years' Exploration Among the Tribes of the Western Sierra Madre; in the Tierra Caliente of Tepic and Jalisco; and Among the Tarascos of Michoacan.* Glorieta, New Mexico: The Rio Grande Press.

Lumsden, A.G.S. (1979). Pattern formation in the molar dentition of the mouse. *Journal de Biologie Buccale* **7**, 77–103.

Lundström, A. (1963). Tooth morphology as a basis for distinguishing monozygotic and dizygotic twins. *American Journal of Human Genetics* **15**, 34–43.

Macintosh, N.W.G. and Larnach, S.L. (1976). Aboriginal affinities looked at in world context. In *The Origin of Australians*, eds. R.L. Kirk and A.G. Thorne, pp. 113–26. Canberra: Australian Institute of Aboriginal Studies.

Mahalanobis, P.C. (1936). On the generalized distance in statistics. *Proceedings of the National Institute of Science, India* **2**, 49–55.

Manabe, Y., Rokutanda, A., Kitagawa, Y. and Oyamada, J. (1991). Genealogical position of native Taiwanese (Bunun tribe) in East Asian populations based on tooth crown morphology. *Journal of the Anthropological Society of Nippon* **99**, 33–47.

Manabe, Y., Rokutanda, A. and Kitagawa, Y. (1992). Nonmetric tooth crown traits in the Ami tribe, Taiwan aborigines: comparisons with other East Asian populations. *Human Biology* **64**, 717–26.

Marks, J. (1995). *Human Biodiversity: Genes, Race, and History.* New York: Aldine de Gruyter.

Masters, D.H. and Hoskins, S.W. (1964). Projection of cervical enamel into molar furcation. *Journal of Periodontology* **38**, 330–4.

Mather, K. (1949). *Biometrical Genetics.* London: Methuen.

Matis, J.A. and Zwemer, T.J. (1971). Odontognathic discrimination of United States Indian and Eskimo groups. *Journal of Dental Research* **50**, 1245–8.

Matsumura, H. (1990). Geographical variation of dental characteristics in the Japanese of the protohistoric Kofun period. *Journal of the Anthropological Society of Japan* **98**, 439–49.

Mayhall, J.T. (1979). The dental morphology of the Inuit of the Canadian central Arctic. *OSSA* **6**, 199–218.

Mayhall, J.T. and Saunders, S.R. (1986). Dimensional and discrete dental trait asymmetry relationships. *American Journal of Physical Anthropology* **69**, 403–11.

Mayhall, J.T., Saunders, S.R. and Belier, P.L. (1982). The dental morphology of North American whites: a reappraisal. In *Teeth: Form, Function, and Evolution*, ed. B. Kurtén, pp. 245–58. New York: Columbia University Press.

Mayr, E. (1942). *Systematics and the Origin of Species.* New York: Columbia University Press.

McKusick, V. A. (1990). *Mendelian Inheritance in Man.* 9th edn. Baltimore: The Johns Hopkins University Press.

Mellanby, H. (1941). The effect of maternal dietary deficiency of vitamin A on dental tissues in rats. *Journal of Dental Research* **20**, 489–503.

Mellanby, M. and Holloway, P.J. (1956). The effect of vitamin A on the growth of tooth germs in tissue culture. *Journal of Dental Research* **35**, 960–1.

Merrill, R.G. (1964). Occlusal anomalous tubercles on premolars of Alaskan Eskimos and Indians. *Oral Surgery, Oral Medicine, Oral Pathology* **17**, 484–96.

Merriwether, D.A., Rothhammer, F. and Ferrell, R.E. (1995). Distribution of the four founding lineage haplotypes in Native Americans suggests a single wave of migration for the New World. *American Journal of Physical Anthropology* **98**, 411–30.

Mizoguchi, Y. (1977). Genetic variability in tooth crown characters: analysis by the tetrachoric correlation method. *Bulletin of the National Science Museum, Series D (Anthropology)* **3**, 37–62.

Mizoguchi, Y. (1978). Tooth crown characters on the lingual surfaces of the maxillary anterior teeth: analysis of the correlations by the method of path coefficients. *Bulletin of the National Science Museum, Series D (Anthropology)* **4**, 25–57.

Mizoguchi, Y. (1981). Variation units in the human permanent dentition. *Bulletin of the National Science Museum, Series D* (Anthropology) **7**, 29–38.

Mizoguchi, Y. (1985). *Shovelling: A Statistical Analysis of its Morphology*. Tokyo: University of Tokyo Press.

Mizoguchi, Y. (1986). Correlated asymmetries detected in the tooth crown diameters of human permanent teeth. *Bulletin of the National Science Museum, Tokyo, Series D* (Anthropology) **12**, 25–45.

Mizoguchi, Y. (1987). Mirror imagery and genetic variability of lateral asymmetries in the mesiodistal crown diameters of permanent teeth. *Bulletin of the National Science Museum, Tokyo, Series D* (Anthropology) **13**, 11–19.

Mizoguchi, Y. (1988a). Statistical analysis of geographical variation in dental size. *Report of the Ministry of Education, Science and Culture, Japan*, pp. 1–124.

Mizoguchi, Y. (1988b). Degree of bilateral asymmetry of nonmetric tooth crown characters quantified by the tetrachoric correlation method. *Bulletin of the National Science Museum, Tokyo, Series D* (Anthropology) **14**, 29–49.

Mizoguchi, Y. (1989). Genetic variability of left-right asymmetries and mirror imagery in nonmetric tooth crown characters. *Bulletin of the National Science Museum, Series D* (Anthropology) **15**, 49–61.

Mizoguchi, Y. (1990). Covariation of asymmetries in metric and nonmetric tooth crown characters. *Bulletin of the National Science Museum, Series D* (Anthropology) **16**, 39–47.

Mizoguchi, Y. (1993). Adaptive significance of the Carabelli trait. *Bulletin of the National Science Museum, Series D* (Anthropology) **19**, 21–58.

Moggi-Cecchi, J. (Ed.) (1995). *Aspects of Dental Biology: Paleontology, Anthropology and Evolution*. Florence: International Institute for the Study of Man.

Møller, I.J. (1967). Influence of microelements on the morphology of the teeth. *Journal of Dental Research* **46**(suppl. to no. 5), 933–7.

Molnar, S. (1972). Tooth wear and culture: a survey of tooth functions among some prehistoric populations. *Current Anthropology* **13**, 511–26.

Molnar, S. (1975). *Races, Types, and Ethnic Groups*. Englewood Cliffs: Prentice-Hall.

Montagu, M.F.A., (Ed.) (1964). *The Concept of Race*. London: Collier-Macmillan Limited.

Moorrees, C.F.A. (1951). The dentition as a criterion of race with special reference to the Aleut. *Journal of Dental Research* **30**, 815–21.

Moorrees, C.F.A. (1957). *The Aleut Dentition: A Correlative Study of Dental Characteristics in an Eskimoid People.* Cambridge: Harvard University Press.

Moorrees, C.F.A. (1962). Genetic considerations in dental anthropology. In *Genetics and Dental Health*, ed. C.J. Witkop, Jr., pp. 101–12. New York: McGraw-Hill.

Morris, D.H. (1965) *The Anthropological Utility of Dental Morphology.* PhD dissertation, Department of Anthropology, University of Arizona, Tucson.

Morris, D.H. (1970). On deflecting wrinkles and the *Dryopithecus* pattern in human mandibular molars. *American Journal of Physical Anthropology* **32**, 97–104.

Morris, D.H. (1975). Bushman maxillary canine polymorphism. *South African Journal of Science* **71**, 333–5.

Morris, D.H. (1981). Maxillary first premolar angular differences between North American Indians and non-North American Indians. *American Journal of Physical Anthropology* **54**, 431–3.

Morris, D.H. (1986). Maxillary molar polygons in five human samples. *American Journal of Physical Anthropology* **70**, 333–8.

Morris, D.H., Glasstone Hughes, S. and Dahlberg, A.A. (1978). Uto-Aztecan premolar: the anthropology of a dental trait. In *Development, Function and Evolution of Teeth*, eds. P.M. Butler and K.A. Joysey, pp. 69–79. New York: Academic Press.

Morton, N.E. (1959). Genetic tests under incomplete ascertainment. *American Journal of Human Genetics* **11**, 1–16.

Morton, N.E. and MacLean, C.J. (1974). Analysis of family resemblance. III. Complex segregation analysis of quantitative traits. *American Journal of Human Genetics* **26**, 489–503.

Morton, N.E., Yee, S. and Lew, R. (1971). Complex segregation analysis. *American Journal of Human Genetics* **23**, 602–11.

Mourant, A.E. (1954). *The Distribution of the Human Blood Groups.* Springfield: C.C. Thomas.

Mourant, A.E., Kopec, A.C. and Domaniewska-Sobczak, K. (1976). *The Distribution of Human Blood Groups and Other Polymorphisms.* Oxford: Oxford University Press.

Nakata, M. (1985). Twin studies in craniofacial genetics: a review. *Acta Geneticae, Medicae, et Gemellologiae* **34**, 1–14.

Neel, J.V. and Schull, W.J. (1954). *Human Heredity.* Chicago: University of Chicago Press.

Nei, M. (1972). Genetic distance between populations. *American Naturalist* **106**, 283–92.

Nei, M. (1992). The origins of human populations: genetic, linguistic, and archeological data. In *The Evolution and Dispersal of Modern Humans in Asia*, eds. T. Akazawa, K. Aoki and T. Kimura, pp. 71–91. Tokyo: Hokusen-Sha Publishing Co.

Nei, M. and Roychoudhury, A.K. (1982). Genetic relationship and evolution of human races. *Evolutionary Biology* **14**, 1–59.

Nei, M. and Roychoudhury, A.K. (1993). Evolutionary relationships of human populations on a global scale. *Molecular Biology and Evolution* **10**, 927–43.

Nelson, C.T. (1938). The teeth of the Indians of Pecos Pueblo. *American Journal of Physical Anthropology* **23**, 261–93.

Nichol, C.R (1989). Complex segregation analysis of dental morphological variants. *American Journal of Physical Anthropology* **78**, 37–59.

Nichol, C.R. (1990). *Dental Genetics and Biological Relationships of the Pima Indians of Arizona*. PhD dissertation, Department of Anthropology, Arizona State University, Tempe.

Nichol, C.R. and Turner, C.G., II (1986). Intra- and interobserver concordance in classifying dental morphology. *American Journal of Physical Anthropology* **69**, 299–315.

Nie, N., Hull, C., Jenkins, J., Steinbrenner, K. and Bent, D. (1975). *SPSS Statistical Package for the Social Sciences*. New York: McGraw-Hill.

Noss, J.F., Scott, G.R., Potter, R.H.Y., Dahlberg, A.A. and Dahlberg, T. (1983a). The influence of crown size dimorphism on sex differences in the Carabelli trait and the canine distal accessory ridge in man. *Archives of Oral Biology* **28**, 527–30.

Noss, J.F., Scott, G.R., Potter, R.H.Y. and Dahlberg, A.A. (1983b). Fluctuating asymmetry in molar dimensions and discrete morphological traits in Pima Indians. *American Journal of Physical Anthropology* **61**, 437–45.

Oehlers, F.A.C. (1956). The tuberculated premolar. *The Dental Practitioner* **6**, 144–8.

Omoto, K. (1984). The Negritos: genetic origins and microevolution. *Acta Anthropogenetica* **8**, 137–47.

Omoto, K. (1992). Some aspects of the genetic composition of the Japanese. In *International Symposium on Japan as a Member of the Asian and Pacific Populations*, ed. K. Hanihara, pp. 138–44. Kyoto: International Research Center for Japanese Studies.

Oöe, T. (1965). A study of the ontogenetic origin of human permanent tooth germs. *Okajimas Folia Anatomica Japan* **40**, 429–37.

Ortner, D.J. (1966). A recent occurrence of an African type tooth mutilation in Florida. *American Journal of Physical Anthropology* **25**, 177–80.

Osborn, H.F. (1888a). The evolution of the mammalian molars to and from the tritubercular type. *American Naturalist* **22**, 1067–79.

Osborn, H.F. (1888b). The nomenclature of the mammalian molar cusps. *American Naturalist* **22**, 926–8.

Osborn, H.F. (1897). Trituberculy: a review dedicated to the late Professor Cope. *American Naturalist* **31**, 993–1016.

Osborn, H.F. (1907). *Evolution of Mammalian Molar Teeth, To and From the Triangular Type*. New York: The Macmillan Company.

Osborn, J.H. (1978). Morphogenetic gradients: fields versus clones. In *Development, Function and Evolution of Teeth*, eds. P.M. Butler and K.A. Joysey, pp. 171–201. New York: Academic Press.

Osborne, R. H. (1963). Respective roles of twin, sibling, family, and population methods in dentistry and medicine. *Journal of Dental Research* **42**(suppl. to no. 6), 1276–87.

Osborne, R. H. (1967). Some genetic problems in interpreting the evolution of the human dentition. *Journal of Dental Research* **46**(suppl. to no. 5), 945–8.

Ossenberg, N.S. (1981). An argument for the use of total side frequencies of bilateral nonmetric skeletal traits in population distance analysis: the regression of symmetry on incidence. *American Journal of Physical Anthropology* **54**, 471–9.

Ossenberg, N.S. (1992). Native people of the American Northwest: population history from the perspective of skull morphology. In *The Evolution and Dispersal of Modern Humans in Asia*, eds. T. Akazawa, K. Aoki, and T. Kimura, pp. 493–530. Tokyo: Hokusen-Sha Publishing Co.

Oster, G. and Alberch, P. (1982). Evolution and bifurcation of developmental programs. *Evolution* **36**, 444–59.

Pal, A. (1972). Double-rooted human lower canine – a rare anomaly. *Journal of the Indian Anthropological Society* **7**, 171–4.

Palomino, H., Chakraborty, R. and Rothhammer, F. (1977). Dental morphology and population diversity. *Human Biology* **49**, 61–70.

Paynter, K.J. and Grainger, R.M. (1956). The relation of nutrition to the morphology and size of rat molar teeth. *Journal of the Canadian Dental Association* **22**, 519–31.

Paynter, K.J. and Grainger, R.M. (1962). Relationship of morphology and size of teeth to caries. *International Dental Journal* **12**, 147–60.

Pearson, K. (1926). On the coefficient of racial likeness. *Biometrika* **18**, 105–17.

Pedersen, P.O. (1949). The East Greenland Eskimo dentition. *Meddelelser om Grønland* **142**, 1–244.

Pedersen, P.O. and Thyssen, H. (1942). Den cervicale Emaljerands Forløb hos Eskimoer. *Odontologisk Tidskrift* **50**, 444–92.

Pedersen, P.O., Dahlberg, A.A. and Alexandersen, V. (Eds.) (1967). Proceedings of the International Symposium on Dental Morphology. *Journal of Dental Research* **46**(suppl. to no. 5), 769–992.

Penrose, L.S. (1954). Distance, size and shape. *Annals of Eugenics* **18**, 337–43.

Perzigian, A.J. (1984). Human odontometric variation: an evolutionary and taxonomic perspective. *Anthropologie* **22**, 193–8.

Peyer, B. (1968). *Comparative Odontology*. Chicago: University of Chicago Press.

Pietrusewsky, M. (1990). Craniofacial variation in Australasian and Pacific populations. *American Journal of Physical Anthropology* **82**, 319–40.

Pietrusewsky, M., Yongyi, L., Xiangqing, S. and Quyen, N.G. (1992). Modern and near modern populations of Asia and the Pacific: a multivariate craniometric interpretation. In *The Evolution and Dispersal of Modern Humans in Asia*, eds. T. Akazawa, K. Aoki and T. Kimura, pp. 531–58. Tokyo: Hokusen-Sha Publishing Co.

Poirier, F.E., Stini, W.A. and Wreden, K.B. (1994). *In Search of Ourselves*. 5th edn. Englewood Cliffs: Prentice-Hall.

Portin, P. and Alvesalo, L. (1974). The inheritance of shovel shape in maxillary central incisors. *American Journal of Physical Anthropology* **41**, 59–62.

Potter, R.H. and Nance, W.E. (1976). A twin study of dental dimension. I. Discordance, asymmetry, and mirror imagery. *American Journal of Physical Anthropology* **44**, 391–6.

Potter, R.H.Y., Yu, P.-L., Dahlberg, A.A., Merritt, A.D. and Conneally, P.M. (1968). Genetic studies of tooth size factors in Pima Indian families. *American Journal of Human Genetics* **20**, 89–100.

Powell, J.F. (1993). Dental evidence for the peopling of the New World: some methodological considerations. *Human Biology* **65**, 799–815.

Radlanski, R.J. and Renz, H. (Eds.) (1995). *Proceedings of the 10th International Symposium on Dental Morphology*. Berlin: 'M' Marketing Services, C. & M. Brünne GbR.

Rampino, M.R. and Self, S. (1993). Bottleneck in human evolution and the Toba eruption. *Science* **262**, 1955.

Reid, C., van Reenan, J.F. and Groeneveld, H.T. (1991). Tooth size and the Carabelli trait. *American Journal of Physical Anthropology* **84**, 427–32.

Reid, C., van Reenan, J.F. and Groeneveld, H.T. (1992). The Carabelli trait and maxillary molar cusp and crown base areas. In *Structure, Function and Evolution of Teeth*, eds. P. Smith and E. Tchernov, pp. 451–66. London: Freund Publishing House.

Relethford, J.H. (1994). Craniometric variation among modern human populations. *American Journal of Physical Anthropology* **95**, 53–62.

Relethford, J.H. and Harpending, H.C. (1994). Craniometric variation, genetic theory, and modern human origins. *American Journal of Physical Anthropology* **95**, 249–70.

Relethford, J.H. and Harpending, H.C. (1995). Ancient differences in population size can mimic a recent African origin of modern humans. *Current Anthropology* **36**, 667–74.

Richards, L.C. and Telfer, P.J. (1979). The use of dental characters in the assessment of genetic distance in Australia. *Archaeology and Physical Anthropology in Oceania* **14**, 184–94.

Robertson, A. and Lerner, I.M. (1949). The heritability of all-or-none traits: viability of poultry. *Genetics* **34**, 395–411.

Robinson, J.T. (1956). *The Dentition of the Australopithecinae*. Pretoria: Transvaal Museum Memoir, Number 9.

Romero, J. (1970). Dental mutilation, trephination, and cranial deformation. In *Handbook of Middle American Indians*, Vol. 9, *Physical Anthropology*, pp. 50–67. Austin: University of Texas Press.

Romesburg, H.C. (1990). *Cluster Analysis for Researchers*. Malabar, Florida: Robert E. Krieger Publishing Company.

Rothhammer, F., Lasserre, E., Blanco, R., Covarrubias, E. and Dixon, M. (1968). Microevolution in Chilean populations. IV. Shovel shape, mesial-palatal version and other dental traits in Pewenche Indians. *Zeitschrift für Morphologie und Anthropologie* **60**, 162–9.

Roychoudhury, A.K. and Nei, M. (1988). *Human Polymorphic Genes: World Distribution*. New York: Oxford University Press.

Ruhlen, M. (1987). *A Guide to the World's Languages*, Vol. 1. Stanford: Stanford University Press.

Ruhlen, M. (1994). *The Origin of Language: Tracing the Evolution of the Mother Tongue*. New York: John Wiley & Sons, Inc.

Russell, D.E., Santoro, J.-P., and Sigogneau-Russell, D. (Eds.) (1988). *Teeth Revisited*. Mémoires du Muséum Nationale d'Histoire Naturelle, Série C, Sciences de la Terre, volume 53.

Saheki, M. (1958). On the heredity of the tooth crown configuration studied in twins. *Acta Anatomica Nipponica* **33**, 456–70.

Sakai, T. (1975). The dentition of the Hawaiians. *Journal of the Anthropological Society of Nippon* **83**, 49–81.

Sakai, T., Sasaki, I. and Hanamura, H. (1967). A morphological study of enamel-dentin border on the Japanese dentition. II. Maxillary canine. *Journal of the Anthropological Society of Nippon* **75**, 155–72.

Sakai, T., Hanamura, H. and Ohno, N. (1969). The dentition of the Pushtun and Tajik in Afghanistan. *Aichi-Gakuin Journal of Dental Science* **7**, 106–37.

Sakuma, M, Irish, J.D. and Morris, D.H. (1991). The Bushman maxillary canine of the Chewa tribe in east-central Africa. *Journal of the Anthropological Society of Nippon* **99**, 411–17.

Saunders, S.R. and Mayhall, J.T. (1982a). Fluctuating asymmetry of dental morphological traits: new interpretations. *Human Biology* **54**, 789–99.

Saunders, S.R. and Mayhall, J.T. (1982b). Developmental patterns of human dental morphological traits. *Archives of Oral Biology* **27**, 45–9.

Schanfield, M. S. (1992). Immunoglobin allotypes (GM and KM) indicate multiple founding populations of native Americans: evidence of at least four migrations to the New World. *Human Biology* **64**, 381–402.

Schulz, P.D. (1977). Task activity and anterior tooth grooving in prehistoric California Indians. *American Journal of Physical Anthropology* **46**, 87–92.

Schwartz, J.H. (1995). *Skeleton Keys: An Introduction to Human Skeletal Morphology, Development, and Analysis.* New York: Oxford University Press.

Sciulli, P.W. (1977). A descriptive and comparative study of the deciduous dentition of prehistoric Ohio Valley Amerindians. *American Journal of Physical Anthropology* **47**, 71–80.

Sciulli, P.W. (1990). Deciduous dentition of a late Archaic population of Ohio. *Human Biology* **62**, 221–45.

Sciulli, P.W., Schneider, K.N., and Mahaney, M.C. (1984). Morphological variation of the permanent dentition in prehistoric Ohio. *Anthropologie* **22**, 211–15.

Scott, G.R. (1971). Canine *tuberculum dentale. American Journal of Physical Anthropology* **35**, 294. (Abstract.)

Scott, G.R. (1972). An analysis of population and family data on Carabelli's trait and shovel-shaped incisors. *American Journal of Physical Anthropology* **37**, 449. (Abstract.)

Scott, G.R. (1973). *Dental Morphology: a Genetic Study of American White Families and Variation in Living Southwest Indians.* PhD dissertation, Department of Anthropology, Arizona State University, Tempe.

Scott, G.R. (1974). A general model of inheritance for nonmetrical tooth crown characteristics. *American Journal of Physical Anthropology* **41**, 503. (Abstract.)

Scott, G.R. (1977a). Classification, sex dimorphism, association, and population variation of the canine distal accessory ridge. *Human Biology* **49**, 453–69.

Scott, G.R. (1977b). Interaction between shoveling of the maxillary and mandibular incisors. *Journal of Dental Research* **56**, 1423.

Scott, G.R. (1977c). Lingual tubercles and the maxillary incisor-canine field. *Journal of Dental Research* **56**, 1192.

Scott, G.R. (1978). The relationship between Carabelli's trait and the protostylid. *Journal of Dental Research* **57**, 570.

Scott, G.R. (1979). Association between the hypocone and Carabelli's trait of the maxillary molars. *Journal of Dental Research* **58**, 1403–4.

Scott, G.R. (1980). Population variation of Carabelli's trait. *Human Biology* **52**, 63–78.

Scott, G.R. (1991a). Dental anthropology. *Encyclopedia of Human Biology*, Vol. 2. pp. 789–804. San Diego: Academic Press.

Scott, G.R. (1991b). Continuity or replacement at the Uyak site: a physical anthropological analysis of population relationships. In *The Uyak Site on Kodiak Island: Its Place in Alaskan Prehistory*, University of Oregon Anthropological Papers No. 44, pp. 1–56.

Scott, G.R. (1994). Teeth and prehistory on Kodiak Island. In *Reckoning with the Dead: The Larsen Bay Repatriation and the Smithsonian Institution*, eds. T.L. Bray and T.W. Killion, pp. 67–74. Washington: Smithsonian Institution Press.

Scott, G.R. and Alexandersen, V. (1992). Dental morphological variation among medieval Greendlanders, Icelanders, and Norwegians. In *Structure, Function and Evolution of Teeth*, eds. P. Smith and E. Tchernov, pp. 467–90. London: Freund Publishing House.

Scott, G.R. and Dahlberg, A.A. (1982). Microdifferentiation in tooth crown morphology among Indians of the American Southwest. In *Teeth: Form, Function, and Evolution*, ed. B. Kurtén, pp. 259–91. New York: Columbia University Press.

Scott, G.R. and Gillispie, T.E. (1997). The dentition of the prehistoric inhabitants of St. Lawrence Island, Alaska. In *St.-Lorenz Insel Studien*, ed. P. Haupt, Geneva: Publications Interdisciplinaires de l'Academie Suisse des Sciences Humaines et de la Societe Helvetique des Sciences Naturelle. (In press.)

Scott, G.R. and Potter, R.H.Y. (1984). An analysis of tooth crown morphology in American white twins. *Anthropologie* **22**, 223–31.

Scott, G.R. and Turner, C.G., II (1988). Dental anthropology. *Annual Review of Anthropology* **17**, 99–126.

Scott, G.R., Halffman, C.M. and Pedersen, P.O. (1992). Dental conditions of medieval Norsemen in the North Atlantic. *Acta Archaeologica* **62**, 183–207.

Scott, G.R., Potter, R.H.Y., Noss, J.F., Dahlberg, A.A. and Dahlberg, T. (1983). The dental morphology of Pima Indians. *American Journal of Physical Anthropology* **61**, 13–31.

Scott, G.R., Street, S.R. and Dahlberg, A.A. (1988). The dental variation of Yuman speaking groups in an American Southwest context. In *Teeth Revisited*, eds. Russel, D.E., Santoro, J.-P., and Sigogneau-Russell, D., *Mémoires du Muséum Nationale d'Histoire Naturelle, Série C, Sciences de la Terre* **53**, 305–19.

Searle, A.G. (1954). Genetical studies on the skeleton of the mouse. XI. The influence of diet on variation within pure lines. *Journal of Genetics* **52**, 413–24.

Sekikawa, M, Kanazawa, E. and Ozaki, T. (1987a). Study of the cuspal ridges of the upper first molars in a modern Japanese population. *Acta Anatomica* **129**, 159–64.

Sekikawa, M, Kanazawa, E., Ito, T. and Ozaki, T. (1987b). Cuspal ridges of the lower first molar in a modern Japanese population. *Japanese Journal of Oral Biology* **29**, 763–9.

Sekikawa, M., Kanazawa, E., Ozaki, T. and Richards, L.C. (1990). Cuspal ridges of deciduous upper second molars in Japanese subjects. *Journal of the Anthropol-*

ogical Society of Nippon **98**, 39–47.

Senyurek, M.S. (1952). A study of the dentition of the ancient inhabitants of Alaca Hoyuk. *Turk Tarih Kuruma Belleten* **16**, 153–224.

Shapiro, M.M.J. (1949). The anatomy and morphology of the tubercle of Carabelli. *The Official Journal of the Dental Association of South Africa* **4**, 355–62.

Sharma, J.C. and Kaul, V. (1977). Dental morphology and odontometry in Panjabis. *Journal of the Indian Anthropological Society* **12**, 213–26.

Shaw, J.C.M. (1931). *The Teeth, the Bony Palate, and the Mandible in the Bantu Races of South Africa*. London: Bale and Danielsson.

Shields, G.F., Schmiechen, A.M., Frazier, B.L., Redd, A., Voevoda, M.I., Reed, J.K. and Ward, R.H. (1993). mtDNA sequences suggest a recent evolutionary divergence for Beringian and northern North American populations. *American Journal of Human Genetics* **53**, 549–62.

Siegel, M.I. and Doyle, W.J. (1975a). The differential effects of prenatal and postnatal audiogenic stress on fluctuating dental asymmetry. *Journal of Experimental Zoology* **191**, 211–14.

Siegel, M.I. and Doyle, W.J. (1975b). The effects of cold stress on fluctuating asymmetry in the dentition of the mouse. *Journal of Experimental Zoology* **193**, 385–9.

Siegel, M.I. and Mooney, M.P. (1987). Perinatal stress and increased fluctuating asymmetry of dental calcium in the laboratory rat. *American Journal of Physical Anthropology* **73**, 267–70.

Siegel, M.I., Doyle, W.J. and Kelley, C. (1977). Heat stress, fluctuating asymmetry and prenatal selection in the laboratory rat. *American Journal of Physical Anthropology* **46**, 121–6.

Simmons, R.T. (1976). The biological origin of the Autralian Aboriginals. In *The Origin of the Australians*, eds. R.L. Kirk and A.G. Thorne, pp. 307–28. Canberra: Australian Institute of Aboriginal Studies.

Simpson, G.G. (1944). *Tempo and Mode in Evolution*. New York: Columbia University Press.

Skrinjaric, I., Slaj, M., Lapter, V. and Muretic, Z. (1985). Heritability of Carabelli's trait in twins. *Collegium Antropologicum* **2**, 177–81.

Smith, B.H., Garn, S.M. and Cole, P.E. (1982). Problems of sampling and inference in the study of fluctuating dental asymmetry. *American Journal of Physical Anthropology* **58**, 281–9.

Smith, P. and Shegev, M. (1988). The dentition of Nubians from Wadi Halfa, Sudan: an evolutionary perspective. *Journal of the Dental Association of South Africa* **43**, 539–41.

Smith, P. and Tchernov, E. (Eds.) (1992). *Structure, Function and Evolution of Teeth*. London: Freund Publishing House.

Snyder, R.G. (1960). Mesial margin ridging of incisor labial surfaces. *Journal of Dental Research* **39**, 361–4.

Sofaer, J.A. (1970). Dental morphologic variation and the Hardy–Weinberg law. *Journal of Dental Research* **49**, 1505–8.

Sofaer, J.A., Niswander, J.D., MacLean, C.J. and Workman, P.L. (1972a). Population studies on Southwestern Indian tribes. V. Tooth morphology as an indicator of biological distance. *American Journal of Physical Anthropology* **37**, 357–66.

Sofaer, J.A., MacLean, C.J. and Bailit, H.L. (1972b). Heredity and morphological variation in early and late developing human teeth of the same morphological class. *Archives of Oral Biology* **17**, 811–16.

Sofaer, J.A., Smith, P. and Kaye, E. (1986). Affinities between contemporary and skeletal Jewish and non-Jewish groups based on tooth morphology. *American Journal of Physical Anthropology* **70**, 265–75.

Sokal, R.R. and Sneath, P.H.A. (1963). *Principles of Numerical Taxonomy*. San Francisco: W.H. Freeman.

Solheim, W.G., II (1968) Early bronze in northeastern Thailand. *Current Anthropology* **9**, 59–62.

Somogyi-Csizmazia, W. and Simons, A.J. (1971). Three-rooted mandibular first permanent molars in Alberta Indian children. *Journal of the Canadian Dental Association* **37**, 105–6.

de Souza-Freitas, J.A., Lopes, E.S. and Casati-Alvares, L. (1971). Anatomic variations of lower first permanent molar roots in two ethnic groups. *Oral Surgery, Oral Medicine, Oral Pathology* **31**, 274–8.

Spencer, B. and Gillen, F.J. (1899). *The Native Tribes of Central Australia*. London: Macmillan and Co.

Spuhler, J.N. (1954). Some problems in the physical anthropology of the American Southwest. *American Anthropologist* **56**, 604–19.

Staley, R.N. and Green, L.J. (1974). Types of tooth cusp occurrence asymmetry in human monozygotic and dizygotic twins. *American Journal of Physical Anthropology* **40**, 187–96.

Staski, E. and Marks, J. (1992). *Evolutionary Anthropology*. Fort Worth: Harcourt Brace Jovanovich.

Stein, M.R. (1934). Polyisomerism of the human dentition. *Journal of Dental Research* **14**, 125–37.

Stein, P.L. and Rowe, B.M. (1996). *Physical Anthropology*, 6th edn. New York: McGraw-Hill.

Steinberg, A.G. and Cook, C.E. (1981). *The Distribution of the Human Immunoglobulin Allotypes*. Oxford: Oxford University Press.

Stewart, T.D. (1942). Persistence of the African type of tooth pointing in Panama. *American Anthropologist* **44**, 328–30.

Stewart, T.D. and Groome, J.R. (1968). The African custom of tooth mutilation in America. *American Journal of Physical Anthropology* **28**, 31–42.

Stringer, C.B. (1993). Understanding the fossil human record: past, present and future. *Revista di Antropologia (Roma)* **71**, 91–100.

Sumiya, Y. (1959). Statistical study on dental anomalies in the Japanese. *Journal of the Anthropological Society of Nippon* **67**, 215–33.

Suzuki, M. and Sakai, T. (1965). Labial surface pattern on permanent upper incisors of the Japanese. *Journal of the Anthropological Society of Japan* **73**, 1–8. (In Japanese, with English summary.)

Suzuki, M. and Sakai, T. (1966). Morphological analysis of the shovel-shaped teeth. *Journal of the Anthropological Society of Nippon* **74**, 202–18.

Swindler, D.R. (1976). *Dentition of Living Primates*. London: Academic Press.

Swindler, D.R., Drusini, A.G. and Ferrando, C.C. (1995). Molar morphology of precontact Easter Islanders. In *Proceedings of the 10th International Symposium on Dental Morphology*, eds. R. J. Radlanski and H. Renz, pp. 354–7.

Berlin: Marketing Services, Christine and Michael Brünne GbR.

Szathmary, E.J.E. (1984). Peopling of northern North America: clues from genetic studies. *Acta Anthropogenetica* **8**, 79–109.

Szathmary, E.J.E. (1985). Peopling of North America: clues from genetic studies. In *Out of Asia: Peopling of the Americas and the Pacific*, eds. R.L. Kirk and E.J.E. Szathmary, pp. 79–104. Canberra: The Journal of Pacific History, Inc. Australian National University.

Szathmary, E.J.E. (1993). Genetics of aboriginal North Americans. *Evolutionary Anthropology* **1**, 202–20.

Szathmary, E.J.E. and Ossenberg, N.S. (1978). Are the biological differences between North American Indians and Eskimos truly profound? *Current Anthropology* **19**, 673–701.

Takei, T. (1990). An anthropological study on the tooth crown morphology in the Atayal tribe of Taiwan aborigines: comparative analysis between Atayal and some Asian-Pacific populations. *Journal of the Anthropological Society of Japan* **98**, 337–51. (In Japanese, with English summary.)

Tegako, L.I. and Salivon, I.I. (1979). Byelorussians. In *Ethnic Odontology of the USSR*, eds. A.A. Zubov and N.I. Khaldeeva, pp. 48–65. Moscow: Nauka. (In Russian.)

Ten Cate, A.R. (1994). *Oral Histology: Development, Structure, and Function*. 4th edn. St. Louis: Mosby.

Thomsen, S. (1955). *Dental Morphology and Occlusion in the People of Tristan da Cunha*. Results of the Norwegian Scientific Expedition to Tristan da Cunha, 1937–38, No. 25. Oslo: Det Norske Videnskaps-Akademi.

Thorne, A.G. and Wolpoff, M.H. (1981). Regional continuity in Australasian Pleistocene hominid evolution. *American Journal of Physical Anthropology* **55**, 337–49.

Tobias, P.V. (1955). Teeth, jaws and genes. *Journal of the Dental Association of South Africa* **10**, 88–104.

Tomes, C.S. (1889). *A Manual of Dental Anatomy: Human and Comparative*. London: J. & A. Churchill.

Tonge, C.H. (1971). The role of the mesenchyme in tooth development. In *Dental Morphology and Evolution*, ed. A.A. Dahlberg, pp. 45–58. Chicago: University of Chicago Press.

Torroni, A., Schurr, T.G., Yang, C.-C., Szathmary, E.J.E., Williams, R.C., Schanfield, M.S., Troup, G.A., Knowler, W.C., Lawrence, D.N., Weiss, K.M. and Wallace, D.C. (1992). Native American mitochondrial DNA analysis indicates that the Amerind and Nadene populations were founded by two independent migrations. *Genetics* **130**, 153–62.

Townsend, G.C. and Brown, T. (1980). Dental asymmetry in Australian aboriginals. *Human Biology* **52**, 661–73.

Townsend, G.C. and Brown, T. (1981a). Morphogenetic fields within the dentition. *Australian Orthodontic Journal* **7**, 3–12.

Townsend, G.C. and Brown, T. (1981b). The Carabelli trait in Australian aboriginal dentition. *Archives of Oral Biology* **26**, 809–14.

Townsend, G.C. and Martin N.G. (1992). Fitting genetic models to Carabelli trait data in South Australian twins. *Journal of Dental Research* **71**, 403–9.

Townsend, G.C., Yamada, H. and Smith, P. (1986). The metaconule in Australian aboriginals: an accessory tubercle on maxillary molar teeth. *Human Biology* **58**, 851–62.

Townsend, G.C., Richards, L.C., Brown, R. and Burgess, V.B. (1988). Twin zygosity determination on the basis of dental morphology. *Journal of Forensic Odonto-Stomatology.* **6**, 1–15.

Townsend, G.C., Yamada, H. and Smith, P. (1990). Expression of the entoconulid (sixth cusp) on mandibular molar teeth of an Australian aboriginal population. *American Journal of Physical Anthropology* **82**, 267–74.

Townsend, G.C., Richards, L.C., Brown, T., Burgess, V.B., Travan, G.R. and Rogers, J.R. (1992). Genetic studies of dental morphology in South Australian twins. In *Structure, Function and Evolution of Teeth*, eds. P. Smith and E. Tchernov, pp. 501–18. London: Freund Publishing House Ltd.

Townsend, G.C., Dempsey, P., Brown, T., Kaidonis, J. and Richards, L. (1994). Teeth, genes and the environment. *Perspectives in Human Biology* **4**, 35–46.

Tratman, E.K. (1938). Three-rooted lower molars in man and their racial distribution. *British Dental Journal* **64**, 264–74.

Tsatsas, B., Mandi, F. and Kerani, S. (1973). Cervical enamel projections in the molar teeth. *Journal of Periodontology* **44**, 312–14.

Tsuji, T. (1958). Incidence and inheritance of the Carabelli's cusp in a Japanese population. *Japanese Journal of Human Genetics* **3**, 21–31.

Turner, C.G., II (1967). Dental genetics and microevolution in prehistoric and living Koniag Eskimo. *Journal of Dental Research* **46**(suppl. to no. 5), 911–17.

Turner, C.G., II (1968). Review of Rupert I. Murrill, Cranial and Postcranial Skeletal Remains from Easter Island. *Science* **162**, 555–6.

Turner, C.G., II (1969). Microevolutionary interpretations from the dentition. *American Journal of Physical Anthropology* **30**, 421–6.

Turner, C.G., II (1970). New classifications of non-metrical dental variation: cusps 6 and 7. Paper presented at 39th annual meeting of the American Association of Physical Anthropologists, Washington, D.C.

Turner, C.G., II (1971). Three-rooted mandibular first permanent molars and the question of American Indian origins. *American Journal of Physical Anthropology* **34**, 229–41.

Turner, C.G., II (1976). Dental evidence on the origins of the Ainu and Japanese. *Science* **193**, 911–13.

Turner, C.G., II (1981). Root number determination in maxillary first premolars for modern human populations. *American Journal of Physical Anthropology* **54**, 59–62.

Turner, C.G., II (1983a). Dental evidence for the peopling of the Americas. In *Early Man in the New World*, ed. R. Shutler, Jr., pp. 147–57. Beverly Hills: Sage Publications.

Turner, C.G., II (1983b). Sinodonty and Sundadonty: a dental anthropological view of Mongoloid microevolution, origin, and dispersal into the Pacific basin, Siberia, and the Americas. In *Late Pleistocene and Early Holocene Cultural Connections of Asia and America*, ed. R.S. Vasilievsky, pp. 72–6. Novosibirsk: USSR Academy of Sciences, Siberian branch.

Turner, C.G., II (1984). Advances in the dental search for native American origins. *Acta Anthropogenetica* **8**, 23–78.

Turner, C.G., II (1985a). Dental evidence for the peopling of the Americas.

National Geographic Society Research Reports **19**, 573–96.

Turner, C.G., II (1985b). The dental search for native American origins. In *Out of Asia*, eds. R.L. Kirk and E. J.E. Szathmary, pp. 31–78. Canberra: Journal of Pacific History, Inc., Australian National University.

Turner, C.G., II (1985c). The modern human dispersal event: the eastern frontier. Review of Out of Asia: Peopling of the Americas and the Pacific, eds. R.L. Kirk and E.J.E. Szathmary, *Quarterly Review of Archaeology* **6**, 8–10.

Turner, C.G., II (1986a). The first Americans: the dental evidence. *National Geographic Research* **2**, 37–46.

Turner, C.G., II (1986b). Dentochronological separation estimates for Pacific rim populations. *Science* **232**, 1140–2.

Turner, C.G., II (1987). Late Pleistocene and Holocene population history of East Asia based on dental variation. *American Journal of Physical Anthropology* **73**, 305–21.

Turner, C.G., II (1989). Teeth and prehistory in Asia. *Scientific American* **260**, 88–96.

Turner, C.G., II (1990a). Major features of Sundadonty and Sinodonty, including suggestions about East Asian microevolution, population history, and late Pleistocene relationships with Australian aboriginals. *American Journal of Physical Anthropology* **82**, 295–317.

Turner, C.G., II (1990b). Origin and affinity of the prehistoric people of Guam: a dental anthropological assessment. In *Recent Advances in Micronesian Archaeology*, Micronesica Supplement No. 2, ed. R.L. Hunter-Anderson, pp. 403–16. Mangilao: University of Guam Press.

Turner, C.G., II (1991). *The Dentition of Arctic Peoples*. New York: Garland Publishing, Inc.

Turner, C.G., II (1992a). Sundadonty and Sinodonty in Japan: the dental basis for a dual origin hypothesis for the peopling of the Japanese Islands. In *International Symposium on Japanese as a Member of the Asian and Pacific Populations*, ed. K. Hanihara, pp. 96–112. Kyoto: International Research Center for Japanese Studies.

Turner, C.G., II (1992b). The dental bridge between Australia and Asia: following Macintosh into the East Asian hearth of humanity. *Perspectives in Human Biology 2/Archaeology in Oceania* **27**, 143–52.

Turner, C.G., II (1992c). Microevolution of East Asian and European populations: a dental perspective. In *The Evolution and Dispersal of Modern Humans in Asia*, eds. T. Akazawa, K. Aoki, and T. Kimura, pp. 415–38. Tokyo: Hokusen-Sha Publishing Company.

Turner, C.G., II (1993). Southwest Indians: prehistory through dentition. *National Geographic Research & Exploration* **9**, 32–53.

Turner, C.G., II (1995). Shifting continuity: modern human origin. In *The Origin and Past of Modern Humans as Viewed from DNA*, eds. S. Brenner and K. Hanihara, pp. 216–43. Singapore: World Scientific Publishing Company.

Turner, C.G., II and Cadien, J.D. (1969). Dental chipping in Aleuts, Eskimos and Indians. *American Journal of Physical Anthropology* **31**, 303–10.

Turner, C.G., II and Hanihara, K. (1977). Additional features of the Ainu dentition. V. Peopling of the Pacific. *American Journal of Physical Anthropology* **46**, 13–24.

Turner, C.G., II and Hawkey, D.E. (1995). Whose teeth are these? Carabelli's trait. *American Journal of Physical Anthropology* (Suppl. 20), 213. (Abstract.)

Turner, C.G., II and Markowitz, M.A. (1990). Dental discontinuity between late Pleistocene and recent Nubians: peopling of the Eurafrican-South Asian triangle I. *Homo* **45**, 32–41.

Turner, C.G., II and Scott, G.R. (1977). Dentition of Easter Islanders. In *Orofacial Growth and Development*, eds. A.A. Dahlberg and T.M. Graber, pp. 229–49. The Hague: Mouton Publishers.

Turner, C.G., II, Nichol, C.R. and Scott, G.R. (1991). Scoring procedures for key morphological traits of the permanent dentition: the Arizona State University dental anthropology system. In *Advances in Dental Anthropology*, eds. M.A. Kelley and C.S. Larson, pp. 13–31. New York: Wiley-Liss.

Turner, E.P. (1967). Early development of the deciduous molar crown in man. *Journal of Dental Research* **46**(suppl. to no. 5), 862–64.

Ubelaker, D.H., Phenice, T.W., and Bass, W.M. (1969). Artificial interproximal grooving of the teeth in American Indians. *American Journal of Physical Anthropology* **30**, 145–9.

Vandebroek, G. (1961). The comparative anatomy of the teeth of lower and nonspecialized mammals. *Koninklijke Vlaamse Academie voor Wetenschappen, Letteren en Schone Kunsten van Belgie* **1**, 215–313.

Van Valen, L. (1962). A study of fluctuating asymmetry. *Evolution* **16**, 125–42.

Varrela, J. (1992). Multirooted mandibular premolars in 45,X females: frequency and morphological types. In *Structure, Function and Evolution of Teeth*, eds. P. Smith and E. Tchernov, pp. 519–26. London: Freund Publishing House, Ltd.

Varrela, J. and Alvesalo, L. (1989). Taurodontism in females with extra X chromosomes. *Journal of Craniofacial Genetics and Developmental Biology* **9**, 129–33.

Vigilant, L.R., Pennington, R., Harpending, H., Kocher, T.D. and Wilson, A.C. (1989). Mitochondrial DNA sequences in single hairs from Southern African population. *Proceedings of the National Academic of Sciences USA* **86**, 9350–4.

Vigilant, L.R., Stoneking, M., Harpending, H., Hawkes, K. and Wilson, A.C. (1991) African populations and the evolution of human mitochondrial DNA. *Science* **253**, 1503–7.

Voronina, V.G. and Vaschaeva, V.F. (1979). Maritime territories. In *Ethnic Odontology of the USSR*, eds. A.A. Zubov and N.I. Khaldeeva, pp. 212–28. Moscow: Nauka. (In Russian.)

Waddington, C.H. (1957). *The Strategy of the Genes*. London: Allen and Unwin.

Weidenreich, F. (1937). The dentition of *Sinanthropus pekinensis*. *Palaeontologica Sinica*, Whole series 101, New series D-1, pp. 1–180.

Weiner, J.S. and Huizinga, J. (Eds.) (1972). *The Assessment of Population Affinities in Man*. Oxford: Clarendon Press.

Weiss, K.M. (1990). Duplication with variation: metameric logic in evolution from genes to morphology. *Yearbook of Physical Anthropology* **33**, 1–23.

Weiss, M.L. and Mann, A.E. (1978). *Human Biology and Behavior*. 2nd edn. Boston: Little, Brown.

Wheeler, R.C. (1965). *A Textbook of Dental Anatomy and Physiology*. 4th edn. Philadelphia: W.B. Saunders Company.

White, T.D. (1991). *Human Osteology*. San Diego: Academic Press.

Williams, B.J. (1973). *Evolution and Human Origins*. New York: Harper & Row.

Williams, R.C., Steinberg, A.G., Gershowitz, H., Bennett, P.H., Knowler, W.C., Pettitt, D.J., Butler, W., Baird, R., Dowda-Rea, L., Burch, T.A., Morse, H.G. and Smith, C.H. (1985). GM allotypes in native Americans: evidence for three distinct migrations across the Bering land bridge. *American Journal of Physical Anthropology* **66**, 1–19.

Wissler, C. (1917). *The American Indian: An Introduction to the Anthropology of the New World*. New York: D.C. McMurtrie.

Witkop, C. (1960). Dental genetics. *Journal of the American Dental Association* **60**, 564–77.

Wolpoff, M.H. (1971). *Metric Trends in Hominid Dental Evolution*. Cleveland: Case Western Reserve University Press.

Wood, B.A. and Abbott, S.A. (1983). Analysis of the dental morphology of Plio-Pleistocene hominids. I. Mandibular molars: crown area measurements and morphological traits. *Journal of Anatomy* **136**, 197–219.

Wood, B.A. and Engleman, C.A. (1988). Analysis of the dental morphology of Plio-Pleistocene hominids. V. Maxillary postcanine tooth morphology. *Journal of Anatomy* **161**, 1–35.

Wood, B.A. and Uytterschaut, H. (1987). Analysis of the dental morphology of Plio-Pleistocene hominids. III. Mandibular premolar crowns. *Journal of Anatomy* **154**, 121–56.

Wood, B.A., Abbott, S.A. and Graham, S.H. (1983). Analysis of the dental morphology of Plio-Pleistocene hominids. II. Mandibular molars – study of cusp areas, fissure pattern and cross sectional shape of the crown. *Journal of Anatomy* **137**, 287–314.

Wood, B.A., Abbott, S.A. and Uytterschaut, H. (1988). Analysis of the dental morphology of Plio-Pleistocene hominids. IV. Mandibular postcanine root morphology. *Journal of Anatomy* **156**, 107–39.

Wood, B.F. and Green, L.J. (1969). Second premolar morphologic trait similarities in twins. *Journal of Dental Research* **48**, 74–87.

Wright, H.B. (1941). A frequent variation of the maxillary central incisors with some observations on dental caries among the Jivaro (Shuara) Indians of Ecuador. *American Journal of Orthodontics* **27**, 249–54.

Wright, S. (1934). The results of crosses between inbred strains of guinea pigs, differing in number of digits. *Genetics* **19**, 537–51.

Wright, S. (1968). *Evolution and the Genetics of Populations*, Vol. 1. *Genetic and Biometric Foundations*. Chicago: University of Chicago Press.

Wright, S. (1969) *Evolution and the Genetics of Populations*, Vol. 2. *The Theory of Gene Frequencies*. Chicago: University of Chicago Press.

Wu, L. and Xianglong, Z. (1995). Preliminary impression of current dental anthropology research in China. *Dental Anthropology Newsletter* **9**, 1–5.

Wu, X. (1992). Origins and affinities of the Stone Age inhabitants of Japan. In *International Symposium on Japanese as a Member of the Asian and Pacific Populations*, ed. K. Hanihara, pp. 1–7. Kyoto: International Research Center for Japanese Studies.

Yamada, E. (1932). The anthropological study of the Japanese teeth. *Journal of the Nippon Dental Association* **25**, 15–46.

Yamada, H. and Kawamoto, K. (1988). The dentition of Cook Islanders. In *People of the Cook Islands – Past and Present*, eds. K. Katayama and A. Tagaya, pp. 143–209. Rarotonga: The Cook Islands Library and Museum Society.

Younes, S.A., Al-Shammery, A.R. and Al-Angbawi, M.F. (1990). Three-rooted permanent mandibular first molars of Asian and black groups in the Middle East. *Oral Surgery, Oral Medicine, Oral Pathology* **69**, 102–5.

Zegura, S. (1978). Components, factors and confusion. *Yearbook of Physical Anthropology* **21**, 151–9.

Zubov, A.A. (1968). *Odontology: A Method of Anthropological Research*. Moscow: Nauka. (In Russian.)

Zubov, A.A. (1977). Odontoglyphics: the laws of variation of the human molar crown relief. In *Orofacial Growth and Development*, eds. A.A. Dahlberg and T.M. Graber, pp. 269–82. The Hague: Mouton Publishers.

Zubov, A.A. and Khaldeeva, N.I. (1979). *Ethnic Odontology of the USSR*. Moscow: Nauka. (In Russian.)

Zubov, A.A. and Nikityuk, B.A. (1978). Prospects for the application of dental morphology in twin type analysis. *Journal of Human Evolution* **7**, 519–24.

Index

Page numbers in **bold** indicate figures and those in *italics* indicate tables.

371